Topics in Medicinal Chemistry

Volume 42

Topics in Medicinal Chemistry (TMC) covers all relevant aspects of medicinal chemistry research, e.g. pathobiochemistry of diseases, identification and validation of (emerging) drug targets, structural biology, drugability of targets, drug design approaches, chemogenomics, synthetic chemistry including combinatorial methods, bioorganic chemistry, natural compounds, high-throughput screening, pharmacological in vitro and in vivo investigations, drug-receptor interactions on the molecular level, structure-activity relationships, drug absorption, distribution, metabolism, elimination, toxicology and pharmacogenomics. Drug research requires interdisciplinary team-work at the interface between chemistry, biology and medicine. To fulfil this need, TMC is intended for researchers and experts working in academia and in the pharmaceutical industry, and also for graduates that look for a carefully selected collection of high quality review articles on their respective field of expertise. This book series supports and advances the UN's sustainable development goals, in particular SDG3 (Good Health and Well Being).

Medicinal chemistry is both science and art. The science of medicinal chemistry offers mankind one of its best hopes for improving the quality of life. The art of medicinal chemistry continues to challenge its practitioners with the need for both intuition and experience to discover new drugs. Hence sharing the experience of drug research is uniquely beneficial to the field of medicinal chemistry.

All chapters from Topics in Medicinal Chemistry are published OnlineFirst with an individual DOI. In references, Topics in Medicinal Chemistry is abbreviated as Top Med Chem and cited as a journal.

Victoria Calzada • Hugo Cerecetto •
Juan Pablo Tosar
Editors

Biotherapeutics

From small to large molecules and cells

With contributions by

N. S. Adler · M. Barros · J. Bayo · P. Cabral · V. Calzada ·
X. Camacho · M. J. Cantero · M. Casadei · N. Ceaglio ·
H. Cerecetto · E. Fiore · M. Fusco · L. G. V. Gelves · R. Glisoni ·
G. González-Aseguinolaza · L. M. Lima · M. Malvicini · P. Moeini ·
M. Oggero · A. S. Peinetti · J. R. Ramírez · J. P. Tosar ·
A. C. L. B. Trindade · C. P. Vaca G · D. Vallés · C. Villadóniga

Springer

Editors
Victoria Calzada
University of the Republic
Montevideo, Uruguay

Hugo Cerecetto
University of the Republic
Montevideo, Uruguay

Juan Pablo Tosar
University of the Republic
Montevideo, Uruguay

ISSN 1862-2461 ISSN 1862-247X (electronic)
Topics in Medicinal Chemistry
ISBN 978-3-031-81024-4 ISBN 978-3-031-81025-1 (eBook)
https://doi.org/10.1007/978-3-031-81025-1

This Springer imprint is published by the registered company Springer Nature Switzerland AG
The registered company address is: Gewerbestrasse 11, 6330 Cham, Switzerland

Preface

Small-molecule drugs appear to have traditionally dominated medicinal chemistry. However, proteins like insulin, blood derivatives or vaccines have also been used in the therapeutic setting for a long time. More recently, there has been a shift in favour of biotherapeutics for the treatment of several diseases such as cancer and inflammatory disorders, probably motivated by the therapeutic success of monoclonal antibodies. Despite the early difficulties, a future where biotherapeutics surpass small-molecule drugs as therapeutic tools can be safely envisioned. The diversification of biological therapeutic entities has been substantially accelerated in recent years due to advances in structural chemistry and biology, making the development of innovative therapies possible. In this sense, thanks to the coming of age of genetic engineering and nanotechnology, the development of chemical and biological methods to selectively inhibit, activate or modify proteins, mRNAs and cells, or even manipulate genomic information, is now possible. As a consequence, the process between basic studies identifying new molecular targets and the design of new effective biotherapeutics is becoming more straightforward, with an expanding toolbox of biomolecules that can be converted into drugs.

Many chapters in this book endeavour to emphasize the contributions made by medicinal chemistry and related disciplines in the development of biotherapeutics while also taking into account other related topics like cell therapy and regulation of biosimilars. The book consists of six parts with 15 chapters. The first part, including Chapters 1 and 2, deals with general aspects of medicinal chemistry and biotherapeutics, introducing basic concepts that will be useful for understanding later sections. The second part, including Chapters 3, 4 and 5, covers proteins as biotherapeutics. Chapter 3 shows emblematic examples of proteins, other than antibodies, used to treat diabetes, stimulate erythropoiesis, control viral infections and treat haemophilia and cancer. Chapter 4 briefly presents monoclonal antibodies, for both therapy and diagnosis. This topic is only presented at an introductory level, since several excellent books on biotherapeutics are centred specifically on therapeutic antibodies. Finally, Chapter 5 presents the production of biotherapeutic proteins in mammalian cells. This chapter includes the production of second-generation biotherapeutics, biobetters and glycoengineering approaches to produce long-lasting agents. Chapters 6, 7, 8, 9 and 10 (i.e. Part 3) cover the use of oligonucleotides and gene therapy. The aim is to address biotherapeutics based on the oligonucleotide structure, from antisense oligonucleotides (Chapter 6) to small interfering RNAs (Chapter 7) and finally aptamers (Chapter 8). Gene therapy is covered in detail in Chapter 9. The current strategies for production and structural modifications used to obtain antisense oligonucleotides, small interfering RNAs and aptamers are described in Chapter 10. In the fourth part, we did not want to leave aside other biotherapeutic tools that, according to current definitions, should be included as biotherapeutics too. Thus, Chapter 11 presents the state of the art and recent advances in cell therapy. Chapter 12 describes low-molecular-mass hormones (peptidic and steroidal) that have been blockbusters in the biopharmaceutical market for a long time. Finally, Chapter 13 covers small peptides used as biotherapeutics. These peptides might have antimicrobial effects, immunomodulation properties and antioxidant effects or might be involved in the regulation of metabolic pathways, etc. Finally, Chapter 14 (i.e. Part 4) covers relevant aspects related to the encapsulation and structural modification of biotherapeutics to enhance their therapeutic index, reduce immunogenicity or prolong their half-life. Part 6 is composed of Chapter 15 and presents regulatory aspects related to the approval of biosimilars.

The authors of each chapter are highly skilled scientists working on different areas of the drug development pipeline. Readers of this book will obtain a general understanding of current challenges and perspectives in the biotherapeutic field and successful examples of biotherapeutics that are pushing the boundaries in disease treatment. In every chapter, the authors provide different viewpoints, using in some cases examples from their own research, to stimulate further research in this topic. We expect that the biomedical community, as well as students in medicinal chemistry-related areas, will find this book of great interest and generally appealing.

We express our sincere gratitude to the authors of each chapter for their excellent work in compiling such a relevant selection of references and personal insights, helping readers to navigate the complex and dynamic landscape of biological medicinal products.

Montevideo, Uruguay Victoria Calzada
Montevideo, Uruguay Hugo Cerecetto
Montevideo, Uruguay Juan Pablo Tosar

Contents

Part I
Basic Concepts and Definitions

Top Med Chem
https://doi.org/10.1007/7355_2024_165

Published online: 28 December 2024

Medicinal Chemistry and Biotherapeutics: Overview and Definitions

Hugo Cerecetto

Contents

Abstract Medicinal chemistry is an old discipline and a fast-evolving interdisciplinary research area that aims to improve human, animal, and vegetal life by creating and developing medicines to fight diseases. The medicines could be small-molecule drugs or biotherapeutics. This chapter includes some definitinions related to small-molecule drugs and biotherapeutics, medicinal chemistry, the pipeline for creating medicines and to the medicines themselves.

Keywords Biotherapeutic, Medicinal chemistry, Small-molecule drug

1 Introduction

This chapter aims to expose and update some concepts related to medicinal chemistry and the research, development, and innovation (R + D + i) of biotherapeutic agents.

H. Cerecetto (✉)
Área de Radiofarmacia-Centro de Investigaciones Nucleares and Grupo de Química Orgánica Medicinal-Instituto de Química Biológica, Facultad de Ciencias, Universidad de la República, Montevideo, Uruguay
e-mail: hcerecetto@cin.edu.uy

1.1 Medicinal Chemistry

According to the International Union of Pure and Applied Chemistry (IUPAC), "***Medicinal chemistry*** concerns the invention, discovery, design, identification and preparation of biologically active compounds, the study of their metabolism, the interpretation of their mode of action at the molecular level and the construction of structure-activity relationships" [1]. This discipline must be nourished by chemistry, such as organic, inorganic, and product natural-chemistry, and life sciences, including biochemistry, biophysics, molecular biology, genetics, immunology, pharmacology (pharmacodynamics and pharmacokinetics), and toxicology, as well as chemistry-based disciplines like theoretical and experimental physical chemistry, crystallography, spectroscopy, and data analysis. Although drugs are the focus of medicinal chemistry, ***bioactive compounds*** (see IUPAC definition below) in general are also in the interests of medicinal chemists [2].

Bioactive compounds that are subjects of ***medicinal chemistry*** are substances used in humans, animals, and plants [1]:

(i) To treat diseases, for example, the protein *etanercept* indicated for reducing signs and symptoms of rheumatoid arthritis [3].
(ii) To cure diseases, for example, the protein *interferon-alpha* for chronic hepatitis C [4].
(iii) To prevent diseases, for example, the *vaccines* for influenza.
(iv) To control, for example, *DvSnf7 dsRNA*-expressing maize for Western corn rootworm via the RNA interference pathway [5].
(v) To restore, for example, the protein *insulin*, which manages diabetes.
(vi) To correct, for example, the RNA aptamer *avacincaptad pegol* for geographic atrophy [6][1].
(vii) To modify, for example, *equine umbilical cord mesenchymal stem cells* for lameness associated with osteoarthritis in horses [7], or.
(viii) To diagnose diseases, for example, the ^{111}*In radiolabeled capromab pendetide* for prostate cancer diagnosis [8].

The discovery and development of compounds, ***medicines***, or ***drug products*** [9] or the portfolio of a pharmaceutical company or research organization is known as a ***pipeline*** [10]. In the flowchart shown in Fig. 1, different concepts appear that are defined in the following sections.

1.2 Definitions (I)

The ***pipeline*** to reach a ***medicine*** or a ***drug product***, that according to World Health Organization (WHO) is a pharmaceutical product type in a defined container

[1] See Chapter "Aptamers".

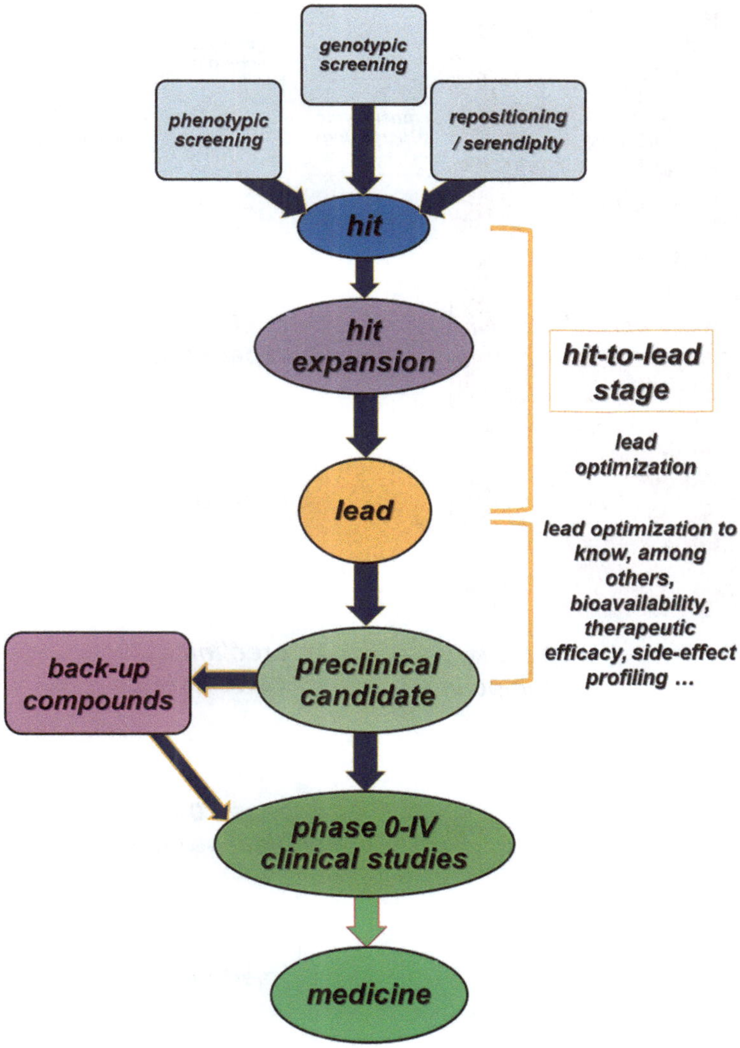

Fig. 1 Drug product pipeline

containing a ***bioactive compound*** in association with excipients [9], involves a series of steps (Fig. 1) characterized by relevant milestones [10]:

1. It begins with the identification of a ***hit***, which according to IUPAC [10] is an agent that produces reproducible activity above a defined threshold in a primary biological assay and whose structural identity has been established. A ***hit*** can be derived through i) ***phenotypic screening*** [10], which is the evaluation in cells, tissues, or organisms of a compound's ability to modify the system in a measurable manner; ii) ***target-based screening*** or ***genotypic screening***, where a

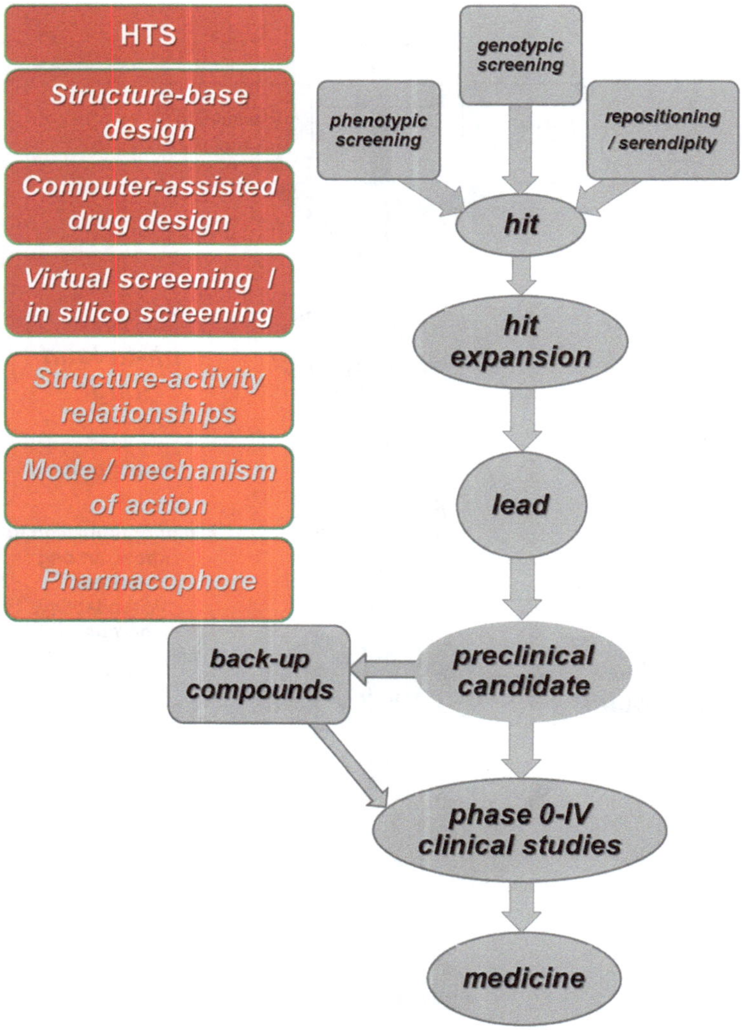

Fig. 2 Contributions (red boxes) and findings (orange boxes) that emerge or are required throughout the drug product pipeline

compound is challenged to modify a defined purified molecular target; iii) ***drug repositioning***, which is a strategy to discover new applications for an existing drug that were not previously referenced and not currently investigated [10]; and iv) ***serendipity*** [11], which refers to the fortuitous discovery of something while searching for another thing.

Phenotypic and ***genotypic screening*** could be performed ***using high-throughput screening*** (***HTS***) [10] (Fig. 2), which is a method for the rapid assessment of the activity of samples from large compound collections using, generally, microplates of at least 96 wells in robots. The term "ultra-high-throughput screening" (UHTS) is employed when the assay rate is at least 10^5 assays per day.

Related tools could support the identification of ***hits*** by ***genotypic screening*** procedures, for example (Fig. 2) [1], i) **de novo** or ***structure-base design***, which is the design of new compounds – new potential ***hits*** – through the incremental construction of a ligand model within the structural model of the bioactive site. When the process is supported by computer-assisted techniques, it is known as ***computer-assisted drug design***. Another example is ii) ***virtual screening*** or **in silico** ***screening*** [10], which is the evaluation of compounds using computational methods either by knowing the macromolecular structure of the biotarget or the physicochemical parameters or by knowing bioactive ligand ***structure-activity relationships*** (see below).

2. Next is the ***hit expansion*** procedure [10], which is the generation of additional compound sets that include structural motifs and scaffolds with activity in the primary screen. ***Hit expansion*** permits the identification of additional hits and new scaffolds and develops ***structure-activity relationships*** [1] (Fig. 2) around existing hits.
3. Following this, there is the ***"hit-to-lead" stage*** [10], by which a ***hit*** compound or series from the primary screens or from the ***hit expansion*** procedure is chemically or biologically improved to generate an adequate ***lead*** or a new series of compounds. The ***lead*** [10] is a compound or compound series that satisfies predefined minimum criteria (for example, adequate activity, selectivity, and handled ***structure-activity relationship***) for further structure and activity optimization. ***Lead generation*** or ***lead discovery*** takes place through a process known as ***lead optimization***, which involves structural modifications to fulfill all stereoelectronic, physicochemical, pharmacokinetic, and toxicologic requirements for clinical usefulness [1].
4. In this step, a ***preclinical candidate*** or ***safety assessment candidate*** [10] is generated. It is an optimized ***lead*** compound that has successfully passed key screening, selectivity (i.e., among others, bioavailability, therapeutic efficacy in an appropriate disease model, and side-effect profiling), and physicochemical criteria sufficient to warrant further detailed pharmacological and pharmacokinetic evaluations in animal models. At this stage of the pipeline, there must also be one or more ***back-up compounds*** [10], which are molecules selected as a replacement for the ***preclinical candidate*** in further preclinical evaluation or in clinical studies if the main drug candidate fails. Ideally, a ***back-up compound*** should be pharmacologically equivalent to the ***preclinical candidate*** but should have significant structural differences. The possession of a distinct core scaffold is optimal.
5. Finally, to obtain the ***medicine***, ***clinical studies*** must be carried out, from phase 0 to phase IV. ***Phase 0 clinical studies*** or ***exploratory investigational new drug studies*** [10] are exploratory first-in-human trials that involve the microdosing of drugs to allow the assessment of pharmacokinetic parameters with limited drug exposure. These trials have no therapeutic or diagnostic intent but are designed

to assist decision-making by providing bioavailability, metabolism, and other limited data from a small number of patients. ***Phase I clinical studies*** [10] are an initial introduction of the investigational new drug into humans, with the intention to determine the metabolic and pharmacologic actions of the drug in humans and the side-effects associated with increasing doses and, if possible, to gain early evidence on effectiveness. ***Phase II clinical studies*** [10] are controlled studies conducted in a limited number of individuals with a disease or condition to obtain some preliminary data on the effectiveness of the investigational new drug for indication. ***Phase IIa trials*** [10] are pilot clinical trials to evaluate efficacy and safety in selected populations of patients, while ***phase IIb*** [10] trials extend ***phase IIa*** to well-controlled trials that evaluate the same parameters in similar patient populations. ***Phase III clinical studies*** [10] are expanded controlled and uncontrolled trials in humans. These studies are performed after preliminary evidence is obtained suggesting the effectiveness of the drug product candidate in ***phase II*** and are intended to gather additional information about the effectiveness and safety needed to evaluate the overall benefit/risk relationship of the medicine. ***Phase IV clinical studies*** [10] expand postmarketing studies in humans. These trials are designed to generate information concerning the treatment risks, benefits, and optimal use of medicine during treatment.

1.3 Definitions (II)

During the ***drug product pipeline***, other aspects must be studied or must arise as a result of research (Fig. 2). Thus, it is relevant to know [1]:

The ***structure-activity relationship*** [1]: this is the relationship between chemical structure and biological activity for a series of compounds belonging to the ***hit expansion*** group. It is the result of an observational process, which will allow obtaining correlations, qualitative or quantitative, between structural changes and their effects on bioresponses. ***Quantitative structure-activity relationships*** (***QSAR***) [1] are mathematical relationships linking chemical structure and pharmacological activity in a quantitative manner for a series of compounds. Methods that can be used in QSAR include various regression and pattern recognition techniques. In this sense, a ***three-dimensional quantitative structure-activity relationship*** (***3D-QSAR***) [1] involves the analysis of the quantitative relationship between the biological activity of a set of compounds and their spatial properties using statistical methods. Knowing the ***structure-activity relationship*** allows i) designing new compounds in order to improve their biological behaviors; ii) contributing to the mechanism of action elucidation; and iii) identifying the pharmacophore.

The ***mode of action*** and the ***mechanism of action*** of the bioactive agents under study [12]: the term ***mode of action*** usually refers to the cellular changes, functional or anatomical, that result from exposure to the agent, whereas a ***mechanism of action*** refers to the specific targets or pathways that the agent modulates. Knowing both about a compound is crucial for various purposes, such as for i) the production of a new generation of more active and selective compounds; ii) the identification of the origin of its toxicity – in this sense, a concept that must be handled is the ***off-target effect*** [10], which is the pharmacological action induced by any molecule at molecular/biological sites distinct from that for which it was designed; and iii) the rationalizing of a phenotypic effect to boost confidence in the ***preclinical candidate*** before a clinical trial.

The ***pharmacophore*** or ***pharmacophoric pattern*** [1]: it is the ensemble of steric and electronic features that are necessary to ensure the optimal supramolecular interactions with a specific biological target structure and to trigger (or block) its bioresponse. A pharmacophore represents not a real molecule or a real association of functional groups but a purely abstract concept that accounts for the common molecular interaction capacities of a group of compounds toward their target structure. The pharmacophore can be considered the largest common denominator shared by a set of active molecules. Some ***pharmacophoric descriptors*** [1] are used to define a pharmacophore, such as hydrogen bonding and hydrophobic and electrostatic interaction sites, defined by atoms, ring centers, and virtual points.

1.4 Small-Molecule Drugs and Biotherapeutics

Currently, more than 90% of the medicines on the market involve, as active components in their formulations, small molecules [13]. ***Small-molecule drugs*** have historically been the pillars of traditional medicine [14]. They are low-molecular-weight (MW) molecules, less than 900 Da, that are made through chemical processes, such as organic and inorganic synthetic chemistry or natural secondary metabolite isolation from ***nonliving biological sources***. In general, the purity, the chemical identity of the active component, and the biological profile of different ***small-molecule drugs***, for example ***generic drugs***, are guaranteed regardless of the chemical process used to obtain them. ***Generic drugs*** are ***medicines*** that contain the same active component but differ, among others, in the manufacturing process, formulation, or excipients. Furthermore, historically, some of the ***biggest blockbusters***[2] in the pharmaceutical industry have been ***small-molecule drugs***, but the market for ***biotherapeutics*** has started to rise rapidly. For instance, three ***biotherapeutics*** (***erythropoietin***, ***insulin***, and ***interferon***) were among the top ten ***blockbusters*** in the early 2000s [15]. By 2023, however, eight ***biotherapeutics*** (not considering COVID-19 vaccines) were among the top ten ***blockbusters***: ***pembrolizumab***, ***adalimumab***, ***semaglutide***, ***dupilumab***, ***ustekinumab***, a combination of ***daratumumab and hyaluronidase-fihj***, ***aflibercept***, and ***nivolumab*** (Fig. 3)

[2]According to IUPAC ***blockbusters*** are drugs that generate annual sales of USD 1 billion or more [10].

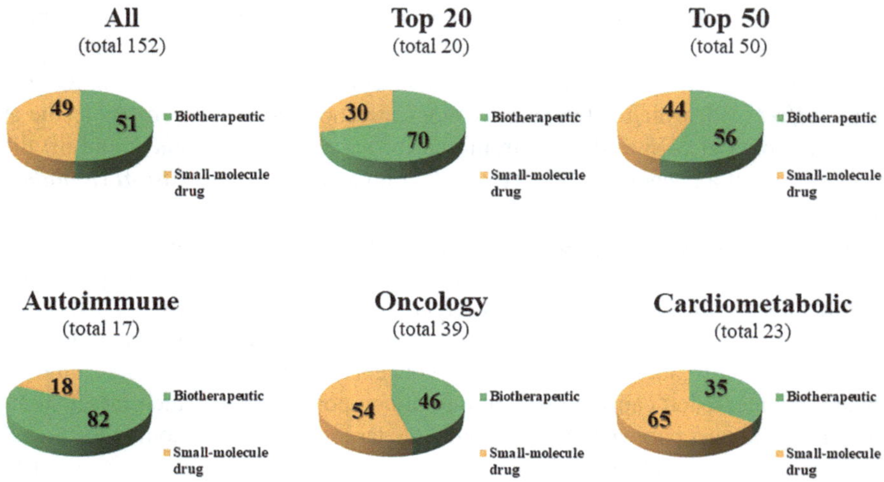

Fig. 3 Distribution of ***biotherapeutics*** in 152 blockbuster drugs in 2023. Distribution of blockbuster ***biotherapeutics*** in the main pathologies [17]

[16]. Several therapeutic areas have been addressed by these ***biotherapeutics*** (Fig. 3) [17].

On the other hand, ***biotherapeutics***, also called ***biodrugs***, ***biopharmaceuticals***, ***biologic drugs***, ***biological agents***, or ***biologics***, are generally large-sized entities, typically MW greater than 1,000 Da, and may be produced from ***living biological sources*** or through biological processes. Structurally speaking, they are, in general, more complex than ***small-molecule drugs*** as they involve many additional levels of structural complexity, for example, the formation of polymeric chains with variable chemical structures and modifications, with each subunit of the polymer being a potential site for structural modifications. Due to their biological source or manufacturing, ***biotherapeutic*** products from different manufacturers, unlike ***generic drugs***, can be substantially different in efficacy and safety profiles [18]. Different definitions of ***biotherapeutics*** have been proposed in the literature, each one providing different aspects. Some of these definitions are as follows:

1. For IUPAC [10], *they are biopolymer-based pharmaceuticals, such as a protein, applicable to the prevention, treatment, or cure of diseases or injuries to man.* ***Biological agents*** *may be any virus, therapeutic serum, toxin, antitoxin, vaccine, blood component or derivative, allergenic product, or analogous products.*
2. For WHO [19], *they are those class of medicines which are grown and then purified from large-scale cell cultures of bacteria or yeast, or plant or animal cells.* ***Biologicals*** *are a diverse group of medicines which includes vaccines, growth factors, immune modulators, monoclonal antibodies, as well as products derived from human blood and plasma. What distinguishes* ***biologicals*** *from other medicines is that these are generally proteins purified from* ***living*** *culture systems or from blood, whereas other medicines are considered as "small molecules" and are either made synthetically or purified from plants.*

Additionally, WHO [9] defines a **biotherapeutic product** as a biological medicinal product with the indication of treating human disease .... including all biologically active protein products (including plasma-fractionated products) which are used in the treatment of human diseases, and those intentionally modified by, for example, fusion proteins,[3] PEGylation[2], conjugation with a cytotoxic drug or modification of rDNA sequences. They also include protein products used for in vivo diagnosis (for example, monoclonal antibody products used for imaging).

3. For the European Union regulations (EUr) [20], a **biological medicinal product** is a protein or nucleic acid-based pharmaceutical substance used for therapeutic or in vivo diagnostic purposes, which is produced by means **other than direct extraction from a native (nonengineered) biological source**.
4. For the US Food and Drug Administration (FDA) [21], **biological products** are used to diagnose, prevent, treat, and cure diseases and medical conditions. **Biological products** are a diverse category of products and are generally large, complex molecules. These products **may be produced** through biotechnology in a **living system**, such as a microorganism, plant cell, or animal cell, and are often more difficult to characterize than small molecule drugs.

The definitions by WHO, the EU, and the FDA allow differentiating between ***small-molecule drugs*** prepared using enzymes and certain animal- and plant-derived natural products and ***biotherapeutics*** since they clarify that the latter must be obtained from ***living biological systems***.

The FDA's definition, indicating that "... the products may be produced ...", opens the door to and fixes several issues with entities that fall within the gray areas without knowing whether they are ***small-molecule drugs***, ***biotherapeutics***, or both. For example, low-molecular steroidal and peptidic hormones, synthetic peptides, aptamers, antisense oligonucleotides, RNA interference, and others can be viewed as ***biotherapeutics*** due to them being similar to, or mimicking, biological molecules, despite being chemically prepared in some cases.

A selection of pertinent data that both resemble and differentiate ***small-molecule drugs*** from ***biotherapeutics*** is presented in Table 1 [22, 23].

Associated with ***biotherapeutics*** is the concept of ***immunotherapy***. ***Immunotherapy*** is the treatment, control, diagnosis, or prevention of diseases by inducing, enhancing, or suppressing the immune response [24]. Immunotherapies, related to the treatment, control, or prevention of diseases, could be divided into i) ***active immunotherapy***, which attempts to boost the host's immune system or a particular immunological response to a disease (e.g., the use of vaccines), and ii) ***passive immunotherapy***, which involves the administration of immune system components or cellular proteins to target and combat a disease (e.g., the use of antibodies).

[3] See Chapter "Encapsulation and Structural Modifications of Biotherapeutics".

Table 1 Some similarities and differences between ***small-molecule drugs*** and ***biotherapeutics***

	Small-molecule drug	Biotherapeutic
Structure	– Low MW – Well-defined structure – Completely characterizable	– High MW – Complex, heterogeneous structure – Not entirely characterizable – Examples of those that have low MW are steroidal and peptidic hormones and small peptides.
Biological behavior	– They can interact with albumin in their transport process. – Due to their low MW, they pass through cell membranes passively. – In general, they have higher stability (affected, among others, by hepatic enzymes and, in some cases, by esterases and other hydrolases). – In general, they are nonimmunogenic.	– Albumin is generally not involved in their transport process. – They involve a specific mechanism of cellular uptake. – In general, they have lower stability (affected, among others, by hepatic enzymes and additionally by esterases, hydrolases, DNA-ases, RNA-ases). – They may be immunogenic.
Manufacture	– Chemical or chemical-biological processes with ***nonliving*** systems – Mostly process-independent	– Mostly by biological processes with ***living*** systems, but they can be chemical processes – Strongly process-dependent (*the process is the product*)
Process scaling capability	Generally low	Depends on the process (for example, in bacteria high but in animal cellular culture low)
Process and product controls	Each batch is tested mainly as to physicochemical quality.	Each batch should be tested extensively as to quality, safety, and biological efficacy.
Storage cost of pharmaceutical products	Generally low	Generally high

1.5 Definitions (III)

Other relevant concepts that need to be considered in the medicinal chemistry of monotherapy with ***biotherapeutics*** include [1, 10] (Fig. 4) the following.

A ***prodrug*** is any entity that undergoes biotransformation before exhibiting its pharmacological effects. ***Prodrugs*** can thus be viewed as drugs containing specialized nontoxic protective groups used in a transient manner to alter or eliminate undesirable properties in the parent entity. For example [25], a receptor-masked interferon alpha-2b prodrug, constituted by active frameworks connected to cleavable tumor-related proteases, has extended serum half-life, reduced systemic toxicity and increased long-term tumor-targeting ability compared to the original unmasked ***biotherapeutic*** (Fig. 5). ***Drug latentiation*** is the structural modification of an active principle to form a new entity, which in vivo will liberate the original entity. ***Drug latentiation*** is synonymous with ***prodrug*** design. On the other hand, a ***bioprecursor***

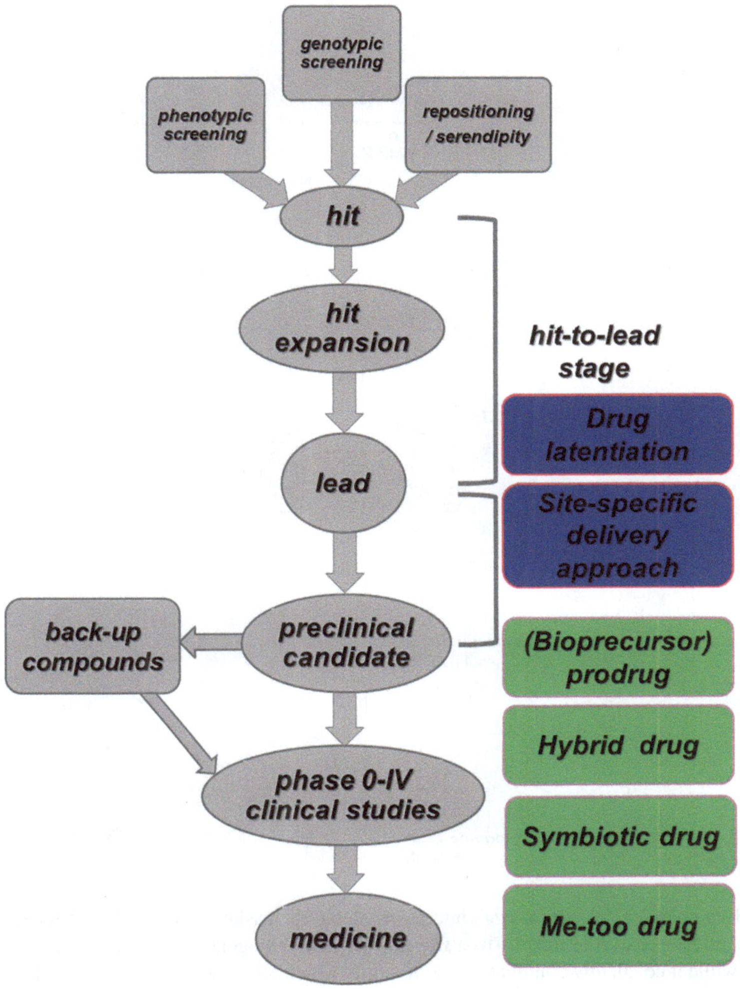

Fig. 4 Other concepts handled in monotherapy with ***biotherapeutics***

prodrug is a ***prodrug*** that does not imply the linkage to a carrier group but results from a molecular modification of the active principle itself. This modification generates a new entity, able to be transformed metabolically or chemically, the resulting product being the active principle.

A related term to the previous one is the ***site-specific delivery*** approach [1], which is a procedure used to target a drug to a specific tissue, using ***prodrugs*** or biomolecule recognition systems. Aptamers are highly effective recognition molecules that can be bound to small-molecule drugs to provide targeted drug delivery as they were demonstrated with the covalent conjugation of the dasatinib framework to a truncated aptamer, Sgc8-c, that recognizes overexpressed PTK-7 tumoral cells (Fig. 5) [26]. After the aptamer binds to a cell surface, PTK-7-receptor, and undergoes

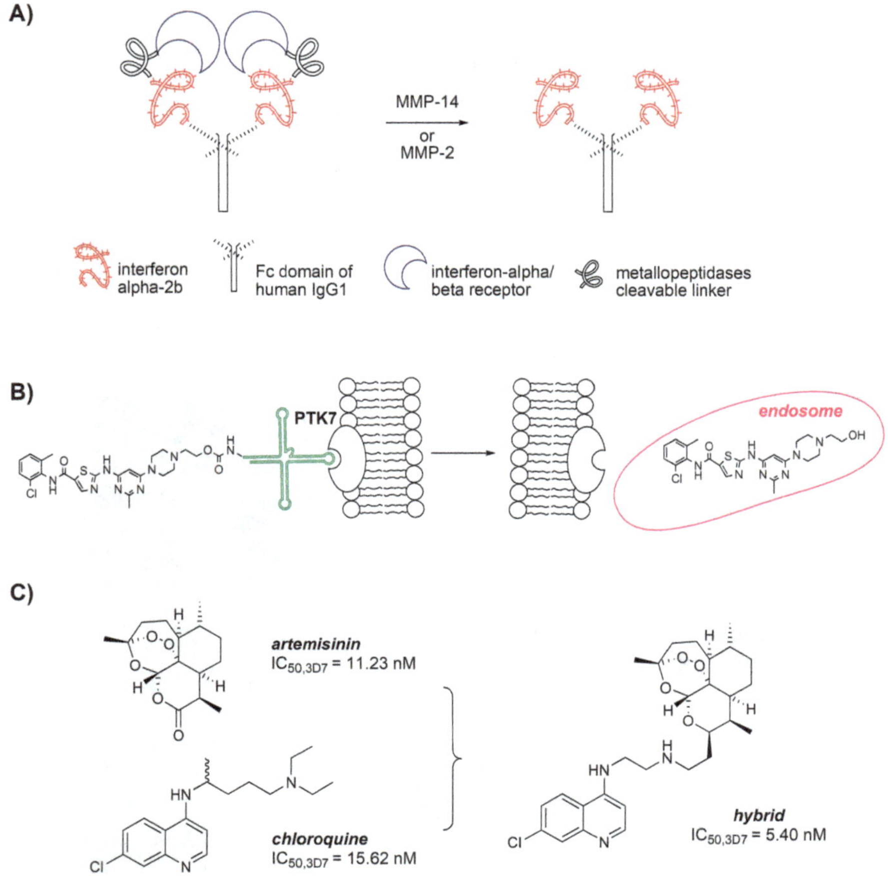

Fig. 5 Examples of (**a**) a ***prodrug***, interferon alpha-2b masked with a natural receptor via a metallopeptidase cleavable linker; (**b**) a ***site-specific delivery*** agent, a PTK7 receptor, recognizing aptamer while it covalently conjugates to a structure, which after endocytosis releases dasatinib; (**c**) a ***dual-action drug***, the hybridization of two anti-*Plasmodium falciparum* pharmacophores in one structure (***hybrid***) with better activity against the chloroquine-sensitive 3D7 strain

receptor-mediated endocytosis, the conjugate is trafficked to the endosome, where the linker is cleaved and dasatinib is released intracellularly.

According to IUPAC, a ***dual-action drug*** [1] is an entity that combines two desired different pharmacological actions at a similarly efficacious dose. Currently, a related idea is being handled, ***hybrid drugs***, which is not necessarily related to identical effectiveness [27]. They are agents with different structural domains with different biological actions, and dual activity means the ***hybrid drug*** acts via two (or more) pharmacophores not necessarily acting on the same biological target. For example, the anti-*Plasmodium falciparum* hybrids developed by Araújo and colleagues (Fig. 5) [28] where in the same structure was incorporated 4-aminoquinolinyl-framework, which increase drug concentration in the parasitic food vacuole, acidic ferrous rich compartment, and endopexoxidyl-unit that produce toxic free radicals to the parasite.

Similar to the previous concept is the idea of entities conceived to act on two or more molecular targets, the ***designed multiple ligands*** [10]. In 2008, Manssour Fraga and Barreiro coined the term ***symbiotic drugs***, which are entities designed to recognize two (or more) distinct biological targets involved in the same pathology and belonging to different biochemical pathways to promote a more efficient bioresponse [29].

Finally, here are two last definitions. On the one hand, a ***me-too drug*** [10] is an entity that is structurally very similar to already known drugs, with only minor pharmacological differences, while an ***orphan drug*** [10] is an entity that has been approved specifically to treat a rare and commercially unfavorable medical condition.

Compliance with Ethical Standards This chapter is a review and no humans or animals were involved in the preparation of the chapter.

References

1. Wermuth CG, Ganellin CR, Lindberg P et al (1998) Glossary of terms used in medicinal chemistry (IUPAC recommendations 1998). Pure Appl Chem 70:1129–1143
2. Imming P (2008) Medicinal chemistry: definitions and objectives, drug activity phases, drug classifications systems. In: Ganellin CR (ed) The practice of medicinal chemistry. Academic Press, San Diego
3. Zhao S, Mysler E, Moots RJ (2018) Etanercept for the treatment of rheumatoid arthritis. Immunotherapy 10:433–445
4. Lau DTY, Kleiner DE, Ghany MC et al (1998) 10-year follow-up after interferon-alpha therapy for chronic hepatitis C. Hepatology 28:1121–1127
5. Khajuria C, Ivashuta S, Wiggins E et al (2018) Development and characterization of the first dsRNA-resistant insect population from western corn rootworm, *Diabrotica virgifera virgifera* LeConte. PLoS One 13:e0197059
6. Kang C (2023) Avacincaptad pegol: First approval. Drugs 83:1447–1453
7. Van Loon VJF, Scheffer CJW, Genn HJ et al (2014) Clinical follow-up of horses treated with allogeneic equine mesenchymal stem cells derived from umbilical cord blood for different tendon and ligament disorders. Vet Q 34:92–97
8. Sarcan ET, Özer Y (2023) Monoclonal antibodies and immune-PET imaging: an overview. FABAD J Pharm Sci 48:165–182
9. World Health Organization (2018) WHO expert committee on biological standardization: sixty-eighth report. World Health Organization, Geneva
10. Buckle DR, Erhardt PW, Ganellin CR et al (2013) Glossary of terms used in medicinal chemistry. Part II (IUPAC recommendations 2013). Pure Appl Chem 85:1725–1758
11. Ban TA (2006) The role of serendipity in drug discovery. Dialogues Clin Neurosci 8:335–344
12. Page SW, Maddison JE, Church DB (2008) Small animal clinical pharmacology. Saunders, Edinburgh
13. Southey MWY, Brunavs M (2023) Introduction to small molecule drug discovery and preclinical development. Front Drug Discov 3:1314077
14. Ngo HX, Garneau-Tsodikova S (2018) What are the drugs of the future? Med Chem Commun 9:757–758

15. Maggon K (2003) The ten billion dollar molecule. In: Pharmaceutical Executive, PharmExec.com. https://www.pharmexec.com/view/ten-billion-dollar-molecule. Accessed 28 Jul 2024
16. Buntz B (2024) Best-selling pharmaceuticals of 2023 reveal a shift in pharma landscape. In: Drug Discovery & Development. https://www.drugdiscoverytrends.com/best-selling-pharmaceuticals-2023/. Accessed 28 Jul 2024
17. Chatsko M (2024) World's bestselling drugs addicted to broken U.S. healthcare system. In: SoltDB. https://www.living.tech/articles/worlds-bestselling-drugs-addicted-broken-us-healthcare. Accessed 28 Jul 2024
18. Declerck P, Mellstedt H, Danese S (2015) Biosimilars – terms of use. Curr Med Res Opin 31:2325–2330
19. World Health Organization (2024) Biologicals. https://www.who.int/health-topics/biologicals#tab=tab_1. Accessed 18 Apr 2024
20. Rader RA (2008) (Re)defining biopharmaceutical. Nat Biotechnol 16:743–751
21. U.S. Food & Drug Administration (2024) Biological product definitions. https://www.fda.gov/media/108557/download. Accessed 19 Apr 2024
22. Declerck PJ (2012) Biologicals and biosimilars: a review of the science and its implications. GaBI J 1:13–16
23. Declerck P, Farouk-Rezk M, Rudd PM (2016) Biosimilarity versus manufacturing change: two distinct concepts. Pharm Res 33:261–268
24. Shahani L, Singh S, Khardori NM (2012) Immunotherapy in clinical medicine; historical perspective and current status. Med Clin North Am 96:421–431
25. Cao X, Liang Y, Hu Z et al (2021) Next generation of tumor activating type I IFN enhances anti-tumor immune responses to overcome therapy resistance. Nat Commun 12:5866
26. Sicco E, Cerecetto H, Calzada V et al (2023) Targeted-lymphoma drug delivery system based on the Sgc8-c aptamer. Cancers (Basel) 15:922
27. Meunier B (2008) Hybrid molecules with a dual mode of action: dream or reality? Acc Chem Res 41:69–77
28. Araújo NCP, Barton V, Jones M et al (2009) Semi-synthetic and synthetic 1,2,4-trioxaquines and 1,2,4-trioxolaquines: synthesis, preliminary SAR and comparison with acridine endoperoxide conjugates. Bioorg Med Chem Lett 19:2038–2043
29. Manssour Fraga CA, Barreiro EJ (2008) New insights for multifactorial disease therapy. The challenge of the symbiotic drugs. Curr Drug Ther 3:1–13

Top Med Chem
https://doi.org/10.1007/7355_2024_166

Published online: 5 January 2025

Biotherapeutics: General Introduction

Hugo Cerecetto, Victoria Calzada, and Juan Pablo Tosar

Abstract As an introduction to the field of biotherapeutics, this brief chapter will discuss some aspects related to the development, challenges, advances and current strategies for their commercialization. The different families of biotherapeutics that make the different sections of this book (i.e., proteins and peptides, nucleic acids, steroids, and cells) are briefly presented. In addition, the chapter briefly describes some biotherapeutics than have gained blockbuster status in recent years.

Keywords Biotherapeutics, Blockbusters, Market

The first ***biotherapeutics*** were ***proteins***, purified from natural sources, which were employed to supplement missing or defective native peptides, for example, the case of ***insulin***, or to resolve specific pathological situations, for example, the case of ***pancreatic enzymes*** [1]. Currently, more than 40 years after the US Food and Drug Administration (FDA) approved recombinant products [2], the pharmaceutical industry has a very powerful therapeutic arsenal based on ***proteins***, including ***monoclonal antibodies***. This last group of ***proteins*** has become, since the approval in the mid-1980s by the FDA of the first ***monoclonal antibody*** as a ***biotherapeutic*** [3], one of the most relevant therapeutic tools for a great number of pathologies.

H. Cerecetto (✉)
Área de Radiofarmacia-Centro de Investigaciones Nucleares and Grupo de Química Orgánica Medicinal-Instituto de Química Biológica, Facultad de Ciencias, Universidad de la República, Montevideo, Uruguay
e-mail: hcerecetto@cin.edu.uy

V. Calzada (✉)
Área de Radiofarmacia-Centro de Investigaciones Nucleares, Facultad de Ciencias, Universidad de la República, Montevideo, Uruguay
e-mail: vcalzada@fcien.edu.uy

J. P. Tosar (✉)
Unidad de Bioquímica Analítica, Centro de Investigaciones Nucleares, Facultad de Ciencias, Universidad de la República, Montevideo, Uruguay

Laboratorio de Genómica Funcional, Institut Pasteur de Montevideo, Montevideo, Uruguay

Table 1 Classification of ***biotherapeutic proteins*** as recommended by Benjamin Leader and colleagues [4]. [a] Selected examples from reference [4]

Group	Subgroup	Examples[a]
Protein therapeutics with enzymatic or regulatory activity	Replacing a protein that is deficient or abnormal	***Insulin*** ***Somatotropin*** ***Amylase***
	Augmenting an existing pathway	***Erythropoietin*** ***Interleukin 11*** ***Trypsin***
	Providing a novel function or activity	***Botulinum toxin type A*** ***Hyaluronidase*** ***Papain***
Protein therapeutics with special targeting activity	Interfering with a molecule or organism	***Bevacizumab*** ***Etanercept*** ***Digoxin immune serum fab***
	Delivering other compounds or proteins	***Ibritumomab tiuxetan*** ***[^{131}I]tositumomab***
Protein vaccines	Protecting against a deleterious foreign agent	***HPV vaccine***
	Treating an autoimmune disease	***Anti-rhesus (Rh) immunoglobulin G***
	Treating cancer	Several phase I/II trials are in progress [5]
Protein diagnostics	–	***Imciromab pentetate*** ***Technetium fanolesomab***

Clinical applications for these ***biotherapeutics*** have been very varied, encompassing a wide range of pathologies and procedures, such as metabolic disorders (e.g., diabetes), supplementation of deficient native peptides (e.g., hemophilia, poor digestion), fertility (e.g., assisted reproduction), diagnostics (e.g., in vivo infectious disease diagnostics, imaging agents), immunological diseases (e.g., hepatitis, rheumatoid arthritis), or age-related diseases (e.g., cancer), among others.

Based on the recommendation of Benjamin Leader and colleagues, ***biotherapeutic proteins*** could be classified according to their pharmacological action (Table 1) [4].

In recent years, ***biotherapeutic proteins*** have surpassed ***small molecules*** as the treatment of choice for certain pathologies. This is evidenced by the profile evolution of the top ten blockbuster[1] drugs in the pharmaceutical industry, which list has been dominated by ***biotherapeutic proteins*** in recent years (Fig. 1) [6–11].

Early in the new millennium, protein therapeutics with enzymatic or regulatory action (Table 1) appeared among the blockbuster ***biotherapeutics*** as a result of the type of pathology for which they were intended, i.e., ***erythropoietin*** for anemia of chronic disease, ***insulin*** for diabetes mellitus, and ***interferon*** for different pathologies. Nevertheless, by the middle of the 2000s, it was evident that the sale of

[1] See Chapter "Medicinal Chemistry and Biotherapeutics. Overview and Definitions".

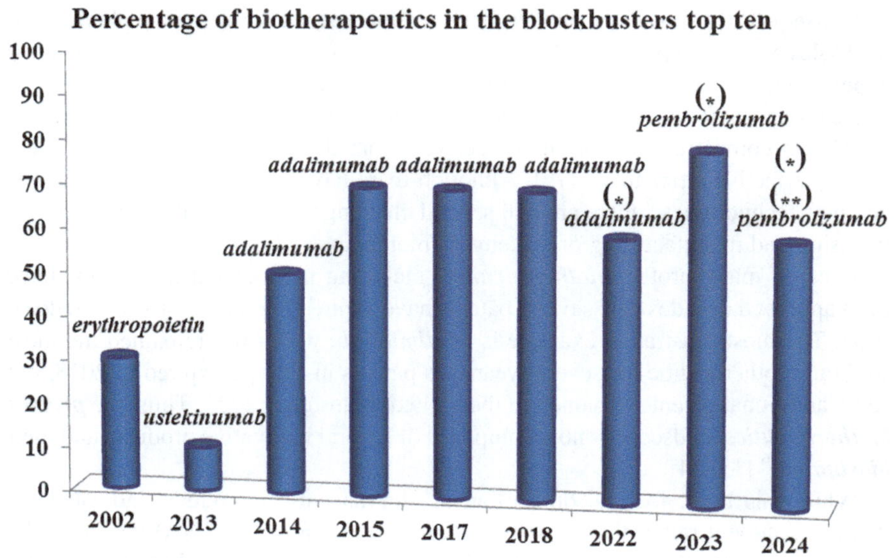

Fig. 1 Percentage of ***protein biotherapeutics*** in the top 10 blockbusters approved by the FDA during the time. The rest of the top 10 blockbusters are small molecules or nonprotein ***biotherapeutics***. In each year is indicated the main ***protein biotherapeutic*** blockbuster. (*) In this case, COVID-19 vaccines were not considered in the top 10. (**) Based on projected 2024 sales

monoclonal antibodies, belonging mainly to the protein therapeutics with a special targeting activity group (Table 1), had increased in response to untreated illnesses for which there was no pharmacological treatment. Consequently, for example, in 2014, in terms of sales, ***insulin*** ranked, among the blockbusters, 15th (***insulin glargine***),[2] 25th (***insulin detemir***),[2] 34th (***insulin aspart***),[2] or 69th (***insulin lispro***),[2] while ***interferons*** ranked 50th (***interferon β-1a***)[2] and 98th (***interferon β-1b***)[2] and erythropoietin 63rd (***epoetin alfa***).[2] ***Monoclonal antibodies*** have taken center stage in more recent times, with the following blockbusters receiving special attention: i) ***adalimumab***,[3] the main ***biotherapeutic*** blockbuster in 2022, for the treatment of rheumatoid arthritis, among other pathologies, and ii) **pembrolizumab**,[3] the main in 2023 and 2024, for different cancer treatment. As a monoclonal antibody, ***pembrolizumab*** belongs to the group of "protein therapeutics with special targeting activity," based on the classification presented in Table 1. It also belongs to the group of ***passive immunotherapies***,[4] based on its mechanism of action, fighting cancer by indirectly boosting the immune system (rather than the conventional mode of targeting cancer cells themselves with ***small-molecule drugs***). This new paradigm is shaping both oncology and the ***protein biotherapeutics*** landscape.

[2] See Chapter "Therapeutic Proteins".

[3] See Chapter "Monoclonal Antibodies as Biotherapeutics".

[4] See Chapter "Medicinal Chemistry and Biotherapeutics. Overview and Definitions".

Consequently, nowadays, ***protein***-derived ***biotherapeutics*** dominate pharmaceutical sales as modern pharmacotherapies (Fig. 1), yet they represent some of the most expensive drugs on the market. These high prices are associated with the complexity of the processes for developing and obtaining these ***biotherapeutics*** (i.e., production costs) and commercial opportunities derived from the lack of less expensive and equally effective alternatives [12]. Although there have been scientific innovations and new technologies,[5] there are still several challenges related to the development, licensing and manufacturing of proteins as biotherapies [12].

Because many protein ***biotherapeutics***, including monoclonal antibodies, have been approved decades ago, several patents have expired in recent years or will do so soon. The most illustrative example is ***adalimumab***, which has remained the most sold out biotherapeutic for several years. Its patents in Europe expired in 2018, but some additional patents remained in the United States until 2022. Thus, the ***protein biotherapeutics*** landscape is now composed of several innovative products and their ***biosimilars***[6] [13, 14].

While this book focuses on biological pharmaceutical products indicated for disease treatment rather than prophylaxis, messenger RNA (mRNA) vaccines like Comirnaty (Pfizer-BioNTech COVID-19 vaccine) and Spikevax (Moderna COVID-19 vaccine) are undoubtedly the most popular examples of nucleic-acid-based pharmaceutical products. However, the coming of age of nucleic acid ***biotherapeutics*** started several years before the pandemic began [15, 16]. And they will continue to grow, perhaps even exponentially, during the next years [17, 18].

Nucleic acid biotherapeutics can be classified into oligonucleotide-based and gene-therapy biotherapeutics.

Oligonucleotide biotherapeutics employ relatively short (15–60 nucleotides) single-stranded or double-stranded RNA or DNA molecules, usually containing several chemical modifications. Depending on their mode of action, they can be further classified into ***antisense oligonucleotides***[7] (***ASOs***), ***small interfering RNAs***[8] (***siRNAs***), and ***aptamers***.[9] ***ASOs*** are single-stranded and unstructured molecules that can recognize specific processed or unprocessed mRNAs and either a) induce their degradation, b) stabilize them, or c) increase the production of their encoded protein products by interfering with the mRNA maturation process (i.e., splicing). Therefore, they can be used to turn up or turn down the expression of a specific gene.

[5] See Chapter "Development and Production of Protein-based Biotherapeutics in Mammalian Cells".

[6] See Chapters "Development and Production of Protein-based Biotherapeutics in Mammalian Cells" and "Evolution of International Standards for the Approval of Biosimilars: A Path to Flexibility".

[7] See Chapter "Antisense Oligonucleotides (ASOs)".

[8] See Chapter "Small Interferring RNAs (siRNAs)".

[9] See Chapter "Aptamers".

Meanwhile, ***siRNAs*** are double-stranded RNA molecules that are exclusively used to degrade specific mRNAs based on the ***RNA interference*** (***RNAi***) process. On the other hand, ***aptamers*** are single-stranded RNA or DNA molecules that work more like antibodies, in the sense that they adopt a tridimensional conformation that enables the recognition of a specific protein antigen. Unlike ***ASOs*** and ***siRNAs***, they do not require cellular internalization for their therapeutic activity. On the contrary, they usually target extracellular proteins (e.g., VEGF) or cell surface receptors.

Gene therapy biotherapeutics[10] are based on the introduction of synthetic genes to provide therapeutic effects (i.e., gene addition) or the manipulation of the patient's own genomic DNA (i.e., gene editing) [19]. Gene addition can be done by introducing a DNA molecule into the nucleus of the target cell population (usually simply called ***gene therapy***) or by introducing a synthetic mRNA encoding for a desired protein product (i.e., ***mRNA therapeutics***). The advantage of the latter is that a) the response is transient as the mRNA will be degraded by cellular nucleases without irreversibly affecting genomic DNA, and b) there is no need to transport the biotherapeutic product into the nucleus. In contrast, DNA-based gene addition offers the attractive possibility of curing a disease forever with a single injection. ***Gene editing biotherapeutics*** include clustered regularly interspaced short palindromic repeats-associated nuclease (***CRISPR-Cas***). Although the ***CRISPR-Cas*** system is usually employed in the laboratory to obtain transgenic organisms by means of germline gene edition, gene editing ***biotherapeutics*** act on somatic cells. Examples include in vivo and ex vivo applications. In the former, the ***CRISPR-Cas*** machinery is transiently introduced into a patient. In the latter, a specific cell type is collected from a patient and gene edited in the laboratory, and the edited cells are introduced back into the patient. Thus, ex vivo gene editing is actually a form of ***cell therapy***.[11] There are also several other ex vivo gene addition biotherapeutics, which, for the same reason, will not be further discussed in this section.

These innovative ***biotherapeutics*** have been expanded to include, currently, over 20 approved medicines, and hundreds more clinical trials are currently being conducted worldwide (Table 2) [20, 21]. Except for COVID-19 vaccines, these ***biotherapeutics*** have not been among the blockbusters[12] in the pharmaceutical industry, especially since they target rare diseases for a limited group of patients. Partly for this reason, they tend to be among the most expensive pharmaceutical products. For example, ***onasemnogene abeparvovec*** (Zolgensma®, a ***gene therapy***) was the most expensive drug on the market, with a list price of $2.1 million. It was surpassed in 2024 by ***atidarsagene autotemcel*** (Lenmeldy®), also a gene therapy, sold at a list price of $4.2 million. However, this tendency might change in the near future, considering the recent approvals of ***biotherapeutics*** like ***inclisiran*** (a ***siRNA***), indicated for the treatment of one of the most popular causes of death worlwide (cardiovascular events).

---

[10] See Chapter "Gene Therapy".

[11] See Chapter "Conventional and Emerging Cell Therapy Strategies".

[12] See Chapter "Medicinal Chemistry and Biotherapeutics. Overview and Definitions".

Table 2 Approved ***biotherapeutics*** derived from ***nucleic acids***

Group	Name (trademark)	Disease indication and comments	Target
ASO	***Fomivirsen*** (Vitravene)	Cytomegalovirus retinitis (first ASO approved for human use by the FDA in 1998)	*IE-2*
	Mipomersen (Kynamro)	Homozygous familial hypercholesteremia	*ApoB-100*
	Eteplirsen (Exondys 51)	Duchenne muscular dystrophy	Dystrophin (exon 51)
	Nusinersen (Spinraza)	Spinal muscular atrophy	*SMN2*
	Inotersen (Tegsedi)	Polyneuropathy of hereditary transthyretin amyloidosis	*TTR*
	Volanesorsen (Waylivra)	Familial chylomicronemia syndrome	*ApoC-III*
	Golodirsen (Vyondys 53)	Duchenne muscular dystrophy	Dystrophin (exon 53)
	Viltolarsen (VILTEPSO)	Duchenne muscular dystrophy	Dystrophin (exon 53)
	Casimersen (Amondys 45)	Duchenne muscular dystrophy	Dystrophin (exon 45)
	Tofersen (Qalsody)	Amyotrophic lateral sclerosis	*SOD1*
	Eplontersen (Wainua)	TTR-meditated polyneuropathy/ TTR-meditated cardiomyopathy	*TTR*
siRNA	***Patisiran*** (Onpattro)	Polyneuropathy of hereditary transthyretin amyloidosis	*TTR*
	Givosiran (Givlaari)	Acute hepatic porphyria	*ALS1*
	Lumasiran (Oxlumo)	Primary hyperoxaluria type 1	*HAO-1*
	Inclisiran (Leqvio)	Heterozygous familial hypercholesteremia, primary hyperlipidemia	*PCSK9*
	Vutrisiran (Amvuttra)	Polyneuropathy of hereditary transthyretin amyloidosis	*TTR*
	Nedosiran (Rivfloza)	Primary hyperoxaluria	Lactate dehydrogenase
Aptamer	***Pegaptanib sodium*** (Macugen)	Neovascular age-related macular degeneration (discontinued)	Vascular endothelial growth factor
	Avacincaptad pegol (Izervay)	Geographic atrophy	Complement component C5
CRISPR-Cas	***Exagamglogene autotemcel*** (Casgevy)—Cell therapy	Sickle cell anemia	*BCL11A*
mRNA	***Tozinameran*** (Comirnaty)	SARS-CoV-2 (mRNA) vaccine	SARS-CoV-2 spike protein
	Elasomeran (Spikevax)	SARS-CoV-2 (mRNA) vaccine	SARS-CoV-2 spike protein
	mRNA-1,345 (mRESVIA)	Respiratory syncytial virus (RSV) vaccine	RSV F glycoprotein

(continued)

Table 2 (continued)

Group	Name (trademark)	Disease indication and comments	Target
DNA	***Voretigene neparvovec-rzyl*** (Luxturna)	Leber congenital amaurosis	*RPE65*
	Onasemnogene abeparvovec (Zolgensma)	Spinal muscular atrophy	*SMN1*
	Etranacogene dezaparvovec (Hemgenix)	Hemophilia B	*Factor IX*
	Beremagene geperpavec (Vyjuvek)	Wound healing in dystrophic epidermolysis bullosa	*COL7A1*
	Delandistrogene moxeparvovec (Elevidys)	Duchenne muscular dystrophy	*Dystrophin*
	Valoctocogene roxaparvovec (Roctavian)	Hemophilia A	*Factor VIII*

Despite the attractiveness of these ***biotherapeutics***, the use of ***nucleic acids*** as drugs has certain problems that must be solved, such as i) immunogenicity, since they can trigger the immune system; ii) off-target organ clearance and kidney excretion; iii) instability, because nucleases easily break them down; iv) unfavorable physicochemical properties, which make cellular internalization and delivery into specific subcellular compartments difficult; v) structural constraints, for an efficient interaction with the intracellular machinery needed to carry out the intended functions; and vi) delivery into specific cell types other than those preferentially targeted by current delivery methods. These problems have been substantially resolved by developments in nucleotide chemistry,[13] mainly by chemical modifications to the oligonucleotides, which promote longer stability and enhanced activity, or by the use of micro- and nanotechnology[14] (e.g., encapsulation inside the lipid nanoparticles of ***siRNA***s and ***mRNAs***) to generate formulations that enhance ***nucleic acid biotherapeutic*** stability, bioavailability, and targeted delivery (e.g., ligand conjugation of ***siRNAs*** and encapsulation inside recombinant viral particles in ***gene therapy***) and avoid self-aggregation and immunogenicity [22].

In this book, we have endeavored to also include other ***biotherapeutic*** entities that do not fully meet the ***biotherapeutic*** definitions previously discussed,[15] e.g., ***cell therapy***, ***low molecular mass (MM) hormones*** (derived from ***steroids*** and ***peptides***), and ***biotherapeutics*** derived from ***small peptides***. On the one hand, it is intended to provide a broad overview of the development and manufacturing of ***cell therapy***.[16] This would be considered biotherapeutic according to the World Health

[13] See Chapter "Nucleic Acids Production".

[14] See Chapter "Encapsulation and Structural Modifications of Biotherapeutics".

[15] See Chapter "Medicinal Chemistry and Biotherapeutics. Overview and Definitions".

[16] See Chapter "Conventional and Emerging Cell Therapy Strategies".

Table 3 Some approved cellular therapies

Name	Disease indication
Lifileucel	Metastatic or unresectable melanoma
Tisagenlecleucel	B-cell acute lymphoblastic leukemia, B-cell non-Hodgkin lymphoma
Axicabtagene ciloleucel	B-cell non-Hodgkin lymphoma, follicular lymphoma
Idecabtagene vicleucel	Multiple myeloma
Allogeneic cultured keratinocytes and fibroblasts in bovine collagen	Mucogingival conditions
Hematopoietic progenitor cell, cord blood	Used in unrelated donor hematopoietic progenitor stem cell transplantation procedures

Organization (WHO) definition: "... biologicals are a diverse group of medicines which includes vaccines, growth factors, immune modulators, monoclonal antibodies, as well as **products derived from human blood and plasma** ...." According to the International Union of Pure and Applied Chemistry (IUPAC), "... [B]iological agents may be any virus, therapeutic serum, toxin, antitoxin, vaccine, **blood component or derivative**, allergenic product, or analogous products ...." On the other hand, it is also described as the art and science of developing ***low MM hormones***[17] and ***small peptides***,[18] which again would fall within the definition of ***biotherapeutics*** as described by WHO: "... they are those class of medicines which are **grown and then purified from large-scale cell cultures of bacteria or yeast, or plant or animal cells** ...."

Cell therapies require complex, advanced, and well-established manufacturing processes. Despite this, a great number of cellular therapies have received approval from health regulatory organisms (Table 3), are awaiting approval, or are in the ***biotherapeutic*** late-stage pipeline development. Additionally, in terms of sales, ***axicabtagene ciloleucel*** (i.e., a CAR-T therapy consisting of manipulated patient-derived T lymphocytes reprogrammed to recognize and attack cancer cells) has reached a blockbuster[1] status [23]. Due to the increased volume of these ***biotherapeutics*** entering the market and their development into such broader medical indications, it is predicted that the use of cellular therapy will expand by more than 207% by 2028.

Low MM hormones and ***small peptides*** have reached a level of development that has allowed many of these biotherapeutics to reach a blockbuster status. For example, some synthetic steroid-like hormones, with mechanisms of action involving the biochemical steps of these hormones, have been found in the top ten blockbusters in recent years, namely, ***abiraterone*** to treat prostate cancer and ***budesonide*** for the long-term management of asthma and chronic obstructive pulmonary disease [24]. On the other hand, the polypeptides ***tirzepatide*** and

[17] See Chapter "Low Molecular-mass Hormones. Peptidic and Steroidal Hormones".

[18] See Chapter "Small Peptides as Biotherapeutics".

semaglutide, acting as an analog of the hormone glucagon-like peptide-1, are indicated for the treatment of type 2 diabetes and/or obesity. However, they are often used off-label for weight loss, making their way into the top ten blockbusters in 2022–2024 [10, 25].

Compliance with Ethical Standards This chapter is a review and no humans or animals were involved in the preparation of the chapter.

References

1. Rosenfeld L (2002) Insulin: discovery and controversy. Clin Chem 48:2270–2288
2. Mathieu C, Martens PJ, Vangoitsenhoven R (2021) One hundred years of insulin therapy. Nat Rev Endocrinol 17:715–725
3. Liu JK (2014) The history of monoclonal antibody development – Progress, remaining challenges and future innovations. Ann Med Surg 3:113–116
4. Leader B, Baca QJ, Golan DE (2008) Protein therapeutics: a summary and pharmacological classification. Nat Rev Drug Discov 7:21–39
5. Hargrave A, Mustafa AS, Hanif A et al (2023) Recent advances in cancer immunotherapy with a focus on FDA-approved vaccines and neoantigen-based vacciones. Vaccines 11:1633
6. Maggon K (2003) The ten billion dollar molecule. In: Pharmaceutical executive, PharmExec.com. https://www.pharmexec.com/view/ten-billion-dollar-molecule. Accessed 28 Jul 2024
7. (2016) Top drugs by sales revenue in 2015: who sold the biggest blockbuster drugs? In: Pharmacompass. http://www.pharmacompass.com/radio-compass-blog/top-drugs-by-sales-revenue-in-2015-who-sold-the-biggest-blockbuster-drugs. Accessed 28 Jul 2024
8. (2017) The top 15 best-selling drugs of 2016. In: Genetic engineering and biotechnology news. https://www.genengnews.com/a-lists/the-top-15-best-selling-drugs-of-2016/. Accessed 28 Jul 2024
9. (2019) Top 15 best-selling drugs of 2018. In: Genetic engineering and biotechnology news. https://www.genengnews.com/a-lists/top-15-best-selling-drugs-of-2018/. Accessed 28 Jul 2024
10. Buntz B (2024) Best-selling pharmaceuticals of 2023 reveal a shift in pharma landscape. In: Drug discovery & development. https://www.drugdiscoverytrends.com/best-selling-pharmaceuticals-2023/. Accessed 28 Jul 2024
11. Chatsko M (2024) World's bestselling drugs addicted to broken U.S. healthcare system. In: SoltDB. https://www.living.tech/articles/worlds-bestselling-drugs-addicted-broken-us-healthcare. Accessed 28 Jul 2024
12. Sauna ZE, Kimchi-Sarfaty C (eds) (2017) Protein therapeutics. Topics in medicinal chemistry, vol 21. Springer, Heidelberg
13. Kesik-Brodacka M (2018) Progress in biopharmaceutical development. Biotechnol Appl Biochem 65:306–322
14. Cordeiro MA, Vitorino C, Sinogas C et al (2024) A regulatory perspective on biosimilar medicines. Pharmaceutics 16:321
15. Rossi JJ, Rossi D (2020) Oligonucleotides and the COVID-19 pandemic: a perspective. Nucleic Acid Ther 30:129–132
16. Anderson EJ, Rouphael NG, Widge AT et al (2020) Safety and immunogenicity of SARS-CoV-2 mRNA-1273 vaccine in older adults. New Engl J Med 383:2427–2438
17. Perry CM, Balfour JAB (1999) Fomivirsen. Drugs 57:375–380
18. Roehr B (1998) Fomivirsen approved for CMV retinitis. J Int Assoc Phys AIDS Care 4:14–16
19. Hofman CR, Corey DR (2024) Targeting RNA with synthetic oligonucleotides: clinical success invites new challenges. Cell Chem Biol 31:125–138
20. Egli M, Manoharan M (2023) Chemistry, structure and function of approved oligonucleotide therapeutics. Nucleic Acids Res 51:2529–2573

21. Belgrad J, Fakih HH, Khvorova A (2024) Nucleic acid therapeutics: successes, milestones, and upcoming innovation. Nucleic Acid Ther 34:52–72
22. Castelli R, Ibarra M, Faccio R et al (2022) T908 polymeric micelles improved the uptake of Sgc8-c aptamer-probe in bearing tumor mice: a co-association study between the probe and preformed nanostructures. Pharmaceuticals 15:15
23. Horlacher ML, Eitelwein O, Jochem L et al (2024) Explore the commercial landscape of cell and gene therapies. Making cell and gene therapies financially sustainable. In: OliverWyman. https://www.oliverwyman.com/our-expertise/insights/2024/may/commercial-viability-gap-in-cell-and-gene-therapy.html. Accessed 28 Jul 2024
24. (2024) Thirteen potential blockbuster drugs to watch in 2024. In: Drug discovery world (DDW). Turning science into business. https://www.ddw-online.com/thirteen-potential-blockbuster-drugs-to-watch-in-2024-27823-202401/. Accessed 28 Jul 2024
25. (2023) Leading drugs worldwide based on projected 2024 sales. In Statista. https://www.statista.com/statistics/973523/top-drugs-by-year-on-year-sales-increase/. Accessed 28 Jul 2024

Part II
Biotherapeutics Derived from Proteins

Top Med Chem
https://doi.org/10.1007/7355_2024_167

Published online: 25 January 2025

Therapeutic Proteins

Hugo Cerecetto

Contents

Abstract Therapeutic proteins were first developed and employed clinically in the last century with the use of insulin to make up for the lack of the natural peptide. In the late 1970s, large-scale protein production was possible because of the introduction of recombinant DNA technology, which also greatly sparked systematic clinical research employing novel therapeutic strategies. This chapter provides an overview of some therapeutic proteins, both in their natural form and as structurally modified analogs, which are used as drugs.

Keywords Arginase, Coagulation factors VIII and IX, Erythropoietin, Insulin, Interferons

H. Cerecetto (✉)
Área de Radiofarmacia-Centro de Investigaciones Nucleares and Grupo de Química Orgánica Medicinal-Instituto de Química Biológica, Facultad de Ciencias, Universidad de la República, Montevideo, Uruguay
e-mail: hcerecetto@cin.edu.uy

1 Introduction

Chains of fewer than 20 amino acids that are linked together by peptide bonds (eupeptide bond [1]) are known as oligopeptides. Polypeptides are longer, continuous, and unbranched peptide chains [2]. Proteins are composed of one or more linked polypeptides and have a molecular weight (MW) of 10,000 Da or more [3]. Therefore, proteins are very large peptides. Oligopeptides, polypeptides, and proteins of a certain type have been recognized as significant regulatory molecules in animals, yeast, and plants, which control a wide range of physiological functions (for example, as hormones) [4].

Historically, proteins have been the first biotherapeutics, used as agents to replenish the lack of the natural peptide absent in a patient. Therefore, for instance, the fermental work on insulin dates back more than a century [5]. Nevertheless, in an effort to optimize their therapeutic profile, for example, reducing the onset times of action, modifying metabolism, increasing residence times in the body, a significant amount of medicinal chemistry is being developed today [6]. Medicinal chemistry procedures on proteins, such as amino acid substitutions, PEGylations,[1] and sialylations, are very common nowadays [7], and biological modifications, among others mutations XTENylations and proteins fusions[1] [8], finding new pharmacological effective biotherapeutics from proteins.

Examples of medicinal chemistry interventions and pharmacological outcomes in each of the five categories of therapeutic proteins—hormones, cytokines, signaling proteins, coagulation factors, and enzymes—will be covered in the sections that follow.

1.1 Insulin

Insulin is the peptidic hormone produced by the beta cells of the pancreatic islets, which is considered the main anabolic hormone of the body, regulating carbohydrate, fat, and protein metabolisms [9]. In the early 1950s, Nobel laureate Frederick Sanger [10] and colleagues determined the primary structure of human insulin by sequencing amino acid structure using four tools to put the puzzle together: (i) the selective cleavage, with performic acid, of disulfide bridges; (ii) the use of 1-fluoro-2,4-dinitrobenzene marker that binds to free NH_2 groups; (iii) partial chemical and enzymatic hydrolyses; and (iv) chromatography. It was not until the 1970s that its exact three-dimensional structure was known [11, 12]. It was found to be a polypeptide hormone (monomer, MW of 5,808 Da) constituted by an A chain with 21 amino acids, which is joined to the B chain, containing 30 amino acids, by two disulfide bridges between Cys A7–B7 and A20–B19. There is an extra intrachain disulfide bridge within the A chain, between Cys 6 and 11 (Fig. 1). The A chain

[1] See Chapter "Encapsulation and Structural Modifications of Biotherapeutics".

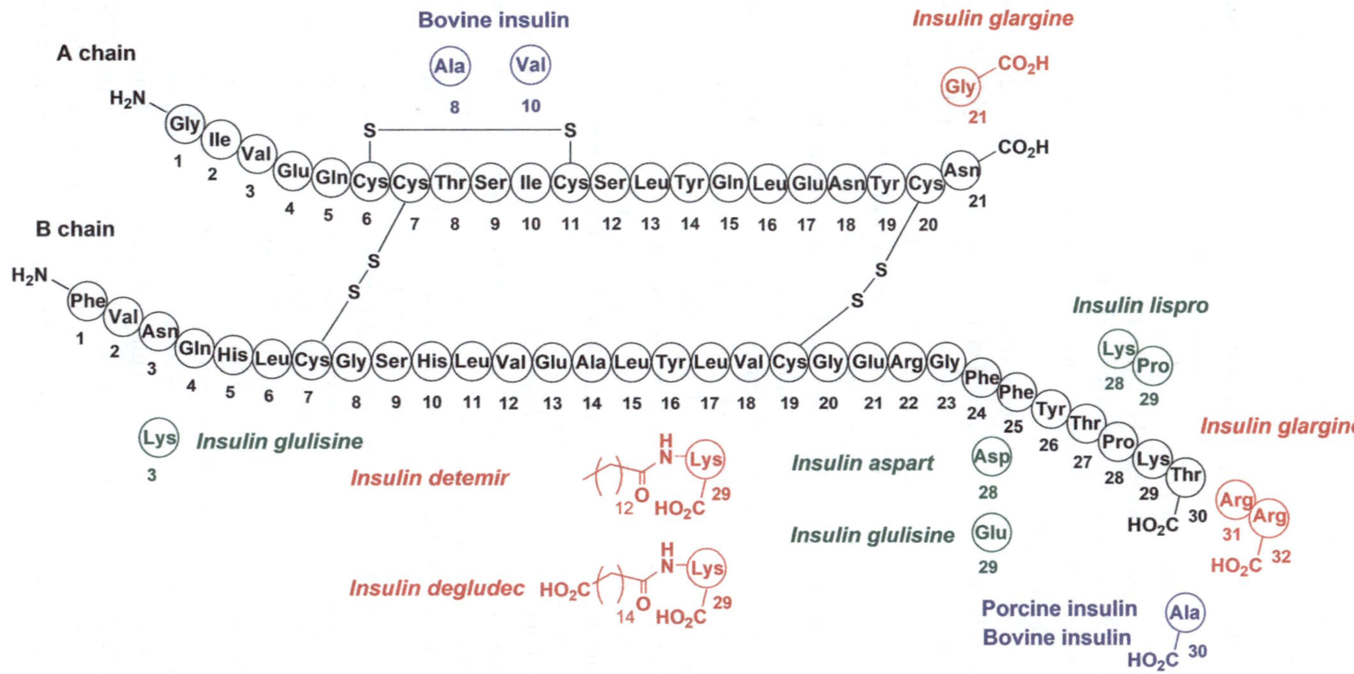

Fig. 1 Primary structure of human insulin (black), animal insulin (blue), and modifications in pharmaceutical insulin analogs (short acting in green and long acting in red)

has two antiparallel α-helical sections, at amino acid1–amino acid8 and amino acid12–amino acid19, while the B chain has two β-sheets (covering residues amino acid7–amino acid10 and amino acid20–amino acid23), a central α-helix (covering residues amino acid9–amino acid19) flanked by both disulfide bridges on each side [13].

The body produces and stores insulin as an inactive hexamer, the monomer being the active form. The six molecules are joined as three dimeric units where Zn^{2+}, surrounded by three His at position 10 of the B chain and three water molecules, is located on the axis of symmetry [13]. Because of its long-term stability, the hexameric form helps preserve the reactive insulin while maintaining its availability, an aspect extremely relevant as a therapeutic tool (see below in the structural modifications of insulin) [14].

With the exception of the sequence found in hystricomorph rodents (such as chinchillas and guinea pigs), both A and B chain sequences are highly preserved in mammals and birds. Porcine insulin and bovine insulin, of great therapeutic relevance in the early years, differ from human insulin in only one or three, respectively, amino acids (Fig. 1). Fish insulin has a significant level of homology, greater than 50%, with mammalian insulin [15]. Insulin-like proteins, which were named glucokinin by James B. Collip [16], were also isolated from plants [17].

Early human clinical trials carried out by pioneers Banting, Best, Collip, Campbell, and Fletcher [18] using animal-derived insulin, for example, extracts from normal adult oxen's pancreas, produced controversial findings [5]. This was mainly due to the purity of the first extracts from animal pancreas, which consisted, in addition to insulin, of salt, water, fats, proteins, and other organic materials, causing abscesses and very painful injections in many patients. However, the years and the improvement in the extractive and purification process, from natural sources, yielded good products and successful therapeutic results.

In this regard, until the beginning of the 1980s, porcine insulin and bovine insulin were the most widely used for diabetes mellitus therapy. Structurally talking, porcine insulin is different from the human protein in the amino acid located at position 30 of B chain, Ala instead of Thr, while bovine insulin in three, Ala replaces Thr at position 8 of A chain, Val replaces Ile at position 10 of A chain, and Ala replaces Thr at position 30 of B chain in human insulin (Fig. 1) [19]. Even though these insulins were utilized extensively for many years, two significant issues were found: (i) complaints from some patients treated with them for a long time, who developed an immune response that compels them to stop administration [20]; however, clinically significant differences were not observed when the pharmacological effects were compared or when the adverse reaction profiles were examined [21]; and (ii) limited availability of pancreases from the aforementioned animals; every year, a diabetic needs the pancreases of 50 pigs. Efficient semi-synthetic procedures to convert porcine to human insulin have been described since the 1980s, for example, transpeptidation, involving enzymatic transformations, which were described by Markussen and colleagues to have a yield of 97% of human insulin ester [22].

Table 1 Pharmacokinetic parameters of different insulin types [30]

Insulin	Onset (h)	Peak (h)	Duration (h)
Insulin lispro	0.25–0.5	0.5–1.5	2–5
Insulin aspart	0.17–0.33	1–3	3–5
Insulin glulisine	0.25	0.5–1.5	1–2.5
Insulin glargine	1.1	–	24
Insulin detemir	0.8–2	–	Up to 24
Insulin degludec	–	–	>25
NPH insulin	1–1.5	6–14	16–24
Regular insulin	0.5–1	2.5–5	8–12

Since the late 1980s, most human insulin was produced biosynthetically, thanks to the development of recombinant DNA technology [17]. Either yeast, *Saccharomyces cerevisiae*, or bacteria, *Escherichia coli*, were used to manufacture the recombinant *insulin*.

Insulin therapy was initially available only as a short-acting form, which required several daily subcutaneous injections, which involved, in order to make insulin therapy easier for patients, efforts related to exploring other administration routes (inhalation, enteral administration) or to developing new formulations to delay subcutaneous absorption (the use of oil suspensions, gum Arabic solutions, lecithin emulsions). In this last sense, the insulin formulates with protamine, a polypeptide arginine-rich (MW of ~4,300 Da) that is able to establish cationic bonds, in the presence of a small amount of zinc and phenol derivatives and at neutral pH, known as Neutral Protamine Hagedorn (NPH) insulin or Isophane Insulin, is still used as a "basal" insulin administering once- or twice-daily, either by itself or in conjunction with a soluble insulin, depending on the situation (Table 1) [23, 24].

Information about the structural aspect of insulin allowed designing and preparing synthetically modified insulin via the DNA recombinant technique, with the desired pharmacodynamic and pharmacokinetic profiles, i.e., short acting and long acting. It could be mentioned the following short-acting modified insulins: *insulin lispro*, *insulin aspart*, and *insulin glulisine* (Fig. 1). In reference to long-acting modified insulins, among others, the following could be brought up: *insulin glargine*, *insulin detemir*, and *insulin degludec* (Fig. 1).

The structure of *insulin lispro* was inspired by the ability of insulin-like growth factor-I (IGF-I) not to self-associate despite the high homology with insulin. In IGF-I, the Pro-Lys sequence that occurs in insulin at locations 28 and 29 of the B chain is reversed (i.e., Lys-Pro), and this change was suggested to be responsible for the prevention of self-assembling, and the modification of the Pro-Lys motif in insulin would produce insulin analogs that are likewise incapable of self-assembling. With this rationale in mind, a large number of synthetic insulins were prepared where position 29 of the B chain was occupied by a Pro and natural and unnatural amino acids were located at position 28, resulting, after a structure-association relationship study, in better *insulin lispro*. *Insulin lispro*, where Lys was located at the 28th position, like IGF-I, has both rapid and short action (Table 1) due to its weakened propensity to self-associate into dimers [25]. *Insulin lispro* exists in its formulation as a hexamer, stabilized with Zn^{2+} and phenols; however, upon injection into the subcutis, the hexamer dissociates to monomers, almost instantly yielding a plasma absorption profile that is identical to that of monomeric insulin.

By substituting the negatively charged Asp for Pro at position 28 of the B chain in insulin, *insulin aspart* is produced. This Asp substitution has a minimal impact on the overall conformation and a few local conformational alterations in the B chain's C-terminus. It has been proposed that this Asp substitution inhibits the formation of dimers by eliminating crucial contacts between the Pro at the 28th position of the B chain and Gly at the 23rd position of the B chain at the monomer-monomer interface [26]. The charge repulsion contributes, similarly to that seen in *insulin lispro*, a decrease in the degree of self-association after subcutaneous injection, having similar potency with human or animal insulins [27].

Insulin glulisine, manufactured by means of *Escherichia coli* recombinant DNA technology, differs from human insulin in that the neutral, polar, and hydrophilic Asn in position 3 of the B chain is replaced by the basic, polar, and hydrophilic Lys, and the basic Lys in position 29 of the B chain is replaced by the acidic, polar, and hydrophilic amino acid Glu. The replacement of Asn by Lys affects the monomer-oligomer equilibria due to the stereoelectronic repulsions between monomers. On the other hand, the substitution of Lys by Glu promotes a decrease in the isoelectric point ($pI_{insuin\ glulisine} = 5.1$, $pI_{insulin} = 5.5$), which leads to an improvement in solubility at physiological pH. Similarly, the significantly lower *pI* of 5.1 found for *insulin aspart* can be attributed to the substitution of the neutral, nonpolar, and hydrophobic Pro for the acidic Asp. Furthermore, the *pI* of *insulin lispro* is the same as that of human insulin ($pI = 5.5$) due to the inversion of amino acids at positions 28 and 29 of the B chain [28]. ***Insulin glulisine*** is equipotent and acts and is absorbed twice as rapidly as regular human insulin (Table 1).

The first once-daily and long-acting insulin, *insulin glargine*, has Gly at the 21st position of the A chain, instead of Asn, and two extra Arg at the C-terminal region of the B chain in the human insulin. As a result of these structural modifications, *insulin glargine* is soluble at a solution of pH 4; consequently, it does not require, unlike NPH, resuspension prior to injection, but when administered subcutaneously, it forms a precipitate from which insulin is very slowly released over the following 24 h (Table 1). Furthermore, *insulin glargine* is a prodrug because following subcutaneous injection, it is rapidly metabolized to i) insulin with Gly at the 21st position of the A chain (A21-Gly-insulin), which is responsible for approximately 90% of the available daily plasma insulin; ii) A21-Gly-des-B30-Thr-insulin; iii) and A21-Gly-B31-Arg-insulin.

Insulin detemir and *insulin degludec* were designed, in addition to their natural strong self-associations, as *insulins* able to bind to albumin, generating a circulating depot from which they are gradually released. Both insulin types do not have Thr at the 30th position of the B chain, and an amide is formed between the ε-amino-moiety of Lys at the 29th position of the B chain and tetradecanoic acid (myristic acid) or hexadecanedioic acid, respectively [29]. Other strategies, for example, PEGylation of insulin, have been studied but still without adequate clinical applications. Some pharmacokinetic parameters of *insulin detemir* and *insulin degludec* are summarized in Table 1.

1.2 Erythropoietin

Human erythropoietin is an acidic glycoprotein cytokine produced by the kidneys, liver, and salivary glands. Its production is controlled by hypoxia-inducible transcription factors in response to cellular hypoxia. Erythropoietin plays an important role in the survival, proliferation, and maturation of erythroid cells. Its MW is between 34,000 and 36,000 Da, according to the glycosylated isoform, and its 165 amino acid residues in a single chain forms four antiparallel α-helices, two β-sheets, and two intrachain disulfide bridges (between Cys7 and Cys161 and between Cys29 and Cys33, Fig. 2). The carbohydrate portion (40% of the MW) comprises three *N*-glycans (at Asn24, Asn38, and Asn83) and one *O*-glycan (at Ser126) [31]. Among the several roles played by glycans-web are i) signal transduction to target cells, ii) modulation of its receptor-binding affinity, and iii) protection against proteases [32]. In this last aspect, *N*-acetylneuraminic acid (sialic

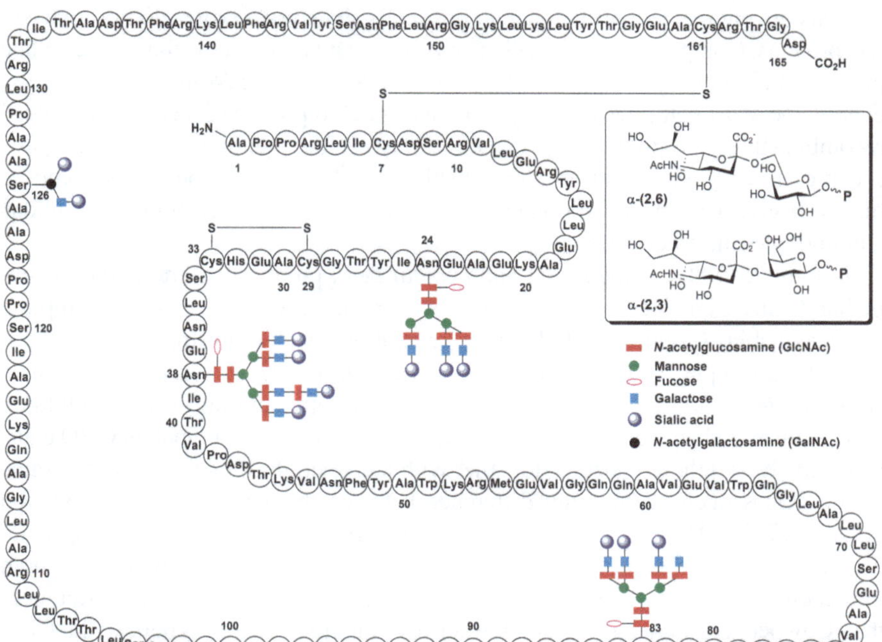

Fig. 2 Primary structure of a glycosylated isoform of erythropoietin. Inset: structures and linkages of some common natural derivatives of sialic acid. It shows sialic acid linked to galactose by either an α-(2,6) or α-(2,3) linkage

acid, Fig. 2), at the glycan terminus, is a relevant component of erythropoietin sugars conferring metabolic in vivo stability. For example, erythropoietin isoforms with a greater number of sialic acids result in decreased receptor-binding affinity but enhanced half-lives and, consequently, biological activities, or nonglycosylated erythropoietin is rapidly cleared from the circulation, while the complete-glycosylated erythropoietin has a half-life of 7–8 h [33]. Glycan synthesis in mammalian cells starts in the cytosol with the addition of unique monosaccharides to dolichol, and then the glycans are transferred to the growing polypeptide in the endoplasmic reticulum. Endogenous human erythropoietins, as well as the recombinants and their analogs, exist as several glycosylated isoforms.

Different erythropoiesis-stimulating proteins have been developed, by recombinant DNA technology, and used in clinics. The term *epoetin* is the nonproprietary drug name for eukaryotic-cell-derived recombinant human erythropoietin, being an amino acidic sequence similar to the native one. When amino acid sequences are different a random prefix is incorporated in the names, for example, in *darbepoetin*; on the other hand, Greek letters, as a second word, are added to indicate that the product varies in glycosylation patterns, for example, *epoetin beta*. It could be mentioned, among other, the following erythropoiesis-stimulating proteins, from recombinant or recombinant/further structural modifications origins: *epoetin alfa*, *epoetin beta*, *epoetin delta*, *epoetin omega*, *epoetin zeta*, *darbepoetin alfa*, and *methoxypolyethylene glycol-epoetin beta*. The majority of therapeutic human glycoproteins are engineered in mammalian host cells since prokaryotes do not glycosylate proteins. Compared to mammalian cells, yeast and filamentous fungi glycosylate proteins differently, adding, for instance, high mannose-type *N*-glycans [34]. However, some yeast cell lines have been "humanized" to produce functional sialylated recombinant erythropoietins by the knockout of four genes to eliminate yeast-specific glycosylation and the introduction of 14 heterologous genes coding human *N*-glycan synthesis, allowing the use of chemically defined culture conditions and high protein production rates.

In the late 1980s, *epoetin alfa* and *epoetin beta*, produced in transfected mammalian Chinese hamster ovary (CHO) cells, were launched for human consumption to treat anemia. Around the 2000s, next-generation epoetin biosimilars were developed, for example, *epoetin delta*, *epoetin omega*, and *epoetin zeta*, produced in *Homo sapiens* fibrosarcoma cells (line R223, a derivative of the HT-1080 cell line), baby hamster kidney cells (BHK, from Syrian hamster), and mammalian CHO cells, respectively. All these epoetins, together with native erythropoietins, have the same amino acid sequence, the main difference between them being the glycosylation pattern [35, 36]. Divergence of therapeutic properties is the result from the production process, is relevant for biosimilars, and the recombinant bioreactor both influencing the glycosylation pattern. For instance, Table 2 lists the key differences that Cowper and colleagues found when examining epoetins produced in CHO cells [37].

When compared to epoetins, *darbepoetin alfa*, produced in mammalian CHO cells, has a different structure in terms of both its amino acid composition and the amount of carbohydrates. It is a hyperglycosylated analog of endogenous

Table 2 Differences in the glycan patterns of endogenous erythropoietin (*Eryt. ref.*) and different recombinant forms of this cytokine [37]

	N-glycans[a]								Ser-126[b]		
	S1	S2	S3	S4	S5	OA	LNA	Z	Ag	Ms	Bs
Epoetin alfa 1[c]	Sv	Sv	Dv (27)	Dv (64)	Dv (1.3)	Sv	Sv	359	Sv	Sv	Dv (20)
Epoetin alfa 2[c]	Sv	Sv	Dv (20)	Dv (69)	Dv (3.0)	Sv	Sv	368	Dv (4)	Dv (14)	Dv (52)
Epoetin beta	Hv (0.9)	Sv	Hv (31)	Lv (59)	Sv	Lv (3)	Sv	351	Lv (1.4)	Sv	Sv
Epoetin zeta	Sv	Hv (10)	Sv	Sv	Sv	Dv (4)	Hv (46)	353	Hv (10)	Hv (44)	Lv (18)
Darb. alfa[d]	Lv (0.0)	Lv (2)	Lv (18)	Hv (76)	Hv (5)	Hv (47)	Lv (13)	383	Ns	Ns	Ns
Eryt. ref.	Sv	Sv	Sv	Sv	Sv	Sv	Sv	359	Sv	Sv	Sv

[a] S1: monosialylated *N*-glycans; S2: bisialylated *N*-glycans; S3: trisialylated *N*-glycans; S4: tetrasialylated *N*-glycans; S5: pentasialylated *N*-glycans; OA: *O*-acetylated sialic acid; LNA: *N*-acetyllactosamine extended antennae; Z: hypothetical *N*-glycan charge [38], defined as the sum of the products of the respective areas in the asialo, monosialo, disialo, trisialo, tetrasialo, and pentasialo regions, each multiplied by the corresponding charge; Sv: similar values between the analyzed groups; Dv: different values between the analyzed groups; Hv: higher value; Lv: lower value; the values in parenthesis indicate the relative peak area in liquid chromatography (some values are expressed approximately)

[b] Ag: aglycosylated; Ms: monosialylated (GalNAc-Gal + Neu5Ac); Bs: bisialylated (GalNAc-(Neu5Ac)-Gal-Neu5Ac); Ns: not studied by the authors

[c] Different commercial origins of biosimilars

[d] *Darb. alfa*: *darbepoetin alfa*

erythropoietin with an MW of 37,100 Da containing two additional *N*-glycans, together with an exchange of five amino acids, i.e., Asn in 30 instead of Ala, Thr in 32 instead of His, Val in 87 instead of Pro, Asn in 88 instead of Trp, and Thr in 90 instead of Pro, as a result of site-directed mutagenesis [33]. A threefold rise in plasma half-life, a decrease in receptor-binding affinity, and an increase in in vivo activity were the results of the sialic acid residues increasing to 22 and the total carbohydrate increasing from 40% to 52% (Table 2) [39].

Methoxypolyethylene glycol-epoetin beta is a third-generation erythropoiesis-stimulating protein chemically synthesized from *epoetin beta*. It is the result of the covalent combination of a long linear methoxypolyethyleneglycolylbutanoic acid of 30,000 Da and *epoetin beta* via an amide bond established by the ε-amino group of one Lys (predominantly Lys in position 52 or Lys in position 45), yielding a product with an MW of approximately 60,000 Da [40]. This structural modification allows *methoxypolyethylene glycol-epoetin beta* to have a long half-life (about 130–140 h on intravenous injection) with different interactions with the erythropoietin receptor compared to other erythropoiesis-stimulated proteins. *Methoxypolyethylene glycol-epoetin beta* is known as a continuous erythropoietin receptor activator due to its constant slower association and higher dissociation to erythropoietin receptors compared with *epoetin beta*. This lower affinity to receptor-binding sites, 50–100 times less, results in the continuous stimulation of erythropoiesis [41].

1.3 Interferons

Interferons are a family of intercellular signaling glycoproteins that are produced by different kinds of eukaryotic cells in response to viruses and other antigenic stimuli [42]. Since they are not a single protein, a specific nomenclature is used that allows distinguishing between each one [43]: i) on the basis of antigenic specificities, interferons are classified as α, β, γ, λ, ω, and τ [44]; ii) on their sequence homology, properties, and ability to bind to cell receptors, they are classified as types I, II, and III. Interferons have MWs that vary between 5,000 and 37,000 Da, amino acid sequences that range from 143 to 166 peptides, and a variety of stabilities at different pHs. The main described biological properties of interferons have been as antiviral and antitumoral being applied in clinical practice enhancing immune responses against infections and to treat autoimmune disorders and cancer [42]. In these cases, the modes of action depend on the control of the gene expression. Some interferons used as biotherapeutics are summarized in Table 3 [45].

Upon administration, most interferons are poorly absorbed in the gastrointestinal tract, being also unstable due to acidic pH and the presence of proteases in these sites. As a result, the majority of interferon formulations are based on parenteral solutions administered via subcutaneous injection. However, after a single subcutaneous administration, serum interferon concentrations drop below the limit of detection in less than 24 h. Many strategies have been developed to improve interferons' pharmacokinetic and clinical utility, for example, encapsulation of them in drug delivery systems or the use of stabilizers/excipients. The chemical coupling of polymers (for example, in *PEGylated interferon α-2a*, *PEGylated interferon α-2b*, or *PEGylated interferon β-1a*, Table 3) and the genetically engineered coupling of fusion proteins (for example, in the *fusion protein of albumin and interferon α-2b*, Table 3) will be commented here.

Different strategies were used in the covalent union of long linear polyethylene glycol moiety to interferons (Fig. 3) [46]. For example [47], one bis-methoxypolyethylene glycol chain, from PEG2-NHS (Fig. 3) with an MW of approximately 40,000 Da, was incorporated into *interferon α-2a*, obtained via recombinant *E. coli* technology and with an MW of approximately 20,000 Da, to produce *PEGylated interferon α-2a* (Table 3) with an MW of approximately 60,000 Da. The process yields a stable amide bond to the interferon via the accessible ε-amine of lysine residues (Lys31, Lys121, Lys131, or Lys134), detecting four major positional isomers and two in minor percentage (by reaction at Lys70 and Lys83). These structural modifications, of any of the six positional isomers, do not disrupt the three-dimensional structure of interferons. The carbamate moieties, in the terminal Lys-amino of the PEG2-NHS reactant, can be site-specific hydrolysis under basic pH. This modification i) modifies renal elimination, due to the cutoff for renal

Table 3 Some biotherapeutic interferons

Interferon	Clinical indication	Production (observations)
Interferon α-2a	Leukemias, sarcoma, lymphomas, hepatitis B and C, melanoma	Recombinant production using an *E. coli*-based system
PEGylated interferon α-2a	Leukemia, hepatitis B, melanoma	Recombinant production using an *E. coli*-based system and further chemical modification
Interferon α-2b	Hepatitis B and C	Recombinant production using an *E. coli*-based system
PEGylated interferon α-2b	Hepatitis C, melanoma	Recombinant production using an *E. coli*-based system and further chemical modification
Fusion protein of albumin and interferon α-2b	Hepatitis C	Fusion genetically engineered using an *E. coli*-based system
Interferon α-2c	Chronic viral hepatitis, HIV infection	Recombinant production using an *E. coli*-based system
Interferon α-n3	Condyloma acuminate	Virus-exposed human leukocytes
Interferon α-n1	Hepatitis B and C, leukemia, HPV infection	Virus-exposed human lymphoblastoid cells
Interferon α-con-1	Hepatitis C	Recombinant production using an *E. coli*-based system
Interferon β-1a	Multiple sclerosis	Recombinant production using CHO cells
PEGylated interferon β-1a	Multiple sclerosis	Recombinant production using CHO cells and further chemical modification
Interferon β-1b	Multiple sclerosis	Recombinant production using an *E. coli*-based system (differs from human interferon in that Cys17 is replaced by Ser)
Interferon γ-1b	Chronic granulomatous disease, osteopetrosis	Recombinant production using an *E. coli*-based system

filtration is a MW of 40,000–50,000 Da, decreasing the renal clearance 100-fold respect to unmodified *interferon α-2a*, giving a half-live of elimination of 77 h versus 9 h for the native interferon; ii) slightly increases stability at higher temperatures; iii) reduces the tendency to aggregation. Additionally, the subcutaneous administration of *PEGylated interferon α-2a* results in prolonged absorption and a limited volume of distribution.

This kind of postproduction modification was also used with *interferon α-2b* reacting to 12,000 Da of methoxypolyethyleneglycolyl succinimidyl carbonate (mPEG-SC, Fig. 3), producing a carbamate-linker product known as *PEGylated interferon α-2b* (Table 3). The commercial *PEGylated interferon α-2b* is a mixture of 14 positional mono-PEGylated interferon isomers. The PEGylation product depends on the reaction pH [48]: (i) at low pH, i.e., 4.5, the reaction occurs at His7 probably as result of interferon oligomer disruption, which exposes this amino acid; (ii) at 6.5 pH, which is the pH at which the commercial product is synthesized, the reaction occurs at His34 and Cys1, in 48% and 13%, respectively, and with lower amounts of

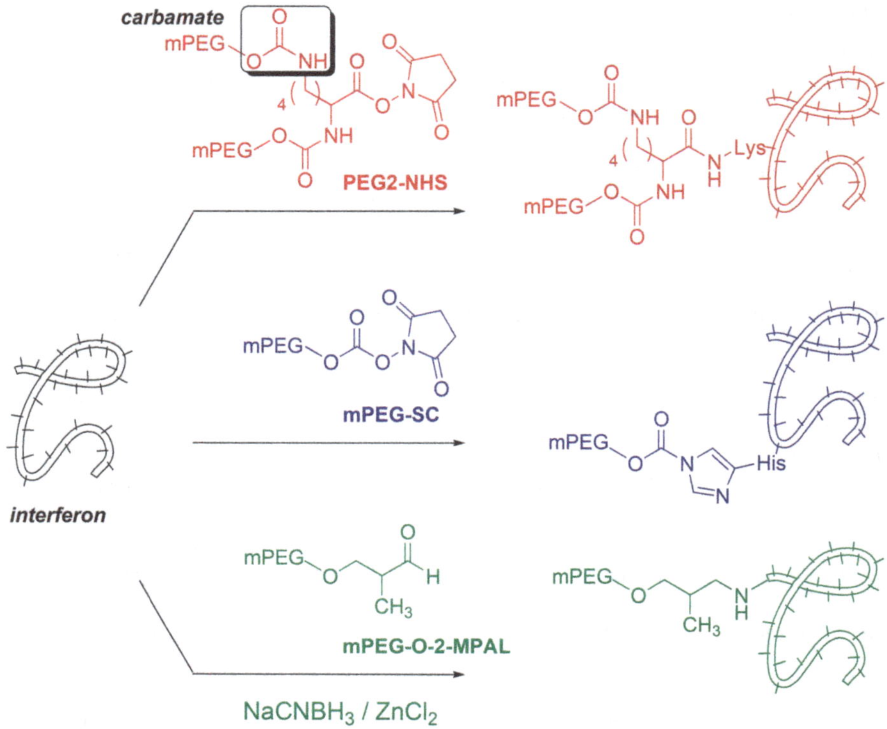

Fig. 3 PEGylation reactions used for the synthesis of interferon derivatives

positional isomers on various Lys residues (Lys31, Lys49, Lys83, Lys112, Lys121, Lys131, Lys 133, Lys 134, and Lys164), Ser163, Tyr129, and His7; and (iii) at 8 pH, the reaction is random and occurs on various Lys residues. These products are the result of the amount of the nucleophilic deprotonated form of amino acids ($pK_{a,His} = 6.0$, $pK_{a,Cys} = 8.2$, $pK_{a,Lys} = 10.5$). Although the His34 reaction product is hydrolytically unstable in basic conditions, *PEGylated interferon α-2b* half-life is sufficiently long in vivo to support a once-weekly dosage regimen.

CHO-recombinant *interferon β-1a* is modified by incorporating a methoxypolyethyleneglycolyl-moiety via reductive amination by reacting with 3-(methoxypolyethyleneglycolyl)-2-methylpropionaldehyde (mPEG-O-2-MPAL, MW of 20,000 Da, Fig. 3) in the presence of $NaCNBH_3$, at 5 pH, to yield *PEGylated interferon β-1a* (MW of 44,000 Da, Table 3) [49]. Because the amino group of the terminal amino acid is substantially less protonated than the ε-amino group of Lys residues, the reaction occurs selectively at this amino acid at 5 pH [50]. *PEGylated interferon β-1a* is prescribed once every 14 days with a subcutaneous dosage, whereas *interferon β-1a* is administered once weekly by intramuscular injection.

Despite the positive effects of polymer coupling, possible immunogenicity problems, together with the low degradability of PEG interferons, have promoted investigation into other serum half-life extenders, such as the genetic fusion of interferons to other proteins. The fusion between a recombinant *interferon α-2b* and human albumin yields the *fusion protein of albumin and interferon α-2b* (Table 3). The *fusion protein of albumin and interferon α-2b*, produced in a genetically modified yeast strain, has an MW of approximately 85,700 Da [51]. This product, from the genetic fusion, increases half-life respect to original interferon, allows dosing at intervals of 2 to 4 weeks, and maintains its stabilization.

1.4 Coagulation Factors VIII and IX

Hemophilia is a bleeding disorder that is usually inherited and causes problems with blood coagulation spontaneously with or after surgeries or injuries. Many proteins, coagulation factors, are found in the blood aiding in halting bleeding through a process known as the coagulation cascade or secondary hemostasis. The coagulation cascade is a sequence of steps taken in response to tissue damage-induced bleeding, where each step initiates the subsequent ones and results in the formation of a blood

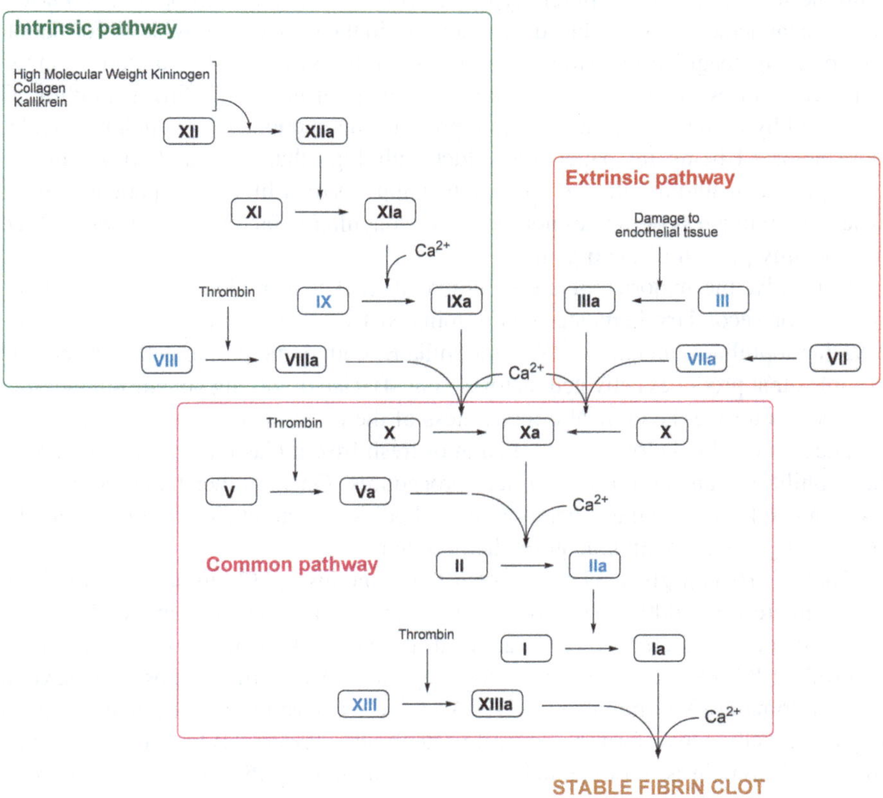

Fig. 4 The steps in the coagulation cascade. In blue biotherapeutics

Table 4 Some biotherapeutic *factor VIII* and *factor IX* [62]

Recombinant protein (name)	Modification/technology	Plasma half-life (h)	Time longer half-life
Factor VIII (rurioctocog alfa)	PEGylated (2 × 20,000 Da)	14.3	1.3–1.5
Factor VIII (turoctocog alfa pegol)	GlycoPEGylated (40,000 Da)	18.4	1.6
Factor VIII (damoctocog alfa pegol)	PEGylated (60,000 Da)	19	1.6
Factor VIII (efmoroctocog alfa)	Fc fusion	19	1.5–1.7
Factor IX (nonacog beta)	GlycoPEGylated (40,000 Da)	93	5.8
Factor IX (Albutrepenanocog alfa)	Albumin fusion	101	5.3
Factor IX (Eftrenonacog alfa)	Fc fusion	82	4.3

coagulum (Fig. 4). The reduction or lack of some coagulation factors is associated with the severity and the different types of hemophilia [7]: (i) hemophilia A, known as classical hemophilia, produced by mutations in the gene that provides information to producing coagulation *factor VIII*; (ii) hemophilia B produced by mutations in the gene that carries instructions for making coagulation *factor IX*; (iii) hemophilia C produced by mutations in the gene that provides instructions to producing coagulation factor XI being for patients a milder pathology than A and B; (iv) acquired hemophilia, an autoimmune hemophilia that appears over lifetime of patients arising when the immune system does not recognize coagulation factors, mainly *factor VIII*, as a healthy protein of the organism.

Currently, the majority of the commercial market to replenish the lack of the natural for secondary hemostasis is dominated by biotherapeutics contains *factor VIII*, hemophilia A, and *factor IX*, hemophilia B, both participating in the proteolytic complex that produces activated factor X (Fig. 4), also involving genetic engineering efforts to improve the clinical effectiveness of them. Usually, drugs that inhibit the breakdown of blood clots or the infusion of fresh frozen plasma are used to manage hemophilia C patients during surgical procedures. On the other hand, controlling bleeding and lowering the generation of antibodies that target clotting factors are the two main goals of treating acquired hemophilia.

Factor VIII is a glycosylated protein that contains 2,332 amino acids (MW of approximately 330,000 Da) composed of three major regions, *N*-terminal 90,000 Da heavy chain connected by a large central region, known as B-domain, to a C-terminal 80,000 Da light chain, both regions are in turn composed of several smaller domains (*N*-terminal-A1-A2-B-A3-C1-C2-*C*-terminal). It is converted into a physiologically active form by several proteolytic cleavages. It has been suggested that the B-domain is not essential for hemostatic activity [52, 53]. Some commercially available *factor VIII* are derived from DNA recombinant technology constructs that eliminate the B domain (Table 4) [54].

Factor IX is a 55,000 Da glycoprotein that circulates as a modular domain structure (*N*-terminal-*gamma*-carboxyglutamic acid (Gla)-EGF1-EGF2-activation peptide protease domain) and numerous posttranslational modifications [55].

In the preparation of *factor VIII* and *factor IX* recombinants (Table 4) with adequate pharmacodynamics/pharmacokinetic behaviors, the expression systems must take into account the fidelity of posttranslational proteolytic processing, γ-glutamyl carboxylation, sulfation, and glycosylation [56]. As previously mentioned in this chapter, for other examples of recombinant proteins, in the case of *factor VIII* and *factor IX*, the following structural postproduction modifications have been studied in order to improve their pharmacodynamics/pharmacokinetic behaviors: (i) conjugation with hydrophilic polymer polyethylene glycol to slow down renal elimination [57]; (ii) fusion with albumin [58]; (iii) fusion with the fragment crystallizable (Fc) region of IgG1; this fragment allows protein binding to the endogenous neonatal Fc receptor, prolonging the half-life of the factor through the natural IgG recycling pathway and delaying the lysosomal degradation of Fc-containing proteins by returning them to the circulation [59]; and (iv) fusion with a hydrophilic unstructured polypeptide, such as XTEN, with low immunogenicity [60]. In general, for the first three technologies, a low increase in half-life between 1.3 and 1.7 times has been observed, compared to native factors, probably due to the chaperone effect of the von Willebrand factor. However, with fusion to XTEN, an increase of three to four, for modified *factor VIII*, and three to five, for modified *factor IX*, has been observed [61]. Some approved *factor VIII* and *factor IX* products are shown in Table 4 [62].

Hemophilias nonfactor biotherapies have also been explored, for example, the use of monoclonal antibodies that mimic cofactors' functions [63] and gene therapy [64]. These types of biotherapeutics and biotherapies will be discussed in Chapters "Monoclonal Antibodies as Biotherapeutics" and "Nucleic Acids Production", respectively.

1.5 Arginase

For more than a century, enzymes have been employed as diagnostic tests, as reagents, and as therapeutic agents. They are especially suited to be biotherapeutics because (i) of their high selectivity, catalyzing a single kind of reaction in physiologic conditions and producing the effect or intended product, and (ii) they can be obtained by a variety of biotechnological techniques using generally straightforward, well-standardized procedures and are formulated in blood-compatible preparations [65–67]. Conversely, actual enzyme applications, such as biotherapeutics, are limited to soluble ones. The use of enzymes as biotherapeutics has been described, among others, to [68, 69] (i) supplement digestive action (for example, pepsin, trypsin, and pancreatin), (ii) combat intestinal nematodes (for example, papain and ficain), (iii) produce fibrinolysis (for example, trypsin, papain, thrombin, plasminogen, and plasmin), (iv) act as anti-inflammatory (for example, chymotrypsin),

(v) increase tissue permeability (for example, hyaluronidase), (vi) combat cardiovascular disorders (for example, agalsidase alfa and plasmin), and (vii) combat cancer. The importance of arginase as a biotherapeutic to treat cancer will be briefly discussed below.

Certain faulty cellular processes, which disrupt normal function and fail to control the development of cells, are the cause of cancer. Targeting cellular metabolism is one explicit strategy for removing cancer cells as they require more energy and nutrients, for example, amino acids, than normal healthy cells for their rapid growth [70]. Depriving cancer cells of amino acid supply is a well-established therapeutic approach that targets the metabolic activity of cells. To this end, biotherapeutics, such as the enzymes asparaginase, methionase, glutaminase, and arginase, have been studied extensively. A considerable correlation has been found between tumor proliferation and Arg levels due to certain facts, such as the emergence of tumors in animals fed a diet rich in Arg and the absence in cancer cells of certain enzymes primordial in the replenishment of the normal pool of Arg, i.e., argininosuccinate synthetase and arginiosuccinate lyase [71, 72].

Some studies have indicated that the enzyme arginase inhibits cancer cell growth without affecting the growth of healthy ones [73]. Arginase is an Mn^{2+}-dependent hydrolase that catalyzes the transformation of Arg to Orn and urea. It is expressed from simple systems, like the mitochondrion form in bacteria, to complex ones, such as arginase 1 in liver cytosol and arginase 2 in kidney mitochondrion in mammals [74]. Arginase 1 and arginase 2 (MW of approximately 105,000 Da), which are closely homologous with identical residues in the Mn-binding region, possess 322 and 354 amino acids, respectively. They consist of three identical subunits with the active site at the bottom of the deep cleft, with two Mn^{2+} ions at the bottom of this cleft. The overall fold of each subunit belongs to the α/β family and consists of a parallel eight-stranded β-sheet flanked on both sides by numerous α-helices. While arginase 1 is involved in removing ammonia produced by amino acid catabolism, arginase 2 is involved in the biosynthesis of polyamines and proline [75]. Clinical studies have indicated that human arginase reacts well at physiological pH without adverse effects in the treatment of solid tumors. The results of different studies established that arginase could be a potential candidate for treating head and neck cancer cells, melanomas, and hepatocellular carcinoma. Additionally, the treatment of hyperarginemia, an autosomal recessive disorder resulting from the lack of arginase 1, could also be managed with human arginase.

The shortcomings observed with human arginase, for example, its short half-life, have been overcome through structural modifications of the recombinant enzyme. PEGylation and protein fusion have been the widely used techniques in successful changes to improve arginase biological behavior, as described in the preceding examples in this chapter [76–79]. This section wants to highlight how the replacement of Mn^{2+} ions by other dications modified the biotherapeutic properties. *HuArgI (Co)-PEG5000* is a PEGylated recombinant enzyme where Mn^{2+} is substituted by Co^{2+}, which also stabilizes the tertiary and quaternary structure but yields tenfold higher catalytic activity in vitro [80]. This biotherapeutic is demonstrated to be cytotoxic to human acute myeloid leukemia, to human glioblastoma, and to ovarian cancer cells [73, 81, 82]. Other examples are *Co-HAI-PEG20L* and *Ni-HAI-PEG20L*, where mono-PEGylated recombinant human arginase 1 was enriched with Co^{2+} and Ni^{2+}, respectively [83]. They demonstrated [84] i) higher catalytic efficiency than Mn^{2+}-recombinant arginase and that the k_{cat}/K_M values of *Co-HAI-PEG20L* and *Ni-HAI-PEG20L* were higher than Mn^{2+}-recombinant arginase by ~8.7-folds and ~ 5.2-folds, respectively; ii) an ability to inhibit the in vitro growth of eight different cancer cells; and iii) that after the intraperitoneal dosage of *Co-HAI-PEG20L* or *Ni-HAI-PEG20L* in mice, the level of Arg is maintained below the detection limit for over 120 h after administration without significant toxicity.

2 Concluding Remarks

This chapter wants to show how different interventions, chemical and biological, can generate biotherapeutic proteins with improved therapeutic performances.

Compliance with Ethical Standards This chapter is a review and no humans or animals were involved in the preparation of the chapter.

References

1. IUPAC-IUB Joint Commission on Biochemical Nomenclature (JCBN) (1983) Nomenclature and symbolism of amino acids and peptides. Eur J Biochem 138:9–37
2. Hamley IW (2020) Introduction to peptide science. Wiley, Chichester
3. International Union of Pure and Applied Chemistry (2006) Proteins. In: IUPAC compendium of chemical terminology, 3rd edn. Online version 3.0.1, 2019. https://doi.org/10.1351/goldbook.P04898. Accessed 16 Mar 2024
4. Ryan CA, Pearce G, Scheer J et al (2002) Polypeptide hormones. Plant Cell 14:s251–s264
5. Rosenfeld L (2002) Insulin: discovery and controversy. Clin Chem 48:2270–2288
6. Zhang Y, Hung-Chieh Chou D (2023) From natural insulin to designed analogs: a chemical biology exploration. ChemBioChem 24:e202300470
7. Sauna ZE, Kimchi-Sarfaty C (eds) (2017) Protein therapeutics. Topics in medicinal chemistry, vol 21. Springer, Heidelberg
8. Khodabakhsh F, Salimian M, Hedayati MH et al (2021) Challenges and advancements in the pharmacokinetic enhancement of therapeutic proteins. Prep Biochem Biotechnol 51:519–529
9. Dimitriadis G, Mitrou P, Lambadiari V et al (2011) Insulin effects in muscle and adipose tissue. Diabetes Res Clin Pract 93(Suppl 1):S52–S59
10. The Nobel Prize in Chemistry 1958 was awarded to Frederick Sanger "for his work on the structure of proteins, especially that of insulin". The Nobel Prize in Chemistry 1958. NobelPrize.org. Nobel Prize Outreach AB 2024. https://www.nobelprize.org/prizes/chemistry/1958/summary/. Accessed 18 Mar 2024
11. Blundell TL, Dodson GG, Hodgkin DC et al (1972) Insulin: the structure in the crystal and its reflection in chemistry and biology. Adv Protein Chem 26:279–402
12. The Peking Insulin Structure Research Group (1974) Studies on the insulin crystal structure: the molecule at 1.8 Å resolution. Sci Sin 17:752–778

13. Fu Z, Gilbert ER, Liu D (2013) Regulation of insulin synthesis and secretion and pancreatic beta-cell dysfunction in diabetes. Curr Diabetes Rev 9:25–53
14. Dunn MF (2005) Zinc–ligand interactions modulate assembly and stability of the insulin hexamer – a review. Biometals 18:295–303
15. Permutt A, Chirgwin J, Giddings S et al (1981) Insulin biosynthesis and diabetes mellitus. Clin Biochem 14:230–236
16. Collip JB (1923) Glucokinin. A new hormone present in plant tissue. J Biol Chem 56:513–543
17. Venâncio TM, Oliveira AEA, Silva LB et al (2003) A protein with amino acid sequence homology to bovine insulin is present in the legume *Vigna unguiculata* (cowpea). Braz J Med Biol Res 36:1167–1173
18. Banting FG, Best CH, Collip JB et al (1922) Pancreatic extracts in the treatment of diabetes mellitus. Preliminary report. Can Med Assoc J 12:141–146
19. McBride SA, Tilger CF, Sanford SP et al (2015) Comparison of human and bovine insulin amyloidogenesis under uniform shear. J Phys Chem B 119:10426–10433
20. Redwan ERM (2009) Animal-derived pharmaceutical proteins. J Immunoassay Immunochem 30:262–290
21. Richter B, Neises G (2005) "Human" insulin versus animal insulin in people with diabetes mellitus. Cochrane Database Syst Rev (1) Art. No. CD003816
22. Markussen J, Damgaard U, Jorgensen KH et al (1983) Human monocomponent insulin. Chemistry and characteristics. Acta Med Scand Suppl 671:99–105
23. Krayenbuhl C, Rosenberg T (1946) Crystalline protamine insulin. Rep Steno Mem Hosp Nord Insulinlab 1:60–73
24. Hagedorn HC (1937) Protamine insulinate: (section of therapeutics and pharmacology). Proc R Soc Med 30:805–814
25. Di Marchi RD, Chance RE, Long HB et al (1994) Preparation of an insulin with improved pharmacokinetics relative to human insulin through consideration of structural homology with insulin-like growth factor I. Horm Res 41:S93–S96
26. Brange J, Vølund A (1999) Insulin analogs with improved pharmacokinetic profile. Adv Drug Deliv Rev 35:307–335
27. Vølund A, Brange J, Drejer K et al (1991) *In vitro* and *in vivo* potency of insulin analogues designed for clinical use. Diabet Med 8:839–847
28. Becker RHA (2007) Insulin glulisine complementing basal insulins: a review of structure and activity. Diabetes Technol Ther 9:109–121
29. Jonassen I, Havelund S, Hoeg-Jensen T et al (2012) Design of the novel protraction mechanism of insulin degludec, an ultra-long-acting basal insulin. Pharm Res 29:2104–2114
30. Hilgenfeld R, Seipke G, Berchtold H et al (2014) The evolution of insulin glargine and its continuing contribution to diabetes care. Drugs 74:911–927
31. Jelkmann W (2013) Physiology and pharmacology of erythropoietin. Transfus Med Hemother 40:302–309
32. Lai PH, Everett R, Wang FF et al (1986) Structural characterization of human erythropoietin. J Biol Chem 261:3116–3121
33. Elliott S, Lorenzini T, Asher S et al (2003) Enhancement of therapeutic protein *in vivo* activities through glycoengineering. Nat Biotechnol 21:414–421
34. Sodoyer R (2004) Expression systems for the production of recombinant pharmaceuticals. BioDrugs 18:51–62
35. Locatelli F, Covic A, Eckardt KU et al (2009) Anaemia management in patients with chronic kidney disease: a position statement by the Anaemia Working Group of European Renal Best Practice (ERBP). Nephrol Dial Transplant 24:348–354
36. Skibeli V, Nissen-Lie G, Torjesen P (2001) Sugar profiling proves that human serum erythropoietin differs from recombinant human erythropoietin. Blood 98:3626–3634
37. Cowper B, Lavén M, Hakkarainen B et al (2020) Glycan analysis of erythropoiesis-stimulating agents. J Pharm Biomed Anal 180:113031

38. Hermentin P, Witzel R, Kanzy EJ et al (1996) The hypothetical *N*-glycan charge: a number that characterises protein glycosylation. Glycobiology 6:217–230
39. Macdougall IC, Gray SJ, Elston O et al (1999) Pharmacokinetics of novel erythropoiesis stimulating protein compared with epoetin alfa in dialysis patients. J Am Soc Nephrol 10: 2392–2395
40. Macdougall IC, Eckardt KU (2006) Novel strategies for stimulating erythropoiesis and potential new treatments for anaemia. Lancet 368:947–953
41. Ohashi N, Sakao Y, Yasuda H et al (2012) Methoxy polyethylene glycol-epoetin beta for anemia with chronic kidney disease. Int J Nephrol Renov Dis 5:53–60
42. Borden EC, Sen GC, Uze G et al (2007) Interferons at age 50: past, current and future impact on biomedicine. Nat Rev Drug Discov 6:975–990
43. Interferon Nomenclature (1980) Nature 286:110
44. Liu L, Leaman DW, Bixby JA et al (1996) A type I ovine interferon with limited similarity to IFN-alpha, IFN-omega and IFN-tau: gene structure, biological properties and unusual species specificity. Biochim Biophys Acta 1294:55–62
45. Castro LS, Lobo GS, Pereira P et al (2021) Interferon-based biopharmaceuticals: overview on the production, purification, and formulation. Vaccine 9:328
46. Kozlowski A, Charles SA, Harris JM (2001) Development of pegylated interferons for the treatment of chronic hepatitis C. BioDrugs 15:419–429
47. Turecek PL, Bossard MJ, Schoetens F et al (2016) PEGylation of biopharmaceuticals: a review of chemistry and nonclinical safety information of approved drugs. J Pharm Sci 105:460–475
48. Wang YS, Youngster S, Grace M et al (2002) Structural and biological characterization of pegylated recombinant interferon alpha-2b and its therapeutic implications. Adv Drug Deliv Rev 54:547–570
49. Baker DP, Lin EY, Lin KC et al (2006) *N*-terminally PEGylated human interferon-â-1a with improved pharmacokinetic properties and *in vivo* efficacy in a melanoma angiogenesis model. Bioconjug Chem 17:179–188
50. Gronke RS, Jaquez OA (2017) Pegylation method. US Patent 0232111 A1, 17 Aug 2017
51. Osborn BL, Olsen HS, Nardelli B et al (2002) Pharmacokinetic and pharmacodynamic studies of a human serum albumin-interferon-alpha fusion protein in cynomolgus monkeys. J Pharmacol Exp Ther 303:540–548
52. Vehar GA, Keyt B, Eaton D et al (1984) Structure of human factor VIII. Nature 312:337–342
53. Wood W, Capon D, Simonsen C et al (1984) Expression of active human factor VIII from recombinant DNA clones. Nature 312:330–337
54. Sandberg H, Almstedt A, Brandt J et al (2001) Structural and functional characterization of B-domain deleted recombinant factor VIII. Semin Hematol 38:4–12
55. Bond M, Jankowski M, Patel H et al (1998) Biochemical characterization of recombinant factor IX. Semin Hematol 35:11–17
56. Kumar SR (2015) Industrial production of clotting factors: challenges of expression, and choice of host cells. Biotechnol J 10:995–1004
57. Ivens IA, Baumann A, McDonald TA et al (2013) PEGylated therapeutic proteins for haemophilia treatment: a review for haemophilia caregivers. Haemophilia 19:11–20
58. Schulte S (2013) Innovative coagulation factors: albumin fusion technology and recombinant single-chain factor VIII. Thromb Res 131:S2–S6
59. McCue J, Osborne D, Dumont J et al (2014) Validation of the manufacturing process used to produce long-acting recombinant factor IX Fc fusion protein. Haemophilia 20:e327–e335
60. Podust VN, Balan S, Sim BC et al (2016) Extension of *in vivo* half-life of biologically active molecules by XTEN protein polymers. J Control Release 240:52–66
61. Ay C, Napolitano M, Hassoun A et al (2024) Classification of recombinant factor VIII products and implications for clinical practice: a systematic literature review. Haemophilia:1–13. https://doi.org/10.1111/hae.15001
62. Mannucci PM (2023) Hemophilia treatment innovation: 50 years of progress and more to come. J Thromb Haemost 21:403–412

63. Voorberg J, Postmus T, Schols S (2022) Next generation FVIII mimetic bispecific antibody for hemophilia A. J Thromb Haemost 20:1301–1305
64. Milani M, Canepari C, Liu T et al (2022) Liver-directed lentiviral gene therapy corrects hemophilia A mice and achieves normal-range factor VIII activity in non-human primates. Nat Commun 13:2454
65. Cooney DA, Rosenbluth RJ (1975) Enzymes as therapeutic agents. Adv Pharmacol Chemother 12:185–289
66. Goldberg DM (1992) Enzymes as agents for the treatment of disease. Clin Chim Acta 206:45–76
67. Farhadi SA, Bracho-Sanchez E, Freeman SL et al (2018) Enzymes as immunotherapeutics. Bioconjug Chem 29:649–656
68. Holcenberg JS (1977) Enzymes as drugs. Annu Rev Pharmacol Toxicol 17:97–116
69. Kumar G, Saini M, Kundu S (2022) Therapeutic enzymes as non-conventional targets in cardiovascular impairments: a comprehensive review. Can J Physiol Pharmacol 100:197–209
70. Lukey MJ, Katt WP, Cerione RA (2017) Targeting amino acid metabolism for cancer therapy. Drug Discov 22:796–804
71. Patil MD, Bhaumik J, Babykutty S et al (2016) Arginine dependence of tumor cells: targeting a chink in cancer's armor. Oncogene 35:4957–9724
72. Riess C, Shokraie F, Classen CF et al (2018) Arginine-depleting enzymes-an increasingly recognized treatment strategy for therapy-refractory malignancies. Cell Physiol Biochem 51: 854–870
73. Tanios R, Bekdash A, Kassab E et al (2013) Human recombinant arginase I(Co)-PEG5000 [HuArgI(Co)-PEG5000]-induced arginine depletion is selectively cytotoxic to human acute myeloid leukemia cells. Leuk Res 37:1565–1571
74. Dizikes GJ, Grody WW, Kern RM et al (1986) Isolation of human liver arginase cDNA and demonstration of nonhomology between the two human arginase genes. Biochem Biophys Res Commun 141:53–59
75. Caldwell RB, Toque HA, Narayanan SP et al (2015) Arginase: an old enzyme with new tricks. Trends Pharmacol Sci 36:395–405
76. De Santo C, Cheng P, Beggs A et al (2018) Metabolic therapy with PEGarginase induces a sustained complete remission in immunotherapy-resistant melanoma. J Hematol Oncol 11:1–5
77. Cheng PNM, Liu AM, Bessudo A et al (2021) Safety, PK/PD and preliminary anti-tumor activities of pegylated recombinant human arginase 1 (BCT-100) in patients with advanced arginine auxotrophic tumors. Invest New Drugs 39:1633–1640
78. Yau T, Cheng PNM, Chiu J et al (2022) A phase 1 study of pegylated recombinant arginase (PEG-BCT-100) in combination with systemic chemotherapy (capecitabine and oxaliplatin) [PACOX] in advanced hepatocellular carcinoma patients. Invest New Drugs 40:314–321
79. Li L, Wang Y, Chen J et al (2013) An engineered arginase FC protein inhibits tumor growth *in vitro* and *in vivo*. Evid Based Complement Alternat Med 2013:423129
80. Stone EM, Glazer ES, Chantranupong L et al (2010) Replacing Mn^{2+} with Co^{2+} in human arginase I enhances cytotoxicity toward L-arginine auxotrophic cancer cell lines. ACS Chem Biol 5:333–342
81. Khoury O, Ghazale N, Stone E et al (2015) Human recombinant arginase I (Co)-PEG5000 [HuArgI (Co)-PEG5000]-induced arginine depletion is selectively cytotoxic to human glioblastoma cells. J Neurooncol 122:75–85
82. El-Mais N, Fakhoury I, Abdellatef S et al (2021) Human recombinant arginase I [HuArgI (Co)-PEG5000]-induced arginine depletion inhibits ovarian cancer cell adhesion and migration through autophagy-mediated inhibition of RhoA. J Ovarian Res 14:13
83. D'Antonio EL, Hai Y, Christianson DW (2012) Structure and function of non-native metal clusters in human arginase I. Biochem 51:8399–8409
84. Chung SF, Kim CF, Tam S et al (2020) A bioengineered arginine-depleting enzyme as a long-lasting therapeutic agent against cancer. Appl Microbiol Biotechnol 104:3921–3934

Top Med Chem
https://doi.org/10.1007/7355_2024_169

Published online: 5 January 2025

Monoclonal Antibodies as Biotherapeutics

Pablo Cabral and Ximena Camacho

Contents

Abstract Monoclonal antibodies (mAbs) have revolutionised modern medicine due to their ability to treat a wide range of diseases, including cancer, autoimmune diseases and infectious diseases. This chapter provides a comprehensive overview of mAbs, covering their selection, general characteristics, therapeutic uses and applications in diagnostics and radionuclide therapy. The rationale for their

P. Cabral (✉) and X. Camacho
Área de Radiofarmacia-Centro de Investigaciones Nucleares, Facultad de Ciencias, Universidad de la República, Montevideo, Uruguay
e-mail: pcabral@cin.edu.uy; xcamacho@cin.edu.uy

development and production, their advantages compared to other treatments and their various clinical applications are explored. In addition, specific examples of mAbs used in current clinical practice and their impact on improving treatment accuracy and efficacy are discussed. This chapter aims to provide an in-depth understanding of mAbs, highlighting their importance and potential in contemporary medicine.

Keywords Contemporary medicine, Diagnostics, Monoclonal antibodies, Radionuclide therapy, Therapeutic uses

1 Introduction

Monoclonal antibodies (mAbs) are laboratory-produced proteins that can bind to specific antigens in the body [1–3]. These antibodies are cloned from a single parent cell, which allows them to have high specificity and affinity for their targets [2]. Since their initial development, mAbs have transformed the way various diseases are treated, providing highly targeted therapies with fewer side effects compared to traditional treatments [2, 4–6].

2 Selection of Monoclonal Antibodies

2.1 Discovery and Development Process

The development of mAbs begins with the identification of a target antigen, usually a specific protein on the surface of pathological cells, such as cancer cells or pathogens [2, 7, 8]. Once the antigen is identified, an animal, usually a mouse, is immunised with this antigen to stimulate an immune response. B cells from the immunised animal, which produce antibodies against the antigen, fuse with myeloma cells to create hybridomas. These hybridomas can be selected and cloned to produce large quantities of specific mAbs [9, 10]. The process of monoclonal antibody discovery and development can be summarised in the following key steps:

(i) Identification of the target antigen.
   - Selection of the antigen is crucial and may include cell surface proteins, receptors, or even specific toxins.
   - The antigen must be relevant to the disease in question and accessible to the antibody.

(ii) Immunisation.
   - An animal (commonly mice) is immunised with the antigen to induce an immune response.

- The immunisation protocol can be varied, using adjuvants to enhance the immune response.

(iii) Fusion of B cells and myeloma cells.

- Antibody-producing B cells are isolated from the spleen of the immunised animal.
- These cells are fused with immortalised myeloma cells to create hybridomas.

(iv) Hybridoma selection.

- Hybridomas are cultured and selected on selection media (e.g., hypoxanthine-aminopterin-thymidine (HAT)).
- Clones that produce specific antibodies against the antigen of interest are selected.

(v) Cloning and expansion.

- Selected hybridomas are cloned to obtain a pure cell line.
- Clonal lines are expanded for large-scale antibody production.

(vi) Purification and characterisation.

- Monoclonal antibodies are purified using techniques such as affinity chromatography.
- They are characterised to determine their specificity, affinity and functionality.

2.2 *Antibody Engineering*

The first mAbs were of murine origin, which limited their use due to their high immunogenicity in humans. To overcome this, genetic engineering techniques were developed that allowed the humanisation of antibodies [11–14]. These techniques include the creation of chimeric, humanised and fully human antibodies:

- Chimeric antibodies: they combine murine variable regions with human constant regions. This reduces immunogenicity while maintaining the specificity of the original antibody.
- Humanised antibodies: only antigen-binding regions (complementarity determining regions (CDRs)) are murine, grafted onto a human framework. This construct further minimises immunogenicity and is suitable for long-term therapeutic applications.
- Fully human antibodies: they are produced using techniques such as mouse transgenesis and phage technology. These antibodies have minimal immunogenicity and are ideal for long-term clinical use.

2.3 Selection Based on Affinity and Specificity

Affinity and specificity are critical factors in the selection of mAbs for therapeutic and diagnostic applications [15]. Selection methods include:

- Enzyme-linked immunosorbent assay (ELISA): to determine the ability of antibodies to bind specifically to the target antigen.
- Surface plasmon resonance (SPR) affinity analysis: to measure the binding strength between the antibody and the antigen.
- Cell-based assays: to assess the efficacy of the antibody in recognising target cells in vitro.

3 General Characteristics of Monoclonal Antibodies

3.1 Structure and Function

Monoclonal antibodies are composed of two light chains and two heavy chains, forming a Y-shaped structure. Each arm of Y contains an antigen-binding site (fragment antigen binding (Fab)), while the tail of Y (Fc) is responsible for effector functions, such as the activation of the immune system. This structure allows mAbs to recognise and bind specifically to their target antigen, marking pathological cells for destruction by the immune system [8, 16, 17].

3.1.1 Functional Regions

Fab consists of a light chain and a heavy chain and contains the hypervariable region (CDR), which binds specifically to the antigen. Additionally, it is responsible for the specificity and affinity of the antibody.

On the other hand, the crystallisable fragment (Fc) is composed of the constant regions of heavy chains and interacts with Fc receptors on immune cells. Furthermore, it mediates effector functions, such as antibody-dependent cell-mediated cytotoxicity (ADCC) and phagocytosis.

3.1.2 Specificity and Affinity

The high specificity of mAbs is due to their ability to recognise and bind to a single epitope on the antigen. This minimises off-target effects and reduces toxicity compared to treatments that affect multiple cellular processes. Affinity refers to the strength with which the mAb binds to its antigen, influencing its therapeutic efficacy [18–22]. To measure affinity, the following assays can be done: (i) affinity – for example, surface plasmon resonance (SPR) and biobilayer interferometry (BLI) are

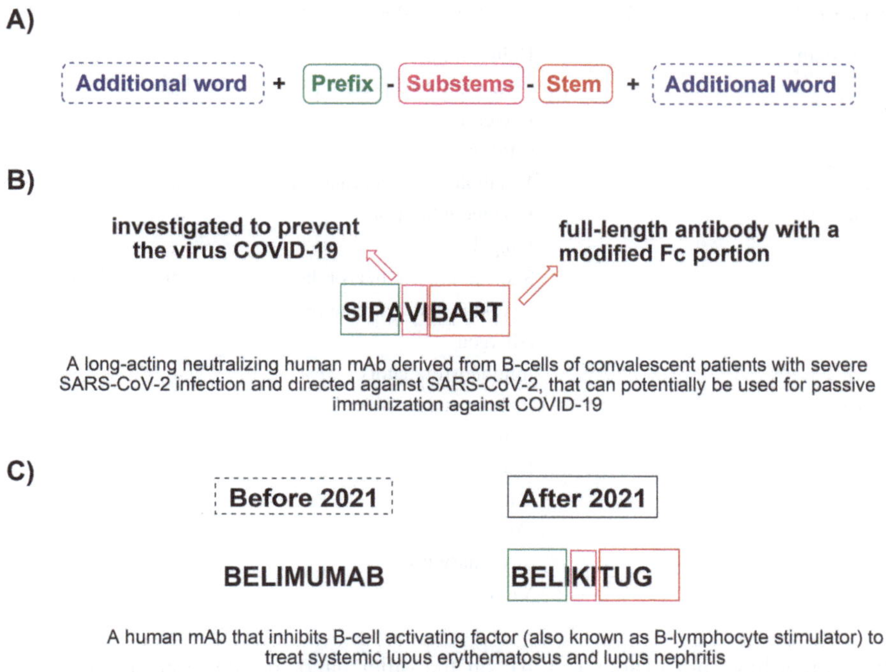

Fig. 1 Naming the scheme of mAbs. **A)** Component and structure to assign generic names to mAbs. **B)** Example. **C)** Name comparison before and after the 2021 recommendations

used to measure the interaction between the antibody and the antigen, and (ii) avidity – the total binding strength when multiple epitopes are present is important in antibodies that bind repetitive antigens.

3.2 *Naming Scheme of MAbs*

The naming scheme allows assigning generic, or non-proprietary, names to mAbs. Since 2021, mAb names have been constituted by different stems, sub-stems (sub-stems for the source and sub-stems for the target), prefixes and, in some cases, additional words (Fig. 1) [23, 24].

Four different groups of stems exist, namely, i) **-tug**: for full-length unmodified immunoglobulins; ii) **-bart**: for full-length antibodies artificial containing at least one engineered mutation; iii) **-mig**: for multi-immunoglobulins of any length, including bispecific and multispecific monoclonal antibodies; and iv) **-ment**: for monospecific antibody fragments without an Fc region.

In regard to sub-stems, i) the source refers to the animal from which the antibody was obtained. Since 2021, some of the previous infixes have been retained, for example, -ba- (when the bacterial origin is) and -fung- (when the fungal origin is),

Table 1 List of sub-stems for mAb nomenclatures adopted in 2021 [23]

Sub-stems	Definition
-ami-	Serum amyloid protein (SAP)/amyloidosis
-Ba-	Bacterial
-ci-	Cardiovascular
-de-	Metabolic or endocrine pathways
-eni-	Enzyme inhibition
-fung-	Fungal
-gro-	Skeletal muscle mass-related growth factors and receptors
-ki-	Cytokine and cytokine receptors
-ler-	Allergen
-sto-	Immunostimulatory
-pru-	Immunosuppressive
-ne-	Neural
-os-	Bone
-ta-	Tumour
-toxa-	Toxin
-vet-	Veterinary use
-vi-	Viral

but others have not (Table 1) [24]; on the other hand, ii) the target refers to the biotherapeutic's target, and similarly, some of the previous ones have been maintained, while others have not (Table 1).

The prefix has no particular meaning contributing to a well-sounding name. For example, for mAbs with the same source and target, sub-stems could be differentiated by their prefixes, such as ***golimumab***[1] and ***adalimumab***, which are both tumour necrosis factor (TNF) inhibitors but have different chemical structures.

Another o other additional words, before or following the name of the mAbs, could be appear indicating that another entity is attached to the biotherapeutic. For example ***technetium (^{99m}Tc) nofetumomab merpentan*** where ***nofetumomab***, the Fab fragment of an IgG2b murine monoclonal antibody NR-LU-10 which is directed against a glycoprotein antigen expressed in a variety of cancers, is attached to radionuclide ^{99m}Tc via the chelator merpentan to produce an imaging agent specific for small-cell lung cancer.

3.3 Production and Purification

The production of mAbs[2] involves the culturing of hybridomas in bioreactors, followed by purification of the antibody by techniques such as affinity

[1] In this chapter, the nomenclature prior to 2021 will still be used.

chromatography. The purity and consistency of the final product are critical to ensuring the safety and efficacy of the mAb in clinical applications [25–27].

The production process consists of:

(i) Cell culture: it is where the hybridomas are cultured in bioreactors under controlled conditions, caring for parameters such as pH, temperature and nutrients, which are optimised to maximise the production of antibodies.
(ii) Purification: using techniques such as affinity chromatography (e.g. protein A/G columns) to isolate antibodies from the culture medium; other methods include ion exchange chromatography and size exclusion chromatography to refine the purity of the product.
(iii) Characterisation: where quality assays are performed to verify the purity, concentration and biological activity of the antibody; additionally, functional and biochemical assays ensure that the antibody retains its specificity and affinity after purification.

4 Therapeutic Uses of Monoclonal Antibodies

4.1 Cancer Treatment

Monoclonal antibodies have been shown to be effective in the treatment of various types of cancer by several modes of action, namely [4, 7, 28, 29]:

- Immunomodulation: mAbs such as ***ipilimumab*** (anti-CTLA-4) and ***pembrolizumab*** (anti-PD-1) work by blocking immune checkpoints, freeing up T cells to attack cancer cells.
- Antibody-dependent cytotoxicity (ADCC): mAbs bind to tumour cells, marking them for destruction by immune cells. Within this group is found the mAb ***trastuzumab*** (Herceptin®), which is targeted against HER2, which is over-expressed in certain types of breast cancer, inhibiting cell proliferation and marking cells for antibody-dependent cellular cytotoxicity (ADCC). ***Trastuzumab*** improves survival in patients with HER2-positive breast cancer. Another example of mAb with this mode of action is ***rituximab*** (Rituxan®), which is targeted against CD20 present in B cells and is used in the treatment of non-Hodgkin's lymphoma and chronic lymphocytic leukaemia. ***Rituximab*** induces B-cell lysis through ADCC, complement-dependent cytotoxicity and apoptosis.
- Toxin delivery: mAbs conjugated with toxins or radioactive agents, such as ***brentuximab vedotin***. ***Brentuximab vedotin*** is a mAb-small-molecule drug conjugate directed to CD30, a transmembrane glycoprotein receptor expressed in

[2] See Chapter "Development and Production of Protein-based Biotherapeutics in Mammalian Cells" for a complete description.

many lymphomas with low expression on normal tissue. It delivers lethal payloads, like the antimitotic agent monomethyl auristatin E, directly to cancer cells.

4.2 Inflammatory and Autoimmune Diseases

Monoclonal antibodies are essential in the treatment of autoimmune diseases such as rheumatoid arthritis, Crohn's disease and multiple sclerosis acting by means of [30–33]:

- Cytokine blockade: mAbs inhibit pro-inflammatory cytokines, reducing inflammation and tissue damage. For example, ***infliximab*** (Remicade®), which is targeted against tumour necrosis factor-alpha (TNF-α), is used in the treatment of Crohn's disease, ulcerative colitis and rheumatoid arthritis, reducing inflammation and improving clinical symptoms. Another example is ***tocilizumab*** (Actemra®), which is used in rheumatoid arthritis and macrophage activation syndrome, reducing inflammation and joint damage, due to its ability to inhibit IL-6 signalling by targeting the IL-6 receptor. The next example is ***adalimumab*** (Humira®), which is the main blockbuster[3] between 2014 and 2022,[4] selected by phage display, and is used to treat, among others, Crohn's disease, ulcerative colitis and rheumatoid arthritis, reducing inflammation and improving clinical symptoms through the inactivation of TNF-α.
- B-cell depletion: for example, ***rituximab***, directed against the CD20 antigen on B cells, is used in the treatment of rheumatoid arthritis and other B-cell-mediated diseases.

4.3 Infectious Diseases

In infectious diseases, mAbs can neutralise toxins and pathogens [34–37]. For example,

- In virus neutralisation, ***palivizumab*** (Synagis®) is used to prevent respiratory syncytial virus (RSV) infections in neonates and young children. It is administered as a prophylaxis to reduce hospitalisation for RSV.
- In bacterial toxin neutralisation, ***obiltoxaximab*** (Anthim®) is used to treat and prevent inhaled anthrax poisoning. It targets the protective antigen of *Bacillus anthracis*, neutralising the toxin and reducing mortality in severe infections.

---

[3] See Chapter "Medicinal Chemistry and Biotherapeutics. Overview and Definitions".

[4] See Chapter "Biotherapeutics. General Introduction".

5 Monoclonal Antibodies Used in Diagnostics and Therapy with Radionuclide

5.1 Diagnostic Imaging

The conjugation of mAbs with radionuclides allows the visualisation of pathological processes in the body by imaging techniques such as positron emission tomography (PET) and single-photon emission computed tomography (SPECT) [38–41]. The commonly used radionuclides include:

- ^{99m}Tc: used in tagging mAbs for SPECT imaging due to its adequate half-life and low toxicity
- ^{18}F: used in PET to provide high-resolution images.

Some examples of these hybrid[5] biotherapeutics are as follows: i) ***^{99m}Tc-arcitumomab***, targeting the carcinoembryonic antigen (CEA), is used for detecting the recurrence of colorectal cancer. It provides clear images of tumours with high CEA expression. ii) HER2-targeted ***^{89}Zr-trastuzumab*** is used in PET to visualise HER2-positive breast cancer, allowing the monitoring of HER2 distribution, drug distribution and the assessment of metastasis.

5.2 Radionuclide Therapy

Therapy with radionuclide-conjugated mAbs offers a form of targeted treatment, delivering therapeutic doses of radiation directly to tumour cells while minimising damage to healthy tissues [42–44].

In this sense, ***ibritumomab tiuxetan*** (Zevalin®) is ^{90}Y conjugated and is used for the treatment of relapsed or refractory non-Hodgkin's lymphoma, providing lethal radiation, due to yttrium beta-emitting radionuclide, to cancer cells. This hybrid biotherapeutic is targeted against CD20, while thiuxetan, a modified version of DTPA (diethylenetriamine-penta-acetic acid), is used to coordinate the metallic radionuclide (Fig. 2).

Another example is ***iodine-131 tositumomab*** (Bexxar®), a mAb covalently conjugated to ^{131}I (Fig. 2), a beta- and gamma emitter, used in the treatment of B-cell lymphomas, especially in patients who do not respond to other treatments. Targeting against CD20, this hybrid biotherapeutic combines immunotherapy[6] with radiotherapy, especially in lymphoma-porting patients who do not respond to other treatments.

[5] See Chapter "Medicinal Chemistry and Biotherapeutics. Overview and Definitions".

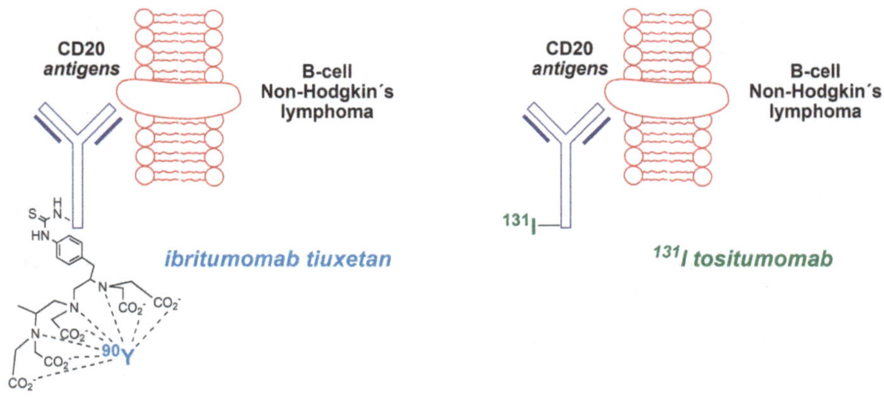

Fig. 2 Schematic representations of two hybrid mAbs-radionuclides developed to treat non-Hodgkin's lymphomas, targeting CD20 antigens

5.3 Study of Cases

5.3.1 Trastuzumab (Herceptin®) in Breast Cancer [41]

Development and mechanism of action: ***trastuzumab*** is a humanised mAb directed against human epidermal growth factor receptor 2 (HER2), which is overexpressed in certain types of breast cancer. By binding to HER2, ***trastuzumab*** inhibits cell proliferation and marks tumour cells for destruction by the immune system.

Clinical applications: used in both adjuvant and metastatic therapies, ***trastuzumab*** has significantly improved disease-free survival and overall survival in patients with HER2-positive breast cancer.

Combination with radionuclides: the use of ***89Zr-trastuzumab*** allows the precise visualisation of mAb distribution in the body by PET, facilitating treatment monitoring and the identification of metastases.

Clinical Efficacy:

- **Clinical trials** have shown that ***trastuzumab***, in combination with chemotherapy, significantly improves progression-free survival and overall survival in patients with HER2-positive breast cancer.
- Monitoring: PET with ***89Zr-trastuzumab*** allows the biodistribution of the drug and its penetration into tumours to be assessed, optimising personalised treatment.

[6]See Chapter "Medicinal Chemistry and Biotherapeutics. Overview and Definitions".

5.3.2 Rituximab (Rituxan®) in Autoimmune Diseases [45]

Development and mechanism of action: ***rituximab*** is a chimeric mAb directed against CD20, a protein expressed on the surface of B cells. By binding to CD20, ***rituximab*** induces B-cell depletion through mechanisms such as ADCC, complement-dependent cytotoxicity and apoptosis.

Clinical applications: in addition to its use in B-cell lymphomas, ***rituximab*** is effective in the treatment of rheumatoid arthritis, granulomatosis with polyangiitis and other autoimmune diseases.

Combination with radionuclides: ***rituximab*** can be conjugated with ^{90}Y to form ***ibritumomab tiuxetan***, providing a combination therapeutic option that uses both immunotherapy[7] and radiotherapy to destroy malignant cells.

Clinical Efficacy:

- **Clinical trials**: ***rituximab*** has shown significant benefits in patients with rheumatoid arthritis who do not respond to conventional treatments, improving joint function and reducing inflammation.
- **Combination therapy**: ***ibritumomab tiuxetan*** has been shown to be effective in the treatment of non-Hodgkin's lymphoma, providing a therapeutic option for relapsed or refractory patients.

6 Future of Monoclonal Antibodies

6.1 Development of New Therapies

Research and development continue to expand the applications of mAbs. Emerging areas include [46–48]:

- **Bispecific antibodies**: they are capable of binding to two different antigens, improving treatment specificity and efficacy. For example, ***emicizumab*** that through simultaneous binding of factor IXa and factor X of the coagulation cascade, bispecific action, mimics the function of factor VIII [49]. ***Emicizumab*** allows subcutaneous dosing of up to once weekly to its long circulatory half-life.
- **Antibody-drug conjugates (ADCs)**: they combine mAbs with cytotoxic agents to target therapy directly to cancer cells, reducing systemic side effects.
- **Gene therapies**: these use mAbs to target viral vectors to specific cells, enabling the delivery of therapeutic genes.[8]

[7] See Chapter "Medicinal Chemistry and Biotherapeutics. Overview and Definitions".

6.2 Personalised Medicine

Advances in genomics and biotechnology are enabling the personalisation of mAb treatments, tailoring them to the specific genetic and molecular characteristics of each patient. This includes the development of specific mAbs for cancer sub-types and the optimisation of doses and therapeutic combinations based on individual biomarker profiles [7, 50, 51].

6.3 Applications in Neurodegenerative Diseases

Monoclonal antibodies are also being explored for the treatment of neurodegenerative diseases such as Alzheimer's and Parkinson's disease [52, 53]. Antibodies targeting pathological proteins such as beta-amyloid and alpha-synuclein are in development, with the aim of halting or reversing the progression of these debilitating diseases.

7 Concluding Remarks

Monoclonal antibodies represent one of the most significant innovations in modern biotherapeutics, offering specific and effective treatment for a variety of diseases. Their ability to be conjugated to radionuclides further expands their application, providing powerful tools for diagnosis and therapy. As research and technology continue to advance, mAbs are likely to play an increasingly important role in personalised medicine and the treatment of complex diseases.

Compliance with Ethical Standards This chapter is a review and no humans or animals were involved in the preparation of the chapter.

References

1. Malik B, Ghatol A (2023) Understanding how monoclonal antibodies work. StatPearls, Treasure Island
2. Quinteros DA, Bermúdez JM, Ravetti S et al (2017) Therapeutic use of monoclonal antibodies: general aspects and challenges for drug delivery. In: Andronescu E, Grumezescu AM (eds) Nanostructures for drug delivery1st edn. Elsevier, Amsterdam
3. Bayer V (2019) An overview of monoclonal antibodies. Semin Oncol Nurs 35:150927
4. Zahavi D, Weiner L (2020) Monoclonal antibodies in cancer therapy. Antibodies (Basel) 9:34

[8] See Chapter "Gene Therapy".

5. Lu RM, Hwang YC, Liu IJ et al (2020) Development of therapeutic antibodies for the treatment of diseases. J Biomed Sci 27:1
6. Fu Z, Li S, Han S et al (2022) Antibody drug conjugate: the "biological missile" for targeted cancer therapy. Signal Transduct Target Ther 7:93
7. Delgado M, Garcia-Sanz JA (2023) Therapeutic monoclonal antibodies against cancer: present and future. Cells 12:2837
8. Sadeghalvad M, Rezaei N (2021) Introduction on monoclonal antibodies. In: Rezaei N (ed) Monoclonal antibodies. IntechOpen, London
9. Mitra S, Tomar PC (2021) Hybridoma technology; advancements, clinical significance, and future aspects. J Genet Eng Biotechnol 19:159
10. Parray HA, Shukla S, Samal S et al (2020) Hybridoma technology a versatile method for isolation of monoclonal antibodies, its applicability across species, limitations, advancement and future perspectives. Int Immunopharmacol 85:106639
11. Ducancel F, Muller BH (2012) Molecular engineering of antibodies for therapeutic and diagnostic purposes. MAbs 4:445–457
12. Doevendans E, Schellekens H (2019) Immunogenicity of innovative and biosimilar monoclonal antibodies. Antibodies (Basel) 8:21
13. Yamashita M, Katakura Y, Shirahata S (2007) Recent advances in the generation of human monoclonal antibody. Cytotechnology 55:55–60
14. Chiu ML, Goulet DR, Teplyakov A et al (2019) Antibody structure and function: the basis for engineering therapeutics. Antibodies (Basel) 8:55
15. Alhazmi HA, Albratty M (2023) Analytical techniques for the characterization and quantification of monoclonal antibodies. Pharmaceuticals 16:291
16. Moorthy BS, Xie B, Moussa EM et al (2015) Structure of monoclonal antibodies. In: Rosenberg A, Demeule B (eds) Biobetters. Protein engineering to approach the curative. Springer, New York
17. Khatib SE, Salla M (2022) The mosaic puzzle of the therapeutic monoclonal antibodies and antibody fragments - a modular transition from full-length immunoglobulins to antibody mimetics. Leuk Res Rep 18:100335
18. Gogesch P, Dudek S, van Zandbergen G et al (2021) The role of Fc receptors on the effectiveness of therapeutic monoclonal antibodies. Int J Mol Sci 22:8947
19. Rabia LA, Desai AA, Jhajj HS et al (2018) Understanding and overcoming trade-offs between antibody affinity, specificity, stability and solubility. Biochem Eng J 137:365–374
20. Dennison SM, Reichartz M, Seaton KE et al (2018) Qualified biolayer interferometry avidity measurements distinguish the heterogeneity of antibody interactions with *Plasmodium falciparum* Circumsporozoite protein antigens. J Immunol 201:1315–1326
21. Murali S, Rustandi RR, Zheng X et al (2022) Applications of surface plasmon resonance and biolayer interferometry for virus-ligand binding. Viruses 14:717
22. Kamat V, Rafique A (2017) Designing binding kinetic assay on the bio-layer interferometry (BLI) biosensor to characterize antibody-antigen interactions. Anal Biochem 536:16–31
23. World Health Organization (2021) New INN monoclonal antibody (mAb) nomenclature scheme: Geneva, November 2021. https://cdn.who.int/media/docs/default-source/international-nonproprietary-names-(inn)/new_mab_-nomenclature-_2021.pdf. Accessed 03 July 2024
24. Guimaraes Koch SS, Thorpe R, Kawasaki N et al (2022) International nonproprietary names for monoclonal antibodies: an evolving nomenclature system. MAbs 14:e2075078
25. Jyothilekshmi I, Jayaprakash NS (2021) Trends in monoclonal antibody production using various bioreactor syst. J Microbiol Biotechnol 31:349–357
26. Li F, Vijayasankaran N, Shen AY et al (2010) Cell culture processes for monoclonal antibody production. MAbs 2:466–479
27. Silva Carvalho L, Bravim da Silva O, Carneiro de Almeida G et al (2017) Production processes for monoclonal antibodies. In: Faustino Jozala A (ed) Fermentation processes. IntechOpen, London

28. Gravbrot N, Gilbert-Gard K, Mehta P et al (2019) Therapeutic monoclonal antibodies targeting immune checkpoints for the treatment of solid tumors. Antibodies (Basel) 8:51
29. Sharma P, Joshi RV, Pritchard R et al (2023) Therapeutic antibodies in medicine. Molecules 28: 6438
30. Atzeni F, Sarzi-Puttini P, Doria A et al (2005) Potential off-label use of infliximab in autoimmune and non-autoimmune diseases: a review. Autoimmun Rev 4:144–152
31. Ogata A, Tanaka T (2012) Tocilizumab for the treatment of rheumatoid arthritis and other systemic autoimmune diseases: current perspectives and future directions. Int J Rheumatol 2012:946048
32. Randall KL (2016) Rituximab in autoimmune diseases. Aust Prescr 39:131–134
33. Ly S, Nedosekin D, Wong HK (2023) Review of an anti-CD20 monoclonal antibody for the treatment of autoimmune diseases of the skin. Am J Clin Dermatol 24:247–273
34. Otsubo R, Yasui T (2022) Monoclonal antibody therapeutics for infectious diseases: beyond normal human immunoglobulin. Pharmacol Therapeut 240:108233
35. Esposito S, Amirthalingam G, Bassetti M et al (2023) Monoclonal antibodies for prophylaxis and therapy of respiratory syncytial virus, SARS-CoV-2, human immunodeficiency virus, rabies and bacterial infections: an update from the World Association of Infectious Diseases and Immunological Disorders and the Italian Society of Antinfective Therapy. Front Immunol 14:1162342
36. Rodriguez-Fernandez R, Mejias A, Ramilo O (2021) Monoclonal antibodies for prevention of respiratory syncytial virus infection. Pediatr Infect Dis J 40:S35–S39
37. Yamamoto BJ, Shadiack AM, Carpenter S et al (2016) Obiltoxaximab prevents disseminated Bacillus anthracis infection and improves survival during pre- and postexposure prophylaxis in animal models of inhalational anthrax. Antimicrob Agents Chemother 60:5796–5805
38. Dammes N, Peer D (2020) Monoclonal antibody-based molecular imaging strategies and theranostic opportunities. Theranostics 10:938–955
39. Morais M, Ma MT (2018) Site-specific chelator-antibody conjugation for PET and SPECT imaging with radiometals. Drug Discov Today Technol 30:91–104
40. National Center for Biotechnology Information (U.S.) (2004) Molecular imaging & contrast agent database: MICAD. NCBI, Bethesda
41. Linders DGJ, Deken MM, van Dam MA et al (2023) ^{89}Zr-Trastuzumab PET/CT imaging of HER2-positive breast cancer for predicting pathological complete response after neoadjuvant systemic therapy: a feasibility study. Cancers 15:4980
42. Gudkov SV, Shilyagina NY, Vodeneev VA et al (2015) Targeted radionuclide therapy of human tumors. Int J Mol Sci 17:33
43. Chamarthy MR, Williams SC, Moadel RM (2011) Radioimmunotherapy of non-Hodgkin's lymphoma: from the 'magic bullets' to 'radioactive magic bullets. Yale J Biol Med 84:391–407
44. Goto H, Shiraishi Y, Okada S (2024) Recent preclinical and clinical advances in radioimmunotherapy for non-Hodgkin's lymphoma. Explor Target Anti-tumor Ther 5:208–224
45. Wang J, Baidoun F, Tun HW et al (2023) ^{90}Yttrium ibritumomab tiuxetan (Zevalin) for the treatment of chronic lymphocytic leukemia/small lymphocytic lymphoma: a report of 5 cases. Blood Lymphat Cancer 13:59–65
46. Singh R, Chandley P, Rohatgi S (2023) Recent advances in the development of monoclonal antibodies and next-generation antibodies. ImmunoHorizons 7:886–897
47. Sun Y, Yu X, Wang X et al (2023) Bispecific antibodies in cancer therapy: target selection and regulatory requirements. Acta Pharm Sin B 13:3583–3597
48. Wang Z, Wang G, Lu H et al (2022) Development of therapeutic antibodies for the treatment of diseases. Mol Biomed 3:35
49. Voorberg J, Postmus T, Schols S (2022) Next generation FVIII mimetic bispecific antibody for hemophilia A. J Thromb Haemost 20:1301–1305
50. Garattini L, Padula A (2019) Precision medicine and monoclonal antibodies: breach of promise? Croat Med J 60:284–289

51. Di Paolo A, Luci G (2021) Personalized medicine of monoclonal antibodies in inflammatory bowel disease: pharmacogenetics, therapeutic drug monitoring, and beyond. Front Pharmacol 11:610806
52. Gklinos P, Papadopoulou M, Stanulovic V et al (2021) Monoclonal antibodies as neurological therapeutics. Pharmaceuticals 14:92
53. Nimmo JT, Kelly L, Verma A et al (2021) Amyloid-β and α-synuclein immunotherapy: from experimental studies to clinical trials. Front Neurosci 15:733857

Top Med Chem
https://doi.org/10.1007/7355_2024_168

Published online: 24 December 2024

Development and Production of Protein-Based Biotherapeutics in Mammalian Cells

Natalia Ceaglio and Marcos Oggero

Contents

Abstract Mammalian-cell-derived-biotherapeutics have revolutionized the treatment of various diseases, with antibodies, fusion proteins, and cytokines being their most significant successes. The so-called innovative or original products, biosimilars, and biobetters have been central pieces of this evolution, alongside the increased capacity of bioprocesses to produce them with better yields and quality. This upward trend, where Chinese hamster ovary cells represent the workhorses for the large-scale production of biotherapeutics, has led to a profound and substantial generation of knowledge, methodologies, and tools applied to research, development, and innovation. Protein-biotherapeutical production is a complex bioprocess that involves a wide range of techniques, from genetic engineering, followed by upstream and downstream stages, up to the final product. In this chapter, we will

N. Ceaglio and M. Oggero (✉)
UNL, CONICET, FBCB, Centro Biotecnológico del Litoral, Ciudad Universitaria, Santa Fe, Pcia. Santa Fe, Argentina
e-mail: nceaglio@fbcb.unl.edu.ar; moggero@fbcb.unl.edu.ar

succinctly outline the key stages in carrying out successful bioprocesses, with emphasis on increasing productivity, and select the appropriate methods for producing protein for therapy. The final section, related to the modification of existing protein biologics to improve their properties, is included, which aims to generate biobetters or second-generation biotherapeutics. A brief description of glycoengineering approaches gives interesting clues to the design of novel long-lasting biodrugs. The main objective of the chapter is to contribute to enhancing the production processes of diverse biotherapeutical proteins for the economically and industrially important sector of biologics.

Keywords Bioprocesses, Downstream, Mammalian cells, Protein biotherapeutics, Upstream

1 Introduction

Almost 40 years have passed since the approval of the first protein-based biotherapeutic produced in mammalian cells (***human tissue plasminogen activator* (*tPA*)**, US Food and Drug Administration (FDA) approval in 1986 [1]). Since then, manufacturing processes have grown exponentially not only in number but also in yield, complexity, and the quality of products. According to Walsh's latest survey of biotherapeutic approval [2], which ranges from 2018 to 2022, mammalian cell systems continue to be the most often used expression system, with almost 70% of approved products. Also, as in previous surveys, mammalian production platforms continue to be dominated by Chinese hamster ovary cells (CHO cells), which were used to produce 90% of biotherapeutic products approved in the mentioned period. During the survey time frame, 197 biotherapeutic products were approved, but only 46% of them can be considered genuinely new, with the rest comprising biosimilars[1] and others. Monoclonal antibodies (mAbs) remain dominant in overall approvals (54% in the last 4 years). Importantly, the market value of biotherapeutics has continued to rise, with total global sales of US$ 343 billion in 2021, including originator recombinant proteins and biosimilars. Altogether, these facts have driven a strong interest for progress in bioprocess development optimization.

This chapter aims to describe the different stages in the bioprocess development of biotherapeutics in mammalian cell platforms, with emphasis on novel strategies for the optimization of the cost-efficiency of manufacturing processes.

[1] Biosimilar is a biotherapeutic that is almost an exact replica of the original product but produced by a different biopharmaceutical company.

2 Generation of Recombinant Cell Lines

2.1 Cell Line Development Process

Whether the recombinant protein emerges as a new biotherapeutic or aims to be approved as a biosimilar, the first step for establishing a production process is cell line development, i.e., the generation of a recombinant cell line able to produce protein in sufficient quantity and adequate quality and can be easily cultured under "low-cost" conditions.

Immortalized cell lines that grow for prolonged periods of time are necessary to produce large amounts of biotherapeutic protein. Also, as will be described in the section *Choice of Host*, the choice of the correct host is crucial for proteins that require posttranslational modifications. Once the appropriate cells are selected, they need to incorporate the expression vector containing the gene of interest (GOI) and other regulatory and accessory sequences (section *Construction of the Gene Expression Cassette*) by means of a transfection or transduction method (section *Methods for Introducing DNA into Mammalian Cells*). The isolation of clones with high productivity, adequate growth profile, stability, and high quality of the produced recombinant protein (section *Screening and Cloning Methods*) represents a time-consuming but essential step to achieving a cost-effective large-scale production process. Finally, clone characterization and the construction of a master and working cell bank are mandatory steps to guarantee the perpetuity of the original recombinant cell line (section *Cell Banking and Cell Line Characterization*).

Figure 1 schematizes the development of a recombinant cell line for the production of a biotherapeutic protein.

2.2 Choice of Host

The expression system used for the production of the recombinant protein is invariably dictated by the posttranslational modification (PTM) requirements of the biotherapeutic [2].

Many human proteins require PTMs, such as phosphorylation, acetylation, glycosylation, and ubiquitination, among others, for proper functioning [3]. Particularly, protein glycosylation plays a crucial role in the efficacy, serum half-life, and antigenicity of a biotherapeutic. Although the expression of many protein-based biotherapeutics has been achieved using bacteria, yeast, insect cells, transgenic plants and transgenic animals, mammalian expression systems are the preferred hosts for the production of most complex therapeutic proteins and nowadays dominate the biotherapeutic market.

Microbial systems, such as *Escherichia coli*, present the advantage of low-cost, short timeline due to rapid growth, easy control, well-known biochemistry and genetics, and high productivity [4, 5] However, the lack of proper PTMs, inclusion

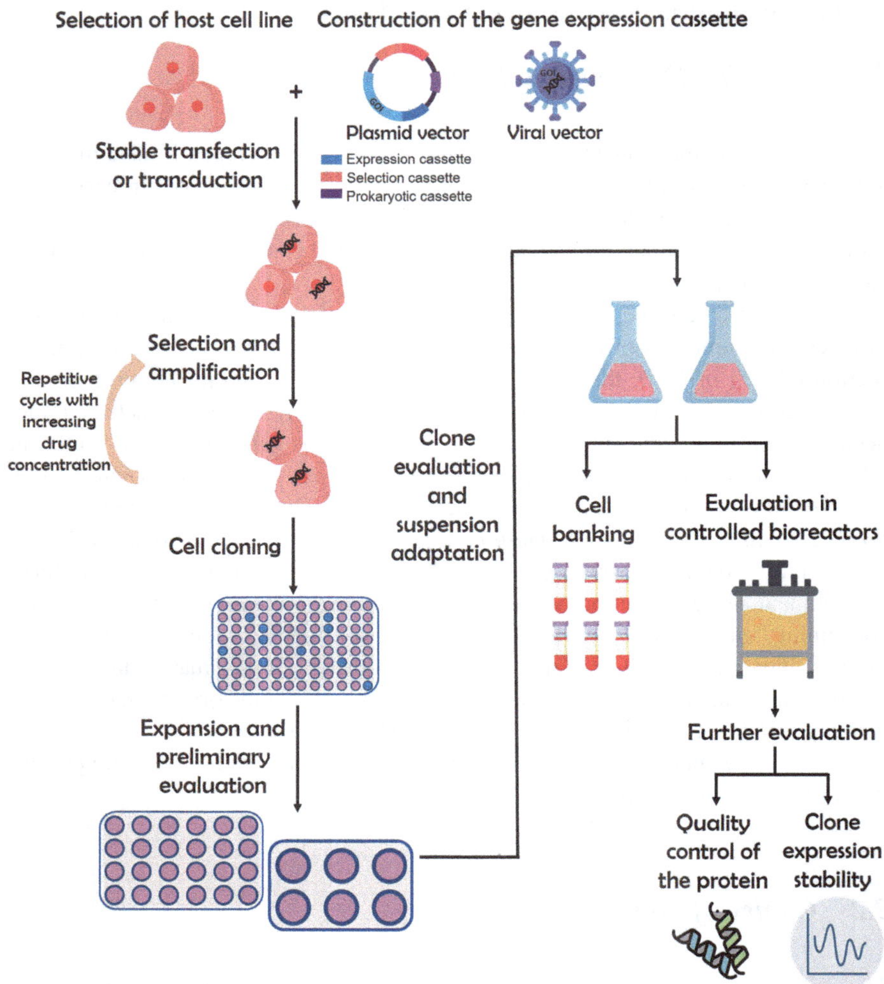

Fig. 1 Schematic representation of the development of a recombinant cell line for the production of a therapeutic protein. This image has been designed using resources from Flaticon.com. Icons were made by Freepik, CreArtive, Iconjam, Paul J., and graphicmall from www.flaticon.com

body formation, codon bias, and endotoxin issues have prevented their use for the expression of large, complex glycoproteins with multiple subunits, cofactors, and disulfide bonds, among other characteristics. Yeast expression platforms such as *Saccharomyces cerevisiae* and *Pichia pastoris* also demonstrate fast growth and high yields, so they can be used to produce certain proteins for which *E. coli* is not appropriate due to folding and glycosylation problems; however, yeast still cannot modify proteins with human glycosylation [6].

The production of recombinant proteins in insect cells is carried out using baculovirus expression vector systems. Insect cells are grown to the desired density and then infected with a recombinant baculovirus containing the GOI. The most common insect cell line is Sf9, but S2, Sf21, Tn-368, and High-Five™ are also employed. All these cell lines share the feature of generating good protein folding and high product yield, but the glycosylation pattern differs from that of mammalian cells since insect cells are not able to carry out complete *N*-glycosylation. Instead, they produce trimmed *N*-glycan precursors that do not develop further into terminal galactose and/or sialic acid residues [7].

Plant-based systems constitute an alternative for cost-effective biotherapeutic production and present some advantages, such as low cost, safety (low risk of contamination with animal viruses and other infectious pathogens to humans), easy scale-up, stability, and ability to produce *N*-glycosylated proteins [8]. However, plants are unable to completely reproduce human *N*- and *O*-glycosylation patterns, and so differences can be found, especially in the structure of complex glycans. Plant-specific *N*-glycans contain α-(1,3)-fucose and/or β-(1,2)-xylose residues, while the *N*-glycans of mammals carry α-(1,6)-fucose attached to the proximal *N*-acetylglucosamine (GlcNAc) and never exhibit xylose [9]. As a consequence, plant complex *N*-glycans are immunogenic to humans, leading to the fast elimination of therapeutic glycoproteins from circulation and low efficacy. Additionally, challenges in biotherapeutic production in plants include contamination with pesticides, herbicides, and toxic plant metabolites and the complexity of the purification process [3].

There are many examples of genetic modification/genome editing of bacteria, yeast, insect cells, and plant cells in order to solve the disadvantages exhibited by each system, especially the lack of PTMs or the differences found compared to human protein's PTMs. The discovery by Szymanski and colleagues [10] of an *N*-linked protein glycosylation pathway of *Campylobacter jejuni* has significantly boosted bacterial glycoengineering. In this way, the entire glycosylation locus of these bacteria has been successfully inserted in *E. coli*, enabling the *N*-glycosylation of various recombinant proteins. However, there is still much work to do to allow this strain to become an industrially competent option for the production of biotherapeutics [4, 11]. A very important example of the glycoengineering of plants is the approval of ZMapp®, a cocktail of three monoclonal antibodies against the Ebola virus, which was granted temporary authorization by the FDA and was administered to humans during the Ebola outbreak of 2014 in different West African states [12]. The ZMapp® antibodies were produced in glycoengineered tobacco plants (*Nicotiana benthamiana*), which were modified by RNA interference (RNAi) technology to target the downregulation of endogenous β-1,2-*xylosyltransferase (XylT)* and α-1,3-*fucosyltransferase (FucT)* genes so that they carried a fucose-free glycosylation profile [13].

Nonetheless, these approaches have not substantially reached the market yet since, as we have mentioned before, mammalian cells are still chosen for the expression of most complex biotherapeutics.

Regarding transgenic animals, at present, the ways of producing recombinant protein-based therapeutics (including mAbs, vaccines, hormones, enzymes, and growth factors) include milk from transgenic mammals and eggs from transgenic chickens. In this system, the natural secretion of proteins takes place, with the correct, human-like PTMs [3]. Recently, coagulation factor VII (coagulation factor VIIA), which is produced in the milk of transgenic rabbits, was approved for human use, but two other products have paved the way: Atryn® (***rh-antithrombin***), produced in the milk of transgenic goats (2006, EU approval, and 2009, US approval), and Kanuma® (***sebelipase alfa***, ***rh-lysosomal acid lipase***), produced in transgenic chicken eggs (EU and US approval in 2015) [2]. Some drawbacks of this technology include ethical concerns, the difficulty of separating human proteins from their animal counterpart, potential zoonotic pathogen contamination, and the fact that some recombinant proteins may be active and deleterious for transgenic animals' health. For example, human erythropoietin produced in rabbit milk affected its health, and less serious side effects were observed with human growth hormones [14]. Last but not least, the generation of transgenic animals still stands as a laborious task [15, 16].

Mammalian cells are therefore the chosen system for the production of complex biotherapeutics since the expressed products are the closest to natural proteins in terms of molecular structure, physicochemical properties, and biological functions [17]. Some drawbacks associated with this system, such as the time-consuming development of the recombinant cell line, high culture cost, a challenging scale-up process, and the risk of viral contamination, are totally compensated by the ability of mammalian cells to perform proper PTMs, mainly glycosylation, which are crucial for stability, half-life in circulation, and immunogenicity, together with their capacity to secrete proteins to culture medium, thus facilitating downstream processes [18, 19]. In addition, low productivity issues are no longer a disadvantage of mammalian cell systems since in the last decades, many improvements in cell line development and process optimization have been made, reaching product yields of up to 20 g.L^{-1} for mAbs under fed-batch and perfusion culture conditions [20, 21].

By far, the most often used cell line for the production of biotherapeutics is the Chinese hamster ovary (CHO), with 89% of the products made in mammalian-based expression systems in the last recorded period (from 2018 to 2022), most of them being monoclonal antibodies [2]. Such popularity reflects the strength that CHO cells exhibit: 1) the ability to grow in serum-free suspension cultures for industrial large-scale production, 2) the capacity of performing PTMs compatible with human use, 3) the easiness of the GOIs to integrate into the CHO genome and reach high levels of expression, 4) the current availability of amplification systems that enable higher gene expression and protein productivity, and 5) the resistance to human viral infections due to the lack of genes responsible for human virus entry [2, 5, 17, 19, 22, 23]. Other features that make CHO cell use attractive are accessibility to a great number of suitable commercial serum-free culture media and the availability of multiple genetic-modified strains for different purposes (the overexpression or knockdown of glycosyltransferases for the optimization of the glycosylation pattern, genome editing for targeted gene integration, the overexpression of anti-apoptotic genes for growth improvement, etc. [24]). For example, the CHO-S cell line has been adapted to grow in suspension cultures in a glutamine-free medium [25]; CHODG44 contains a full deletion of the two dihydrofolate reductase (DHFR) loci on chromosome 2, and it is used for the methotrexate amplification system, while CHO-GS cells are the host for the methionine sulfoximine (MSX) selection system based on glutamine synthetase (GS) [26–28]. Finally but not less importantly, the uninterrupted use of CHO cells as hosts for the production of biotherapeutics during the past three decades has demonstrated the safety of this cell line, making approval from regulatory agencies easier.

BHK (baby hamster kidney) is another cell line also derived from hamster, whose use is especially common for factor VIIa, factor VIII, and vaccine production. Similarly to CHO, these cells can express the nonhuman and potentially immunogenic *N*-glycolylneuraminic acid (Neu5Gc) so that a strict quality control of the product must be performed [19, 29].

Murine myeloma cell lines, such as NS0 and Sp2/0, were originated from tumor cells and initially used for the production of biotherapeutic antibodies for the pharmaceutical industry. NS0 cells have the advantage of lacking endogenous GS activity, making them suitable for the MSX selection and amplification system, which allows the achievement of high antibody productivity [30]. However, mAbs produced in both NS0 and Sp2/0 cells contain not only Neu5Gc but also another glycan, Galα1–3Gal (αGal epitope), both of which are not naturally present in humans and thus are potentially immunogenic when used as biotherapeutics [31]. For this reason, murine cell line usage has been limited for biotherapeutic protein production in the last few years. In the period 2018–2022, seven mAbs produced in NS0 cells and one mAb obtained in Sp2/0 cells have been approved, probably due to stringent clone selection to get a desirable glycan profile [2].

The main advantage of the use of human cell lines is their capability to produce biotherapeutic proteins that are most similar to the ones naturally synthesized by humans. This strategy would guarantee that proteins bear, if not the ideal glycosylation pattern, at least nonimmunogenic glycans. The major concern regarding human hosts refers to the risk of viral infection [32, 33]. Nowadays, HEK293, HKB11, PER.C6, and CAP cells are the most popular human hosts for biotherapeutic production.

The most commonly used human cell line to manufacture glycoprotein biotherapeutics is HEK293 (human embryo kidney 293), which has been generated by transfection with adenovirus type 5 DNA, and so it overexpresses the adenovirus E1A and E1B genes [34]. Like CHO cells, several strains of HEK293 cells with different or improved characteristics are available (HEK293T, HEK293H, HEK293F, etc.). Biotherapeutics that have been approved by the FDA in HEK293 cells include Elocta® (***recombinant factor VIII Fc fusion protein***) in 2014 and Vaxzevria® (***replication-deficient chimpanzee adenovirus encoding SARS-CoV-2 spike protein***) in 2021. The main advantage of these cell lines is their ease of cultivation and, especially, their high transfectability, but the formation of large aggregates in suspension cultures has been an obstacle in the development of bioprocesses [5].

To circumvent aggregation, a hybrid cell line, HKB-11, was developed through the fusion of HEK193S and human 2B8 (human suspension cell line derived from Burkitt's lymphoma) cell lines. HKB-11 retains several attractive properties for biotherapeutic production but has shown an unstable expression of EBNA1 during long-term cultivation [35].

HT-1080 is a human cell line derived from human fibrosarcoma that shows an epithelial-like phenotype. HT-1080 cells have been used to produce Dynepo® (***epoetin delta***) using a gene activation technology proprietary of Shire, which was found to have better homogeneity of its tetra-antennary glycans, higher sialic acid content, and no Neu5Gc, compared to the CHO-produced hormone [32, 36].

PER.C6 cells consist of human retinoblasts immortalized by transfection with adenovirus type 5 E1A- and E1B-encoding sequences [37, 38]. They were originally developed to produce adenovirus vectors for vaccine development, but their ability to grow in suspension cultures at high density and to produce high titers of immunoglobulin G (IgG) has turned them into hosts for recombinant protein production.

The CAP (CEVEC'S amniocyte production) system consists of human amniocytes cotransfected with the E1-expressing plasmid containing the E1A, E1B, and pIX function and the plasmid expressing the target protein. The novelty of this system is that only cells that were transfected can grow and divide, while nontransfected cells cease to grow at early passages, allowing the use of a nontumorigenic cell line and not requiring antibiotics for selection. Other advantages of this system include the high yields and human-like glycosylation profiles of the proteins [29, 33].

Finally, HuH-7 human hepatoma cells have been demonstrated to produce PTM patterns similar to those found in human glycoproteins, such as plasma-derived factor IX (FIX), and improved profiles regarding recombinant FIX produced in CHO cells [19].

2.3 Construction of the Gene Expression Cassette

As is universally known, the central dogma of biology states that information encoded in the DNA is transferred to messenger RNA (mRNA) through a transcription process, and then it directs the synthesis of protein via the translation pathway. The set of instructions that allow the mRNA nucleotide sequence to be decoded into the amino acid sequence is known as the genetic code. This theory postulates that genetic information flows in a single direction. However, viral enzymes that enable reverse transcription, i.e. the conversion of mRNA to DNA, have been largely described and used in molecular biology.

In accordance with these principles, it can be deduced that, for a gene to be expressed in a eukaryotic cell, it is not enough to count on the DNA sequence alone, but other sequences that regulate the whole gene expression process must be present. Together, these elements are called the gene expression cassette. Thus, knowing the key control points in the recombinant protein production pathway is essential to get a better protein expression.

Regarding the DNA level, efforts should be made to get an optimum expression vector design and to achieve the correct transgene genome integration, being aware that the chromosomal context will strongly influence gene expression levels ("position effect" [39]). The stability and processing rate of RNA is essential. Although information that governs RNA transcription is encoded in DNA, primary RNA transcript contains the signals that determine the nature and velocity of processing to give rise to mature RNA. Finally, protein must bear the necessary elements to guarantee adequate secretory event progression and the signals to allow posttranslational modifications.

In the next section, the elements required for gene expression in eukaryotic cells will be reviewed.

2.3.1 Elements Required for Gene Expression

Transcriptional Control Elements

As has been mentioned, a typical eukaryotic transcription unit consists of the GOI flanked by upstream and downstream sequences. So the first step toward gene expression involves obtaining the GOI, which can be isolated from either genomic DNA (gDNA) or mRNA, the latter requiring a previous reverse transcription reaction to obtain the complementary DNA (cDNA).

Genomic DNA contains its own regulatory elements so that the success of gene expression depends on the recognition of such sequences by the host machinery. The main disadvantage of the use of gDNA sequences is their length since, unlike cDNA, they contain introns and other elements that make them large, thus complicating their manipulation and reducing the efficiency of their introduction into the cell. In the case of mRNA, the reverse transcription step might introduce errors, although this drawback has been considerably reduced due to the development of reverse transcriptase enzymes with proofreading activity, i.e., with the ability to check the synthesized sequences thoroughly and correct them in the case of finding a mistake [40]. In addition, cDNA does not contain regulatory elements, which must be supplied by the expression vector. However, the short length of mRNA/cDNA sequences stands as their main advantage since it facilitates the cloning and cell introduction processes.

Among the upstream and downstream regions flanking the GOI, they can be classified according to the level at which they operate in transcriptional or translational regulating elements.

Since transcription regulation in eukaryotic organisms comprises a complex process that requires the interplay of different genetic elements, the resulting expression efficiency of a GOI is determined by several factors. The most important ones include the promoter/enhancer sequences, the number of copies of the GOI integrated into the host's genome, and the "position effect," which has been previously mentioned, referring to the structure and elements neighboring such integration sites. Besides, cotranscriptional modifications such as splicing, polyadenylation, capping, and transport to the cytoplasm determine the stability, turnover rate, and translational capability of mRNA [41].

Gene expression starts with transcription by RNA polymerase II (Pol II), which is initiated at the transcription start site (TSS) situated at the 5'end of a gene. The TSS is embedded within a core promoter, one of the most important transcriptional control elements. The core promoter is a short sequence comprising ~50 base pairs (bp) upstream and ~ 50 bp downstream of the TSS, and its main function is to initiate the transcription of the adjacent gene. The core promoter serves as a binding platform for the transcription machinery, formed by Pol II and the general transcription factors (GTFs). In mammals, promoters can be classified into two distinct classes: those that contain the highly conserved TATA-box motif and other upstream sequences, such as the CCAAT box and GC box, and initiate at a single TSS and those with variable CpG-rich regions containing multiple TSS [42].

Promoters used for the expression of recombinant proteins in mammalian cells can be from distinct origins. Viral promoters such as human cytomegalovirus immediate early gene (CMV) promoters have proven high-level expression in a variety of mammalian cells [43], while the simian virus 40 (SV40) promoter allows constitutive expression, which used to be higher in cells that express the large T antigen due to the incorporation of a high vector copy number [44]. However, stable gene expression through time is not always achieved since viral promoters are vulnerable to gene silencing in many cell lines, including CHO [45]. Cellular promoters generally yield lower expression levels than viral ones, but they tend to show higher stability and are used in cell lines that downregulate viral promoters. The human elongation factor-1α subunit gene promoter (EF-1α) is one of the most employed promoters since it permits strong constitutive expression in mammalian cells [46]. Recently, some approaches have been performed to identify and isolate new endogenous CHO cell promoters based on high expression levels in RNA-seq data of CHO-K1 cells cultured in high density [47] and/or histone marks and chromatin states [48], which turned out to be promising alternatives to viral promoters and also constitute the starting point for the design of synthetic promoters for industrial recombinant protein production.

Additionally, promoters can induce constitutive, inducible, tissue-specific, or time-specific gene expression. When the recombinant product is cytotoxic or affects cell growth, the use of a constitutive promoter is not recommended; instead, an inducible promoter needs to be incorporated. Also, inducible systems are effective strategies for the decoupling of the growth and production phases of the cell culture [49]. An ideal inducible promoter should show specificity (no response to an endogen activator and no interference with host physiology), efficiency (no expression or low levels of expression in normal conditions and rapid response after induction), and dose dependency (homogeneous and reliable expression) [41]. Many pairs of sequences/ligands that regulate the promoters of animal cells have been used and cloned into expression vectors, triggering a response to external stimuli (heat, hormones, cytokines, and hypoxia). However, the pleiotropism of these factors has prompted the use of the regulatory elements of evolutionary distant species, such as *E. coli*-derived lactose- (Lac), tetracycline- (Tet), or erythromycin- (E) operon/repressor systems, all of which showing advantages and disadvantages, such as toxicity and considerable basal activity, among others. The tetracycline-controlled Tet-Off and Tet-On gene expression systems are the most commonly used approaches to regulate the activity of genes in eukaryotic cells. They employ DNA elements that regulate the activity of Tet-resistance operon in bacteria. In these systems, the induction of transcription is triggered by the binding of a recombinant transactivator to a synthetic promoter. The transactivator/promoter interaction is modulated by tetracycline or its derivatives, for example, doxycycline (Dox), so that the addition or removal of Tet or Dox results in the control of transcription. In the Tet-On system, when Tet or Dox is added to the culture medium, the transactivator can bind to the synthetic promoter and induce the expression of the regulated gene. Contrarily, in the Tet-Off system, the transactivator can bind and activate the promoter only in the absence of Tet or Dox, activating the expression of the downstream-positioned transgene. One drawback of this method is the basal level or leakage caused by the undesired binding of the transactivator to the promoter even when Dox is not added [22, 50]. This system has been used not only to express toxic proteins but also to control mAb *N*-glycosylation through the inducible expression of key glycosyltransferase genes, achieving a precise fine-tuning of fucosylation and galactosylation [51].

Although core promoters can efficiently direct transcription initiation, they generally have low basal activity, which can be further suppressed by the chromatin state or activated by distant regulatory elements called enhancers. Enhancers bind regulatory proteins, known as transcription factors, and recruit transcription cofactors, modifying the local structure of the chromatin and/or stabilizing the transcriptional complex. In this way, enhancers increase the transcription rate independent of their relative orientation, position, and distance regarding the regulated gene [52]. Enhancers present a variable size (50–1500 bp), and they do not exhibit a conserved sequence; hence, their presence is difficult to predict. Their activity is restricted to certain cellular types or developmental stages [41].

Introns are noncoding DNA sequences that are transcribed into premessenger RNA but are removed by a specific cleavage during a process called splicing to generate the mature mRNA. It is suggested that introns can enhance the efficiency of transcription through different mechanisms, including the stabilization of the RNA structure, the increase of the rate of polyadenylation (see below) and the nuclear export of mRNA, and a less well-understood mechanism termed "intron-mediated enhancement," which increases the processivity of the transcription machinery at the elongation stage [42]. A genome-wide functional analysis of human genes found that approximately 35% of human genes contain introns in the 5'UTR, while, in contrast, 3'UTRs were found to have relatively few introns (5%). For these reasons, their inclusion into the expression vectors at the 5'end of the transcription unit is recommended [53].

Last but not least, there are control elements that are not specifically related to the transcription process itself, but they affect gene expression due to the effect that they exert over the chromatin structure of the site of insertion. When the gene is inserted into the genome, positional effects arise so that the integration site of the transgene is a key factor to the level and stability of the expression. Positional effects include factors associated with the structural organization of the chromatin that influence the transcription of a heterologous gene, such as:

- The condensation state of the chromatin
- The orientation and localization of an exogenous gene regarding other endogenous genes
- The structural elements of the chromosomes of the host cell

The condensation state of the chromatin where the gene is inserted is a crucial factor for its expression. If the region of insertion corresponds to heterochromatin, with highly condensed DNA, there is a high risk of silencing and thus less probable productivity. Contrarily, when the gene inserts in a euchromatin region, with a less compact structure and higher transcriptional activity, then gene expression may be influenced only by other structural elements, which are explained below. The level of histone acetylation determines the condensation state of the chromatin and corresponds to epigenetic regulation, i.e., not directed by genetic sequences.

Regarding chromosomal elements that contribute to the expression of gene cassettes once integrated into the host's chromosomal DNA, they act in *cis* by mechanisms that are different from classical enhancer/silencer effects and comprise distinct sequence motifs, including scaffold/matrix attachment regions (S/MARs), locus control regions (LCRs), insulators, and ubiquitously acting chromatin-opening elements (UCOEs) [54].

S/MARs are AT-rich sequences of about 300 to 3,000 bp and are spaced about every 30 kb in mammalian genomes (*cis*-acting). They associate with nuclear scaffold proteins and form chromatin loops, thus establishing separate chromatin domains and promoting transcription. Many studies have shown that S/MARs can not only increase transgene expression upon the random integration of expression cassettes but also achieve a more stable expression through time [49, 55, 56].

Chromatin insulators are DNA-protein complexes that are able to establish different transcriptional states by controlling the tridimensional organization of the genome [57]. They do not exhibit a transcriptional activity of their own and must be present between the promoter and enhancer/heterochromatin. They allow the isolation of a DNA region through a steric effect (for example, the formation of high-order structures such as hairpin loops), inhibiting the recruitment of histone methyltransferases and thereby affecting DNA methylation, which in turn alters gene expression due to the activation or repression of chromosomal regions. The main drawback of their use in vector design is their large size (about 2.5 kb) [54].

UCOEs are chromosomal elements found upstream of housekeeping genes, which offer an open chromatin conformation in most cellular types due to histone acetylation, making them continuously transcriptionally active. They are defined as CpG-free and methylation-free DNA sequences, also of large size (16 kb), and their efficiency is highly dependent on the UCOE-promoter combination. The inclusion of UCOEs in expression vectors allows achieving a larger number of high-level expression clones. Indeed, the use of UCOEs for enhancing the expression of recombinant proteins in many production cell lines used in industry, such as different CHO lineages, has been largely demonstrated [58].

LCRs are chromosomal elements involved in the regulation of an entire locus (for example, developmentally regulated genes) and are cell-type specific. They are constituted by different *cis*-acting motifs, including insulators, enhancers, and S/MARs, and therefore they exhibit a large size. LCRs are mainly situated upstream of the GOI, and their main activity seems to be their ability to increase transcriptional activity as it was studied in detail for the human developmentally regulated ß-globin locus [54, 59].

Altogether, the inclusion of chromosomal elements into gene expression vectors offers an attractive strategy to generate a favorable transcriptional environment, ameliorating the position effects and reducing the time and cost associated with upstream biomanufacturing stages. Another strategy to reduce the position effects is the addition of factors such as sodium butyrate or trichostatin to the culture medium, which block histone deacetylation and thus induce chromatin structural relaxation, generating transcriptionally accessible genome sites [60].

However, the latest and more desirable approaches to mitigating position effects consist in the development of different targeted integration strategies of the expression cassette to direct the GOI to a transcriptionally active site of the genome, which is known to support a high level of expression (so-called "hotspots"). These strategies not only guarantee a high level of expression of the GOI but also help deal with the high heterogeneity of gene expression upon the random integration of cassettes into the host's chromosomes. Targeted integration strategies include the use of heterologous recombinases such as CRE/LoxP, FLP/FRT, etc. and homologous recombination through double-strand breaks (DSBs), such as zinc fingers and transcription activator-like effector nucleases (TALENs) or the CRISPR/Cas9 system.

Site-specific recombinases that exclusively bind to and recombine short DNA motifs (recombination targets (RTs)) have been identified in some species and could be transferred to heterologous organisms to exploit them for the highly efficient and precise targeted manipulation of genomes. Most of the recombinases can be classified as serine or tyrosine recombinases. Among serine recombinases, the most described are Cre, which derives from P1 bacteriophage and can bind to an RT site named loxP, and Flp, identified in *Saccharomyces cerevisiae*, which recognizes and binds to the RT site named FRT. Tyrosine recombinases can catalyze recombination reactions between two identical recombination targets, and depending on the position of the target sites and their orientation, recombination can result in excision, integration, and inversion, all the reactions being completely reversible. ΦC31 integrase belongs to the family of serine recombinases and catalyzes recombination between the RT sites attB and attP. When a cassette flanked by two attB sequences recombines with a pair of attP sites, the intervening DNA sequence is exchanged in a nonreversible manner. Site-specific recombinases have been used in many ways to find a "hotspot" and direct the GOI to this site in mammalian cells. One of the most employed strategies is called RMCE (recombinase-mediated cassette exchange) and consists of two steps. In the first step (tagging), a DNA sequence, preferably a reporter gene, flanked by a set of specific RTs, is integrated into the genome of a cell by transfection, transduction, or other methods, which will be described later. Upon the random integration of this cassette, exhaustive screening is made to identify clones that exhibit an appropriate expression pattern in terms of level and stability. In the second step (targeting), the cassette containing the GOI also flanked by RT sites is introduced into the selected clones, together with a vector encoding the recombinase. Then a "cut and paste" mechanism occurs, involving the recombinase-mediated excision of both DNA fragments and the posterior exchange of sequences. A prerequisite for RMCE is the use of a set of heterologous RTs that cannot recombine with each other, such as RTs of ΦC31 or mutated Cre and Flp, which have been developed [22, 54, 61].

Homologous recombination consists of a natural process that occurs during meiosis and mitosis with the aim of generating genetic diversity or repairing double-strain breaks that can appear due to cell stress. Although homologous recombination has been evaluated for the targeted integration of external DNA, its low efficiency has limited its application. Nonetheless, the development of methods to specifically induce double-strand breaks (DSBs), which dramatically increase the frequency of homologous recombination, has brought this methodology back to the ring. The first approaches to DSB-induced homologous recombination were based on the use of nucleases such as zinc finger protein-linked nuclease (ZFN) and transcription activator-like effector nuclease (TALEN), which are artificially designed proteins in which a nuclease domain is targeted to a specific DNA sequence. Upon DSB, two repair mechanisms are exploited, including nonhomologous end joining (NHEJ), which consists in the elimination or random addition of nucleotides and homology-directed repair (HDR) that integrates cassettes, resulting in the correction or addition of genes. For more information and details, please see references [62] and [63].

The major inconvenience of the ZFN- and TALEN-based strategies is that they require the de novo design of each DNA-binding protein, resulting in a high cost and a time-consuming task. After the emergence of and the remarkable advancements in the knowledge of the CRISPR-Cas9 system during the last years, this approach has been demonstrated to be the first preference for genetic manipulations because of its high targeting efficiency, cost-effectiveness, and simplicity of design [64]. CRISPR-Cas9 is the most widely used variant and consists in an RNA-guided DNA editing tool comprising two key components: an effector protein (Cas9) and a guide RNA (gRNA). When gRNA binds to Cas9, the complex will search its target, and upon binding to its complementary region in a sequence-specific manner, it will introduce a DSB at a specific region of DNA. The complementary region needs to be flanked by a protospacer adjacent motif (PAM) to be recognized. In this way, endogenous genes can be deleted or transgenes can be integrated [65, 66]. Using this technology, both NHEJ and HDR DNA repair mechanisms in combination with using prevalidated genomic integration sites with high and stable transcriptional activity (hotspots) can lead to the achievement of consistent productivity and cell line stability, especially for CHO cells. Moreover, the CRISPR-Cas9 system can be used for the optimization of recombinant protein production, for example, by knocking out/in different types of genes, such as apoptosis genes, genes that inhibit protein production, or genes that modulate glycan structures [64, 66]. For detailed information on CHO cell line developement and engineering through site-specific integration, it is recommended to refer to [67].

Other events that must occur for transcription to successfully proceed and that require a signal sequence include capping, polyadenylation, and transcription termination. Capping consists of the addition of 7-methylguanosine to 5'mRNA end and protects the transcript from degradation. In general, the promoter provides the capping signal. Polyadenylation is a cotranscriptional process in which a 3′ site-specific cleavage is followed by the polymerization of a long extension of approximately 200 adenosine molecules, known as polyA tail, which is crucial for the formation, stability, and translatability of mRNA. In most eukaryotic organisms, the polyadenylation signal is a highly conserved hexamer motif (AATAAA) located between 10 and 30 bp downstream from the stop codon and immediately upstream of a less-defined U- or GU-rich region [68]. In expression vectors, the most commonly used polyA signals are those corresponding to the bovine growth hormone (bGH) and murine β-globine genes, among others. Transcription termination signals are found between the polyadenylation signal and the 3'end of mRNA. The precise mechanism of termination remains unclear, but it is known that the termination sequence is required to prevent transcriptional interference, a phenomenon that can impair the expression of downstream genes by reducing the accessibility of the promoter to RNA Pol II [69, 70]. Well-characterized prokaryote transcription terminator sequences are usually incorporated in expression vectors.

Translational Control Elements

Up to this point, many transcription control elements have been revised. But, as has been mentioned before, the regulation of expression can also be achieved at the translational level, i.e., through the control of protein synthesis. Among translational regulation elements, two regions can be identified: 5'untranslated region (5'UTR) and 3'untranslated region (3'UTR).

5'UTR is a regulatory DNA region placed at the 5'end of all protein-coding genes that is transcribed into mRNA but not translated into protein. It contains many regulatory sequences that cooperate to control translation initiation, a limiting step in protein synthesis. It has been observed that when the start codon (AUG) is embedded in the consensus sequence GCC(A/G)CCAUGG, known as Kozak's sequence, translation efficiency is increased; also, the presence of other upstream AUG codons severely impairs translation. In addition, the 5'cap structure described above is essential for the efficient translation of mRNA and serves as an interaction site for different eukaryotic initiation factors and to promote the binding of 40S ribosomal subunits and other proteins, which together make up the preinitiation complex. In general, the 5'structure and nucleotide content of the 5'UTRs that enable efficient translation are short, have low GC content, and are relatively unstructured [42].

Regarding 3'UTR, it corresponds to a region placed downstream of the protein-coding sequence, i.e., at the 3'end. Many regulatory processes, including transcript cleavage, stability and polyadenylation, translation, and mRNA localization, are influenced by this region. 3'UTR includes sequences able to alter mRNA stability, AU-rich elements (AREs) being the most well-characterized ones. AREs binding proteins (ARE-BPs) generally promote the decay of mRNA in response to a variety of intra- and extracellular signals so that highly unstable mRNAs have been found to contain multiple ARE motifs. However, other ARE-BPs that modulate translation have also been either positively or negatively described [42, 71].

Importantly, the sequence must bear a stop codon, which binds to the "release factor," promoting the disassociation of the translational complex and, thus, the end of the polypeptide synthesis. Three natural stop codons are generally employed (UAA, UAG, UGA), and the existence of certain adjacent nucleotides can influence the efficiency of the process.

Finally, it is worth mentioning that, for high-level expression genes, an important degree of preference for certain codons has been reported. In these cases, codon optimization constitutes a common strategy to increase protein levels. The goal of codon optimization is to enhance the degree of translation elongation that is prevented due to differences in codon usage and transfer RNA (tRNA) abundance among different species. Due to the degeneracy of the genetic code, which implies that the majority of amino acids can be encoded by more than one codon, this approach can modulate codon usage to best fit the host species. It has been demonstrated that codon usage can have a marked effect on protein expression levels, and accordingly, codon optimization has been reported to increase protein expression by 1000-fold [72]. For example, if a human DNA gene needs to be expressed in CHO cells, it is advisable to check the codon usage frequency in hamsters and optimize it for this species.

Expression of Multiple Genes

The majority of mRNA in animal cells is monocistronic; i.e., it encodes for a single protein. In some circumstances, it might be necessary to express two or more heterologous genes. For instance, it may be desirable to express the GOI together with a selection marker or to produce complex proteins formed by more than one subunit, such as antibodies (light and heavy chains), glycoprotein hormones (heterodimers), etc.

In these situations, some different strategies are available. Multiple genes can be expressed in the same vector using a different promoter for each gene. In this case, the expression of each protein will be independent. Also, fusion proteins can be designed by using a linker under the control of a single upstream promoter. The disadvantage of this approach is that proteins will be expressed not as independent entities but as a unique construction. Multiple expression vectors can be employed, each one bearing a different gene under the influence of its own promoter and a different selection marker. Under this circumstance, the expression of each gene can be considerably different since from the beginning of the strategy, the amount of each vector that is introduced into the cell can be different (although the same method for transfection/transduction is used, the efficiency can vary between plasmids or events) so that a different copy number can be inserted in the genome and in different sites of the genome. For that reason, when two subunits of a heterodimeric protein are expressed, one of them can be overproduced and secreted as a free subunit to the culture supernatant, bringing about subsequent issues in the purification process. Another strategy consists of the expression of large monocistronic transcripts containing the cDNA of each gene separated by sequences that encode proteolytic cleavage sites, for example, of furin. Thus, a single protein is expressed, and after protease digestion, the two proteins are separately produced [41].

Finally, a widely used methodology is the utilization of internal ribosomal entry sites (IRES), which are inserted between the two genes to be expressed. The transcription of both genes is directed by the same promoter, generating a single mRNA. While the translation of the upstream gene is 5'cap-dependent, the presence of the IRES sequence on the mRNA permits the 5′ cap-independent translation initiation of the downstream gene. In this way, two different proteins can be translated from a single mRNA. The advantages of applying IRES include the fact that a single promoter can be used to guide the transcription of a polycistronic mRNA, achieving a more homogeneous expression ratio of both genes, for example, for monoclonal antibody production. Also, if the approach consists in the insertion of the selection marker gene as the downstream gene, the successful transcription of the upstream gene is guaranteed by the selection strategy. This reduces the frequency of selection marker expression that is not accompanied by GOI expression, a situation that might occur after gene fragmentation when a vector containing two genes driven by two different promoters is used [23].

Targeting Signal Sequences and Tags

Proteins' amino acid sequences frequently contain localization signals, whose function is to guide the newly synthesized polypeptide through the different intracellular routes to reach their final localization. These signals are *N*-terminal sequences between 15 and 60 amino acids that are removed when proteins reach their final destination, including secretion to extracellular media, cytosol, the nucleus, peroxisomes, mitochondria, the endoplasmic reticulum, the Golgi apparatus, or the cellular surface.

When expressing a recombinant protein in mammalian cells, especially if it is aimed to be used as a biotherapeutic, it needs to be secreted to the culture supernatant for it to be soluble and easily purified. For that purpose, signal peptides are employed. If the GOI does not have a signal peptide of its own, for example, if it corresponds to an intracellular protein, it is possible to insert a well-characterized sequence from another protein. It has been observed that certain signal peptides are more efficient than others to drive the secretion of the protein of interest, improving global productivity. For example, the use of *Gaussia* luciferase signal peptide allows for achieving highly efficient luciferase levels, as well as the one corresponding to human chymotrypsinogen, compared to other signal peptides, such as the one from human albumin, which shows poor relative activity [73]. Recently, by using bioinformatics-led and synthetic design approaches, a toolkit of signal peptide elements for use in CHO cell vector engineering has been developed [74].

Other sequences can be added to proteins with different objectives. These tags can comprise a few amino acids (for example, His-, FLAG-, c-myc-tags) to large polypeptides (BSA, protein A, GST, GH, GFP, etc.). Tags can be fused to both *N*- and *C*-ends of the protein, and a proteolytic cleavage site (for example, from thrombin, enterokinase, etc.) can be added for removal after use. These tags may have different purposes. One of them is to confer a novel property, such as a higher expression, solubility, stability, and half-life; in this case, tags must not be removed since they will be a part of the new protein with improved properties. Also, tags are of great applicability to confer specific affinity to a certain ligand in order to simplify protein isolation and purification and to enable its detection or selection. The DNA sequence of His-tag, which encodes for a string of six to nine histidine residues, is the most widely used tag in vectors to produce recombinant proteins. The result is the expression of the protein of interest fused to 6xHis in its *N*- or *C*-terminus, which can be easily purified and detected due to its binding to several types of immobilized metal ions (including nickel, cobalt, and copper), under specific conditions, using a methodology called IMAC (immobilized metal-affinity chromatography) [75]. In addition, anti-His-tag antibodies are commercially available for use in several assay methods. Some tags can gather the two mentioned properties; i.e., both serve to improve the properties of the protein of interest and simultaneously confer specificity to facilitate its purification and detection. For example, reference [76] describes the development of two bifunctional peptide tags that are exhibited between four and six *O*-glycosylation sites, a posttranslational modification that confers increased solubility, stability against physicochemical factors and proteases, and improved pharmacokinetics to glycoproteins, and simultaneously, they contain a 4-amino acid epitope that is recognized by a monoclonal antibody, facilitating the purification and detection of fusion proteins.

2.3.2 Transient and Stable Expression

Once introduced into the cell, a gene can be expressed transitorily, during a limited time span, and then eliminated from the population due to successive cell divisions. The ability to express a heterologous gene during a short time period is called transient expression. However, a small fraction of the transfected cells can incorporate the exogenous DNA by random recombination, leading to a stable expression of the gene. Stable expression refers to the fact that every time those cells enter into mitosis, their genome will be replicated, and so the integrated heterologous gene will be transferred to the next generations.

Thus, there are many differences between the transient and stable expression of a gene, and the choice of one or another method will depend on many factors, including the purpose of gene expression and the availability of vectors and reagents, among others.

The main difference between transient and stable expression is that in transient expression, the protein of interest is produced between 16 and 96 h from DNA incorporation but is not transferred to subsequent generations, while in stable expression, the gene is inserted into the cell's genome and transmitted to the next generations. Transient expression allows the fast production of protein for characterization or to evaluate the integrity and functionality of different expression vectors or the efficiency of distinct transfection methods. In contrast, stable expression requires a long period of time for product generation; however, once a producer clone is isolated, an unlimited source of protein is obtained. The transient expression process is simple and does not demand a selection marker (see below), while it is mandatory to employ a selection/amplification marker to select and isolate those cells that have integrated the gene into their genome. An advantage of transient expression is that there is no need to perform clone screening, a tedious and time-consuming task that is essential for stable expression. Importantly, no "position effects" occur during transient expression since the vector does not integrate into the genome; contrarily, a strong "position effect" can be observed in stable expression, together with a different copy number integrated into the genome. A disadvantage of transient expression is that large amounts of circular DNA are needed to produce a significant mass of the recombinant protein (since cells should be transfected simultaneously), while stable expression can be achieved with a small quantity of circular or linear DNA (to promote genome insertion) and the posterior isolation and amplification of the producer clone. In general, low expression levels are attained in transient expression experiments (between 1 and 100 $mg.L^{-1}$), reaching higher expression levels in stable expression (between 5 and 5,000 $mg.L^{-1}$). To sum up, while transient expression represents a low-cost, short process for transitory protein expression, the development of stable recombinant cell lines involves a long process comprising several months and implies a higher cost. In general, the selection reagent is added to the cells about 2 days after transfection, and the selection phase normally lasts for some days to weeks. Those cells that survive the selection pressure, i.e., which contain the selection marker and probably the GOI, are then isolated and characterized. Although the process of stable cell line generation is time-consuming and labor-intensive, it is the recommended practice to produce a large amount of therapeutic protein with consistent product quality [5, 19, 41].

A selection system can thus separate the cell population that incorporated the expression vector into its genome (stable recombinant cells) from those that did not. The selection cassette contains a gene that confers the cells a new property that will be absent in the nonstable cells. A common strategy consists of using a strong promoter to drive the expression of the GOI and a weak promoter to direct the expression of the selection gene so that, theoretically, those cells that show high resistance to the selection pressure might strongly express the GOI. Generally, selection markers constitute sequences that confer resistance to cytotoxic drugs (commonly antibiotics but also other compounds). The most common antibiotics/resistance genes include the pairs neomycin/neomycin phosphotransferase II, hygromycin/hygromycin-B phosphotransferase, puromycin/puromycin *N*-acetyltransferase, and zeocin/Sh ble gene, among others [77].

Other selection markers not only allow the isolation of cells that have stably incorporated the GOI into their genome but also permit gene amplification to achieve.

a higher expression level. These systems are called gene coamplification markers, with dihydrofolate reductase (DHFR) (technology that originated in the 1980s [78]) and Lonza's glutamine synthetase (GS) [79] being the most popularly used ones in the biotherapeutical industry. DHFR catalyzes the reduction of dihydrofolate to tetrahydrofolate (TF4), which is required for the biosynthesis of many essential molecules, including glycine, purine, and thymidylic acid. The DHFR system can be used in the mutant strains of CHO cells in which the *dhfr* gene was mutated or deleted, such as DXB11 and DG44. These auxotrophic mutants need a culture medium containing glycine, hypoxanthine, and thymidine (GHT) to survive. In the DHFR system, a GOI is transfected into the host cells, together with the dhfr gene. When the transfected cells are cultured in media without GHT, only cells bearing the gene construction in their genomes can survive. In addition, the DHFR system can be used as an amplification methodology by using methotrexate (MTX), a DHFR inhibitor. When cells are incubated with increasing concentrations of MTX, a DHFR gene amplification phenomenon that extends along 130–1,300 kbp occurs so that the adjacent GOI is also amplified. Using this gradual selection procedure, cell lines with hundreds and up to thousands of copies of plasmid constructions have been obtained, with the subsequent high productivity of the protein of interest. It is important to perform step-by-step selection process since it has been reported that a single selection step can result in MTX-resistant or MTX-altered transport mutants.

The GS system uses the *gs* gene as a selectable marker to produce the enzyme GS, which catalyzes the conversion of ammonia and glutamate into glutamine. MSX irreversibly binds to GS and prevents the formation of glutamine. Cells that overexpress GS are resistant to high doses of MSX through a gene amplification system. Unlike the DHFR system, the GS system can reach high expression levels after a single round of selection and amplification, shortening the timeline required for cell line generation. Moreover, an advantage of this system is that it reduces toxic ammonia accumulation in the culture medium, and it continuously provides glutamine as an energy source.

In conclusion, when a GOI is coexpressed with a selection marker, such as an antibiotic, the generation of cell clones with high productivity will depend on gene integration into transcriptionally active genome regions, while when the GOI is coexpressed with an amplification marker, highly producing clones will be obtained due to the achievement of a high GOI copy number.

Reporter markers are constituted by genes that encode for proteins that are easily monitored by simple and sensitive methods. These reporter genes or markers are used in many applications, including to monitor DNA delivery methods, to study promoters and regulatory elements or characterize transcription factors, to analyze cellular trafficking and compartments of the protein of interest, to develop high-throughput screening (HTS) methods, etc.

The most widely used reporter markers include bacterial chloramphenicol acetyltransferase (CAT), which is detected using a radioactive or fluorescent substrate (chloramphenicol); *E. coli*-derived beta-galactosidase, whose activity can be detected by using an X-gal substrate (which is transformed into a colored product) or FDG (fluorescein di-beta-D-galactopyranoside) (which releases fluorescein after enzymatic reaction); secreted alkaline phosphatase (SEAP), which also employs fluorescent or chemiluminescent substrates; luciferase (isolated in unicellular seaweed, marine bacteria, and fish), whose substrate (luciferin) is oxidated producing oxyluciferin, CO_2, and a photon, with the intensity of light emitted during the reaction being measured with a luminometer; and, finally, green fluorescent protein (GFP) or other fluorescent proteins, which can be directly determined by measuring their intrinsic fluorescence [41].

2.3.3 Heterologous Expression Systems in Mammalian Cells

All expression systems comprise two components: a vector and a host. The most commonly used mammalian hosts have been described in the section *Choice of Host*. Vectors are defined as DNA or RNA molecules that have been genetically manipulated to contain the GOI and other elements that regulate its expression.

To sum up what has been described until now, a mammalian expression vector should contain the following elements:

- A eukaryotic expression cassette
- A eukaryotic selection cassette
- A prokaryotic cassette

As displayed in Fig. 2, the eukaryotic expression cassette is formed by the enhancer (Enh), the promoter (Prom), and the structural elements (SE), located upstream of the multiple cloning site (MCS), where the open reading frame (ORF) of the GOI will be inserted. The polyA and transcription termination (TT) signals must be added downstream of the MCS. After the promoter, the 5'UTR is positioned together with the translational enhancer (TE) or Kozak's sequence, in which the ATG start codon is embedded. In case of need, the targeting signal (TS) or signal peptide (SP) must be positioned just upstream of the MCS, where the ORF of the GOI is cloned (which can be fused to a fusion protein or tag (TAG), preceded or not by a protease cleavage site). Immediately after the MCS, the STOP codon and the 3'UTR sequence are inserted. To develop a stable cell line, a eukaryotic selection cassette is required, comprising a selection marker, such as zeocin, neomycin, hygromycin, or puromycin, whose expression can be driven by its own promoter and needs a polyA and a termination sequence. Finally, the prokaryotic cassette is needed to perform all the genetic steps required during the construction of the plasmids and their amplification for subsequent transfections. These procedures are generally performed in *E. coli* so that a bacterial replication origin and a selection marker as ampicillin need to be incorporated into the prokaryotic cassette.

Vectors can be classified into plasmid vectors and viral vectors, and accordingly, methods to introduce them into cells are referred to as transfection and infection/transduction.

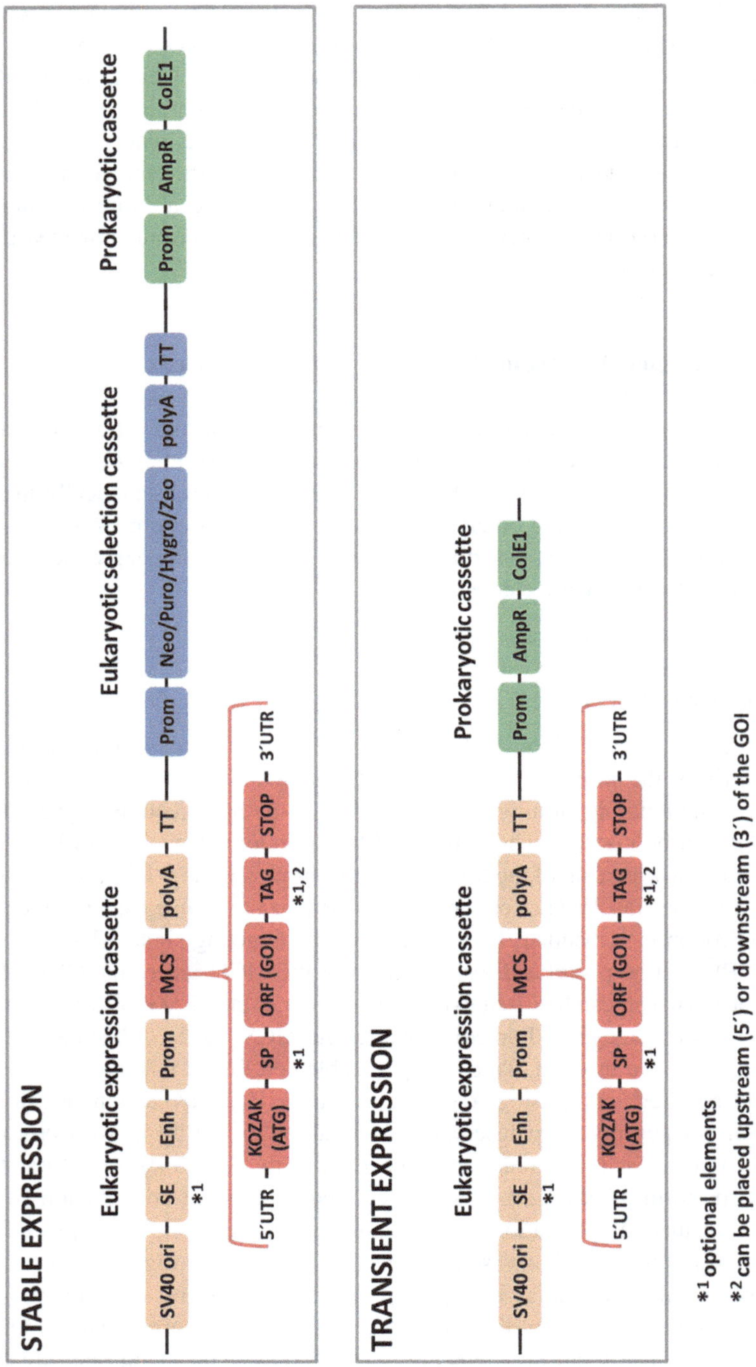

Fig. 2 Essential and optional elements of a mammalian expression vector

2.4 *Methods for Introducing DNA into Mammalian Cells*

2.4.1 Transfection Versus Infection/Transduction Methods

DNA or RNA can be introduced into cells mainly through two methods: transfection and infection/transduction.

Transfection makes use of physical or biochemical methods through which the gene is introduced into the cells, while infection/transduction is a biological process mediated by a virus by binding to cell receptors. Such a virus bears the GOI in its genome.

Transfection is a simple and fast process that requires few reagents, while the generation of the viral vector and the subsequent transduction constitute a more laborious and time-consuming process. This is because it requires two steps: a first step comprising the assembly of the viral particle and a second step involving transduction with the arranged viral vector. Transfection efficiency depends on the interaction or combination between the method and the cell line and does not require specific receptors. Transduction is highly efficient in cell lines containing the viral receptors, but it is restricted to such cells. A major drawback of biological methods is the risk of immunogenicity and the requirement for higher levels of safety containment such that a nonviral gene transfer is sometimes preferred by therapeutic protein manufacturing companies [5, 80].

2.4.2 Transfection Methods

Transfection methods are usually employed to introduce plasmid vectors into cells. Plasmid vectors are double-stranded circular naked DNA molecules of about 2 to 20 kbp in length, which can stay as episomes or integrate into the host's genome. They contain all the abovementioned cassettes to guarantee the transient or stable expression of a GOI in mammalian cells. Linearization of plasmid vectors is sometimes performed to promote the integration of the cassette into the genome.

There are many approaches that can be used to introduce DNA into cells, and, in general, there is not a universal method that functions best for all cell lines and applications. The choice of a transfection method is influenced by several factors, including cost, reproducibility, toxicity, mechanism of delivery, ease of use, availability, and efficiency. Transfection methods can be classified into physical methods, which use physical strength to enhance the permeability of the cell membrane but can injure cells during the process (electroporation, microinjection, biolistic, among others), and chemical methods, which are based on the formation of complexes with the DNA material to facilitate its transport inside the cells (phosphate-calcium coprecipitation, cationic polymers, and lipofection) [81].

Physical Transfection Methods

Electroporation is based on the application of an electric field that induces the generation of transient pores in the membrane through which DNA can enter the cell. In detail, after the application of the electric pulse, the plasma membrane is permeabilized and the DNA is electrophoretically pushed onto the membrane side that faces the cathode. DNA aggregates are inserted into the membrane and can be internalized through endocytosis (DNA in vesicles) and/or through electropores (free DNA). Finally, the introduced DNA needs to cross the cytoplasm and reach the nucleus in order to achieve gene expression. It is important to note that the electroporation procedure also contributes to this traffic since it produces the disassembly of the cytoskeleton and enhances the motility of DNA molecules. The method is appropriate both for transient and stable expressions. The transfection efficiency depends on the cell line, but due to its high performance, it is generally recommended for suspension or it is hard to transfect cell lines. Although a high initial investment is required to acquire an electroporation device, no other delivery reagents are needed. It is important to take into account that a high voltage can induce high cell mortality, so the pulse time and the intensity of the electrical field must be determined empirically to achieve the highest efficiency without damaging cells. Suitable electroporation equipment for the transfection of up to 200 billion cells is available in the market, with very good transfection rates and the preservation of high cell viability [82, 83].

Another physical transfection method is microinjection, which consists of positioning the target cell under a microscope and fixing it with a pipette connected to a vacuum pump. DNA solution is directly injected into the cellular nucleus with a glass capillary microneedle using a microinjector. This method is highly efficient and is used when high transfection efficiency is needed or to control the copy number. It is generally restricted to the generation of transgenic animals in order to inject the DNA of interest into one of the pronuclei of a zygote. The procedure is highly precise, but it is time-consuming and very costly and requires some skills and much training [81].

Biolistic or particle bombardment is a method that employs DNA-coated gold or tungsten particles for cell transfection with a gene gun. A high-pressure helium discharge gives the particles the velocity needed for crossing any physical barriers and for reaching the nucleus. The method is suitable for tissues, cells, and organelles, and results are simple, fast, and efficient, despite the high cost of the equipment. Unlike electroporation, after cell bombardment, cell viability remains above 90%; i.e., the method does not harm the cells.

Other physical transfection methods include laser-assisted transfection, elevated temperature, ultrasound, and hydrodynamic gene transfer [81].

Chemical Transfection Methods

Calcium phosphate coprecipitation condensing DNA into particles is one of the first nonviral techniques for the introduction of plasmid DNA into mammalian cells [34]. Basically, the technique consists in mixing the purified DNA with calcium chloride and phosphate buffer at neutral pH. Under these conditions, a fine, visible precipitate is produced and then incorporated by the cells through phagocytosis. The transfection efficiency can be influenced by variables such as pH, precipitate quality, incubation time, DNA concentration, temperature, and cellular state. Indeed, for CHO cells, it was demonstrated that the phase of the cell cycle has a significant influence on transfection efficiency and that cellular arrest at GO/G1 and S phase resulted in a higher productivity of the GFP reporter gene [84], particularly when combined with osmotic shocks. The main advantages of this method are its low cost, the nonnecessity of sophisticated equipment, and its application to a wide range of adherent and suspension cell lines. However, a crucial drawback involves its low transfection efficiency (near 10%), which demands exhaustive optimization and standardization of the reaction conditions. Besides, there is a risk of cytotoxicity and a high DNA mutation rate [19, 41].

Another chemical transfection method involves the use of cationic polymers, such as diethylaminoethyl-dextran (DEAE-dextran), polyethylenimine (PEI), and polybrene-dimethyl sulfoxide (DMSO). These cationic polymers form complexes with the negatively charged nucleotide backbone, which are then taken up by cells, mainly by endocytosis. If no lysosomal degradation occurs, DNA can be released to the endosome and reach the nucleus, resulting in transgene expression [82].

Finally, one of the most commonly used methods for both stable and transient transfections, evidenced by the wide range of commercial reagents available, is lipid-mediated gene transfer (lipofection). The incubation of plasmid DNA with cationic lipids results in the formation of small unilamellar liposomes, which are actively incorporated into cells by endocytosis. In general, cationic liposomes contain an amphiphilic cationic lipid (2,3-dioleyloxy-*N*-[2-(sperminecarboxamido)ethyl]-*N*,*N*-dimethyl-1-propanaminium trifluoroacetate (DOSPA); 1,2-di-O-octadecenyl-3-trimethylammonium-propane (DOTMA); etc.) and a neutral helper lipid, generally 1,2-dioleoyl-sn-glycero-3-phosphoethanolamine (DOPE), which stabilizes the formation of the liposome. Different lipid-based transfection reagents have been compared, and their performance was shown to highly depend on the cell line used for transfection, as well as other process parameters. The benefits of lipofection include high transfection efficiency (between five and 100 times higher than coprecipitation with calcium phosphate) and the great versatility of applications: it can be used for transient and stable transfections, for continuous or primary cell lines, and for adherent and suspension cultures. Using this methodology, DNA fragments of up to 130 kbp can be introduced into the cells. The lipid formulation and concentration, the DNA:lipid ratio, and the incubation time should be optimized for each cell line. Unfortunately, all lipid-based transfection reagents commonly come along with very high costs, which limits their usage to small-scale transfections. In fact, for the large-scale transient expression of recombinant proteins, other less expensive transfection methods are recommended, such as the use of PEI [19, 82].

For both physical and chemical methods, some considerations should be taken into account before starting the transfection experiments. It is important to characterize the host cell line in terms of growth properties, susceptibility to different transfection methods, age of culture, and nutritional requirements. Besides, to select the appropriate transfection method, it should be defined if the expression needs to be transient or stable and to consider the size and quality of DNA. If necessary, methods for extracting DNA from prokaryotic cells should be optimized. In addition, the use of some particular cell culture medium additives or supplements should be taken into account since serum or antibiotics might interfere with the transfection process, and thus exhaustive washing steps should be included. Finally, an adequate method should be available for the detection and/or quantification of the protein of interest.

2.4.3 Transduction: Viral Vector Systems

Viral vectors take advantage of the natural ability of a virus to deliver its genetic material into a target cell while being genetically modified to remove its ability to replicate [65, 85]. Thus, they have become a valuable tool for obtaining recombinant cell lines with high productivity. The first attempts to express heterologous genes in mammalian cells were performed using viruses as vectors, and for that reason, the process that uses a complete virus to insert a GOI into the host's genome has been called infection. Although viruses can trigger an immune response in the host, they also display evasion mechanisms, allowing viral replication. In this way, if nonessential viral genes are replaced by exogenous genes, they can be used to transduce cells and generate recombinant cell lines.

Four main types of recombinant viruses have been developed: adeno-associated vectors (AAVs), adenoviral (Ad) vectors, retroviral vectors (RVs), and lentiviral vectors (LVs). These viruses belong to different virus families, and they differ in the nature of their nucleic acid (DNA or RNA, single or double stranded) and in their size. Besides, viruses can be distinguished according to their ability to transduce dividing and/or nondividing cells; this influences transduction efficiency, which will be higher if viruses can transduce both types of cells and not only the dividing ones. Another way viral vectors are categorized is as integrating or nonintegrating vectors, based on their ability to insert into the host's genome. In general, nonintegrating vectors transduce quiescent or slowly dividing cells to avoid rapid loss after cell division. Vectors that integrate into the genome can generate stable recombinant cell lines, while nonintegrating vectors can be candidates for vaccine production, producing the antigen of interest transiently and then being removed after successive cell divisions, bringing about higher safety. Viral vectors' tropism can be altered by different methods, including the modification of capsid proteins for naked viruses or the alteration of surface glycoproteins for enveloped viruses. Anther factor to consider when selecting a viral vector is the vector packaging capacity [65].

Adeno-associated natural virus is composed of a 4.7 kbp single-stranded DNA packaged in a nonenveloped capsid. The viral genome consists of three genes, *Rep*, *Cap*, and *AAP*, flanked by inverted terminal repetitions (ITRs) that function as the viral replication origin and the packaging signal. The *Rep* gene encodes for four nonstructural proteins that are required for genome replication, transcriptional regulation, and packaging. The *Cap* gene encodes for three structural proteins (VP1–3) that are arranged to set up the viral capsid, while the *AAP* gene encodes for the assembly-activating protein. AAV is a helper-dependent virus, which enhances AAV production or directly provides the proteins necessary for its replication. In the absence of a helper virus, AAV persists in the host as an episome. To generate the recombinant AAV, the expression cassette is inserted between the ITRs, replacing the *Rep* and *Cap* genes, which are provided *in trans*, together with the helper virus, in the production vector step. Different serotypes can be produced by just changing the capsid protein in the virus helper and maintaining the same expression cassette between the ITRs. In this way, tropism to different tissues or cell types can be altered, a strategy called pseudotyping. The advantages of AAVs include the absence of pathogenicity or cytotoxicity at doses used for transduction. Besides, when administered in vivo, for example, for gene therapy trials, AAV capsids are less immunogenic than those of AV. AAVs can persist as episomes in nondividing cells for years, and their low genomic integration rate reduces the risk of insertional mutagenesis. In addition, AAVs have broad cell tropism. The main disadvantage of this system is its limited packaging capacity (less than 5 kbp), which restricts its use for large genes. The process consists of two steps: first, the cotransfection of the packaging cell line with the AAV DNA, together with a plasmid containing *Rep* and *Cap* genes, and infection with a helper virus, for example, Ad, and, second, the harvest of assembled AAVs and the transduction of the producer cell line [65, 81].

An Ad virus is a nonenveloped virus exhibiting an icosahedral capsid that contains the double-stranded DNA genome, which ranges from 26 to 45 kbp and is flanked by two ITRs. Ad infection is mediated by the binding of the adenovirus fiber protein to the coxsackie and adenovirus receptor (CAR). The most used serotype (Ad5) employs the cell-surface integrin $\alpha V\beta 5$. Once endocytosed, the capsid is partially disassembled and the viral genome is translocated to the nucleus, where it remains episomal. Despite the existence of more than 50 Ad serotypes, most Ad vectors have been produced from the Ad5 virus. In the first Ad vector generation, E1A and E1B genes were suppressed, allowing the insertion of up to 7.5 kbp transgene. The expression of the Ad vector in vivo is generally transient due to the expression of nondeleted viral genes, eliciting immune responses to the expressed viral proteins. For these reasons, Ad vectors can be used for vaccine generation. The principal advantage of Ad includes its large packaging size (about 30 kbp) and its broad tropism, including dividing and quiescent cells. The main drawback of this system is the strong immune innate response to the Ad capsid; however, this "disadvantage" has been exploited for the development of vaccine vehicles since Ad vectors are highly efficient for introducing exogenous DNA into cells and eliciting intrinsic immune responses in the host. A very recent example is the development of ***anti-SARS-CoV-2 vaccines*** based on Ad platforms. Current adenovirus-vector-based vaccines against SARS-CoV-2 infection are JNJ-78435735 by Johnson & Johnson (Janssen), along with the Beth Israel Deaconess Medical Center; AZD1222 by Oxford-AstraZeneca; Sputnik V and Sputnik Light by the Gamaleya Research Institute of Epidemiology and Microbiology; and CanSino vaccine by Can-Sino Inc. As an example, the unique feature of Sputnik V is the use of two types of replication-deficient adenovirus (Ad26 and Ad5) to clone Spike proteins, which are used for the first and the second vaccine doses, respectively [86].

Retroviruses are single-stranded viruses formed by a capsid and an envelope containing two copies of a positive RNA strand. The RNA genome encodes for *gag*, *pol*, and *env* genes, which are flanked by long terminal repeats (LTRs), which work as enhancers and promoters. The *gag* gene encodes for the structural protein (capsid protein), the *pol* gene encodes for the reverse transcriptase and the integrase, and the *env* gene encodes for envelope proteins. Lentiviruses are a subcategory of retroviruses that contain six additional genes that encode for accessory proteins *tat*, *rev*, *vpr*, *vpu*, *nef*, and *vif*, which are important for viral replication, binding, infection, and release. To produce the infection (natural virus) or transduction (recombinant vector), envelope glycoproteins bind to cellular receptors, and the virus is released into the cell cytoplasm. Then viral reverse transcriptase generates a double-stranded DNA genome from the viral original DNA. Depending on the type of retrovirus, double-stranded DNA can enter the nucleus of nondividing cells through nuclear pores (as, for example, lentiviruses) or wait until the nuclear membrane is disseminated during cell division (as gamma retroviruses). In both cases, once in the nucleus, double-stranded DNA integrates into the host's genome [65].

One of the most employed retrovirus vectors is lentiviruses because of their ability to transduce both quiescent and dividing cells. The most widely used lentiviral vectors derive from human immunodeficiency virus (HIV) and are capable of transducing a broad variety of cell types to integrate into the genome and to produce a long-term expression of the transgene. To minimize the risk of generation of replicative competent lentiviral particles (RCLs), different generations of lentiviral vectors have been developed. Initially, lentiviruses were packaged using three plasmids: one of them encoding the transgene flanked by LTRs from HIV; another one containing *gag*, *pol*, and accessory genes; and a third one encoding the envelope protein. To increase safety, the last lentiviral vector generation, called third-generation lentivirus, uses four plasmids with the aim of minimizing recombination events that would lead to the development of a replication-competent virus:

- The transference vector, which contains the GOI, together with all the *cis-acting* sequences needed to produce the genomic RNA and its packaging. These include the LTRs, the packaging signal Psi, a short sequence of the *gal* gene (which has been demonstrated to generate high vector titers), the rev response element (RRE, needed to bind Rev. protein, which promotes RNA exportation from the nucleus to the cytoplasm), a central polypurine tract (cPPT, which promotes viral retrotranscription), a promoter (which can be, for example, the viral promoter CMV), and the woodchuck hepatitis virus postranscriptional regulatory element (WPRE, which stabilizes mRNA).

– The packaging system, constituted of three plasmids that contain the *trans-acting* factors, including the pMDL or packaging plasmid (which contains HIV *gag* and *pol* genes), the pRev plasmid (which contains a rev sequence), and the envelop plasmid (expressing the envelop protein; the native HIV-1 glycoprotein is generally restricted to CD4 positive cells; thus, to promote the uptake of a wider range of cell types, alternative envelopes, such as the vesicular stomatitis virus G glycoprotein (VSV-G), are used).

Packaging cell lines are then cotransfected with the four plasmids, HEK-293 being one of the most popular cells due to their high transfectability, which allows the achievement of high lentivirus titers. Culture supernatants are then harvested and incubated with the host cell line, for instance, CHO cells, for transduction.

Compared to AAVs, the packaging size of lentiviruses is larger (up to 9 kbp). Lentiviral vectors pseudotyped with VSV-G protein exhibit broad tropism and can offer a long-term expression of the GOI. The main disadvantage of this system is the risk of insertional mutagenesis at the site of insertion. Applications of lentiviral vectors comprise not only the generation of cell lines for recombinant protein production but also the ex vivo delivery of DNA sequences for gene therapy [87].

2.5 Screening and Cloning Methods

Once the recombinant cell line is generated using the vectors that contain the expression cassettes for the recombinant protein and the selection and/or coamplification markers, a selection process is performed with the selection drug, for example, the antibiotic or MTX or MSX, in order to increase cell line protein productivity [23].

Besides being a regulatory requirement, in practice, single-cell cloning is necessary because the protein productivities of the transfected, selected, and gene-amplified cell populations are diverse. The heterogeneity of clones is generally attributed to the random integration of the GOI into the genome since, as we have thoroughly explained above, the chromosomal context exerts a strong influence on the transcription rate. Clonal variations also occur because of different transgene copy numbers and the disruption of the genome by gene amplification, thus necessitating considerable screening efforts to select production clones with the desired attributes [30].

Undoubtedly, the most frequently used method to screen for, identify, and isolate high-producer clones is the process named limiting dilution cloning. In this method, transfected or transduced cells are diluted to a low concentration and then plated in multiwell plates such that, on average, only one cell is present per two or three wells and expands to form a colony from a single cell. This method is generally chosen due to its simple operation and no sophisticated equipment requirement, but it has some disadvantages, which limit its application. It is a time-consuming and laborious method that requires multiple rounds of subcloning to guarantee monoclonality.

Also, it demands additional steps related to cell culture and protein tittering (for example, by enzyme-linked immunosorbent assay (ELISA)) to determine the productivity of the clones. Furthermore, for cells expressing nonsecreted proteins, it is not feasible to use ELISAs, so additional downstream processing is required to determine protein productivity. In conclusion, due to a lot of work required to fully characterize each clone isolated by limiting the dilution method, the entire process might take up to eight months [23, 88]. For these reasons, alternative high-throughput screening (HTS) methods that reduce the time and effort in attempting to find rare high-producing cell clones have been developed and preferred by industry, including fluorescence-activated cell sorting (FACS) and semisolid matrix-based systems such as ClonePix™ (Genetix) and Cell Xpress™ systems (Cyntellect) [5]. FACS is the most widely used HTS method, and it is performed after staining cells with antibodies conjugated with fluorescent molecules. The ClonePix™ system is an automated colony picker used to screen a large number of clones in a semisolid medium. Proteins secreted from the colony are captured by antibodies conjugated with fluorescent molecules so that high fluorescent intensity colonies are then picked. Cell Xpress™ not only identifies high-producing clones in a semisolid medium by live cell imaging with a detection reagent but also eliminates non- or low-producing clones by laser-induced apoptosis [5, 30]. For a detailed description of automated HTS methods, refer to reference [23].

Finally, selected clones with a high recombinant protein titer are chosen for progressive expansion and evaluation in controlled bioreactors before cell banking and further evaluations, such as the production stability of clones and the quality of the recombinant protein. It is important to highlight that the productivity levels of a "superproducer" clone should be between 5 and 30 times higher than the average productivity of the original cell line.

2.6 Cell Banking and Cell Line Characterization

An important consideration is to ensure that the chosen clone is able to generate the product in a consistent manner over time. For that, the cells used for processing must be held in a cell bank to guarantee a continuous supply during the product's life span. According to the FDA, a master cell bank (MCB) is defined as a large number of aliquots of a single pool of cells prepared from the clone of origin under defined conditions and conserved under defined conditions (generally in liquid nitrogen, −196 °C). The MCB is used to prepare all the working cell banks (WCBs) used for the manufacture of production batches. The documentation accompanying cell banks should include the inventory of all vials, their identity, and their location. The cell banks must be generated according to good manufacturing practices (GMP), and the manufacturer is responsible for establishing the maximum cultivation period for each particular cell (passage number).

Cell lines need to be characterized in terms of origin (organ, tissue, age, and gender), the history of passages, and culture medium. The characterization of a cell line is vital not only in determining its functionality but also in proving its authenticity, i.e., a confirmation that it is not cross-contaminated or misidentified. Information should include karyotype, isozyme profile, growth rate, morphology, tumorigenicity, and/or genetic markers (DNA fingerprinting). Besides, all the data regarding the recombinant product should be documented, such as the complete sequence of the genetic construction, including the GOI, regulatory elements, and selection markers; the cloning method; and the gene copy number, among other details. Finally, the absence of contaminants like bacteria, mycoplasmas, and viruses must be verified [89, 90].

3 Process Optimization for the Production of Protein-Based Biotherapeutics

3.1 Upstream Operations

3.1.1 Growing Mammalian Cells

In mammalian cell culture, it is necessary to consider the main biological properties of animal cells, whose morphology and functionality have evolved over millions of years. Animal cells have a fragile membrane, a high degree of differentiation, and a natural environment that allows them to reach high cell density (up to 5×10^8 cells per cm^3 of tissue). Also, tissues have a complex circulatory system for supplying nutrients and removing metabolic products. Therefore, for in vitro culturing, they require the effective control of environmental conditions (temperature, pH, etc.) and systems for external regulation through hormones, growth factors, and nutrients.

After establishing the cell line and subsequent clones, a high number of cells are required to generate the necessary amount of the correct recombinant protein for industrial or medical applications. At this point, bioreactors arise as devices or systems that provide a biological environment using the functions of enzymes or organisms to carry out biochemical reactions in vitro.

Their design for animal cell culture fundamentally depends on the type of cell growth; i.e., it should fulfill the requirements of the two main forms of growth: suspension culture (anchorage-independent culture) or adherent culture, depending on the attachment to a specific surface or substrate (anchorage-dependent culture).

Most animal cells derive from solid tissues of their organisms and grow in adherent conditions, in which cell adhesion to the extracellular matrix or to the neighboring cells is achieved through a complex molecular network (mainly proteins and carbohydrates). The most known contributors are proteins synthesized and secreted by cells, including integrins, cadherins, and selectins. Their role does not only imply mechanical consequences for the cells (such as attachment or resistance to compression or stretching forces), but they also act as mechanotransducers [91]. For that reason, when adherent cells cover the surface (confluent state), they can be detached using a protease like trypsin, which aids in their release from the support. Cells are resuspended in a culture medium that allows their subculturing into another device. With a different aim, cells can be grown to a confluent state, and then a production phase can be started by changing the growth medium and replenishing the support with a fresh or not-conditioned one as the production medium. If cells remain attached to the plastic surface, subsequent cycles of harvest and medium replacement can be carried out over time.

The cell and production yield depends on the surface size. Thus, distinct culture devices, such as roller bottles or multitray flasks, are used to increase the surface extent in order to allow the growth of a higher cell mass. Nevertheless, medical applications require a very high number of cells, which would need an incredible quantity of those devices to satisfy the industrial demand.

To overcome the limitation of 2D systems and generate high cell yields, suspension culture methods have been developed [92]. In suspension cultures, cells grow freely suspended in a liquid medium without forming clumps. Only cells derived from the circulatory system (red blood cells or white blood cells) exhibit this behavior spontaneously when cultured in vitro. However, many adherent cell lines of industrial interest, such as CHO, BHK, and HEK cells, can be adapted to grow in suspension. If cells cannot be adapted to grow as individual entities and they need to anchor to a surface to survive, microcarriers can be used to achieve a high final cell concentration. Raising bead concentration in the culture helps increase cell mass; however, a too-high microcarrier concentration leads to higher contact frequency during agitation, as well as concomitant clumping.

After decades of using in vitro 2D monolayer cell cultures, more complex culture systems have arisen to emulate the structure and function of mammalian cell tissues. For example, 3D-organized cell cultures, often named "organoids," are becoming increasingly useful in providing researchers with models for studying organ development and diseases [93]. However, traditional modes of culture using 2D monolayers, suspension cultures of individual cells, or carrier-attached cells are highly utilized to produce recombinant proteins, mainly those with therapeutic applications.

3.1.2 Upstream Processing Development

Upstream processing is a very important stage of the bioprocess that involves the preparation and manipulation of biological material before subjecting it to downstream processing. This includes the development of recombinant cell lines, either mammalian (the focus of this chapter), microbial, or insect cells, that produce therapeutic proteins, vaccines, and other biologics of interest.

Upstream process development (UPD) is crucial to establish the foundations for the manufacturing process, which leads to the regulatory approval of a biotherapeutic. For this reason, the large-scale production of biotherapeutics stands as the main objective of UPD. Nowadays, a growing number of biopharmaceutical companies need to reduce the costs of upstream process development. For this reason, a common practice to shorten process development starts with small-scale procedures in the laboratory [94].

The different variants of a protein of interest, together with the differences between cell line or clone productivities and the cultivation environment or conditions, result in a vast number of candidates being evaluated simultaneously to identify the best under optimal conditions. Therefore, a bottleneck is established where the challenge is to select the most suitable condition or candidate. Establishing a reliable platform using high-throughput (HTP) cultivation systems becomes a highly effective mechanism for early-stage process development [95].

HTP has evolved rapidly in recent years. Since approximately 2006 up to now, HTP cultivation platforms have tried to reduce the culture volume, incorporating real-time monitoring with full system automation [96]. Initiating UPD requires intensive and arduous work to establish the experimental conditions. Simple batch procedures in microtiter plates, shake flasks, or spinner flasks usually play a key role because of their relatively lower prices compared to bioreactors, aimed to be used for HTP. Nevertheless, it has been demonstrated that performance at a small scale does not necessarily extrapolate to a large industrial scale [97]. Scaling-up studies using benchtop bioreactors (BR) of 0.5 to 2 L and pilot-scale bioreactors (at least 100 L) are also limited in their HTP ability and exhibit more difficulties in selecting the best clone and culture medium for the final production stage. Thus, the greatest challenge in this area has been the development of miniaturized bioreactors or microbioreactors (MBR) that emulate large BRs' conditions with the benefit of analyzing many samples in parallel [97, 98]. These devices are currently on the market, and their aptitudes and descriptions can be found on different commercial websites [99, 100]. Briefly, all of them consist of a cell culture platform, including single-use recipients in an automated workstation and a proper software program. They operate in an aseptic environment because they are installed in a biosafety cabinet controlling 24 or 48 MBR in parallel.

3.1.3 Bioreactors for Animal Cells/Operation Modes

The selection of a bioreactor depends on the type of growth and the production mode of the cell culture. These factors primarily depend on the cell growth features (anchorage-dependent or anchorage-independent cells) and on the growth phase, during which the product is secreted (exponential or stationary).

There are four main modes of operation for a bioreactor:

(a) **Batch Mode**

A unit is initially loaded with cells in a starting volume of medium. The culture typically continues until the nutrients are exhausted or waste products accumulate to toxic levels, inhibiting cell growth and causing cell death. No fresh medium is added during cultivation, nor is the used medium removed, keeping the volume constant throughout the entire process.

Once the culture begins, a lag phase is initiated, whose duration will depend on the age of the inoculum and the cells' adaptation to the new environment. Then cells enter a log phase, in which the highest growth rate is displayed for 4–6 days. During this phase, nutrient concentration gradually decreases, whereas toxic metabolites accumulate so that cell growth is eventually inhibited by nutrient limitation (glucose and glutamine) and/or product toxicity (ammonia, lactate). The maximum cell density that can be achieved in a batch culture is about 1–3 × 10^6 cell.mL^{-1}, depending on the anchorage-dependency condition of the cells. The maximum cell density is maintained during a short stationary phase (hours to a couple of days), after which the culture enters the dead phase.

(b) **Fed-Batch Mode**

The bioreactor is filled with a volume of culture medium lower than the maximum working volume and is started as a batch culture. The culture is fed continuously or recurrently with a fresh medium or with a concentrated solution containing the most important nutrients due to its consumption. This procedure increases the volume of the culture until the maximum working volume is reached. Similarly to batch cultures, the spent medium is not withdrawn so that toxic waste metabolites (mainly lactate and ammonia) are accumulated and the environment becomes detrimental to cell growth. Nevertheless, feeding allows more nutrients to be metabolized by the cells, increasing the biomass and the recombinant product yields. Compared to the batch cultivation mode, the fed-batch one is characterized by:

- A longer growth phase (important for cell clones, in which the production phase is associated with the growth phase)
- Higher viable cell density and a longer stationary phase (important to increase the productivity of nongrowth-associated products)
- A higher product concentration

One of the major challenges of the fed-batch culture lies in reaching an appropriate nutrient formulation and feeding strategy with the aim of achieving high titers for the product of interest [21]. The limiting substrates should be identified so as to be supplied at correct levels in critical stages of the culture; in this way, medium feeding can avoid nutrient limitation, thereby promoting higher production of the recombinant protein [101]. Consequently, fed-batch cultures require knowing cell growth features and cell metabolism and the key process parameters that affect the quality and yield of the recombinant protein [102].

(c) **Continuous Culture Without Cell Retention: Chemostat**

This mode is accomplished by continuously removing the entire culture (cells and medium) while simultaneously adding a fresh medium to the reactor. The removal rate is based on the cells' growth rate. In this setup, cells can remain in the exponential growth phase, with the cell density and reactor volume staying constant. However, due to the slow growth rate of mammalian cells, this method is technically challenging and is primarily used in research to study growth parameters. In this way, the chemostat culture is an interesting strategy to improve the understanding of cell responses to culture variables [103]. This is possible because the chemostat enables the control of the specific growth rate by defining the dilution rate when it reaches the steady state. This situation maintains the cells in a defined physiological state and thus provides biologically reliable and homogeneous data on the cell population [104].

(d) **Continuous Culture with Cell Retention: Perfusion**

Like the previously mentioned chemostat system, continuous culture with cell retention is also an open system, but it uses a cell retention device that allows the removal of the spent medium without losing the cells. A fresh medium is simultaneously added, maintaining the reactor volume. This approach addresses the main limitations of the chemostat: low product yield due to low cell densities (since proliferating cells are lost in the outflow) and the limited dilution rate needed to prevent bioreactor washout. In the perfusion mode, cells are retained or recycled back into the culture vessel, allowing for much higher dilution rates [105]. As a result, greater availability of nutrients, higher cell concentrations (1×10^7 - 1×10^8 cell.mL^{-1}), and, consequently, higher productivities are attained [106].

Regarding methods for cell separation, spin filters were first used for perfusion, but they have lost popularity due to scale-up difficulties related to filter fouling [19]. Therefore, the use of porous membranes has become more frequent. Membrane systems include tangential filtration flow (TFF) and alternating tangential flow (ATF) [107].

The productivity of recombinant cell lines has dramatically increased over time. In the 1980s, the maximum cell concentration achieved in a 7-day batch culture in a stirred tank was about 2×10^6 cell.mL^{-1}, with productivity reaching 10 pg $\times$ cell^{-1} $\times$ day^{-1} and recombinant glycoprotein titers of around 0.01–0.05 g.L^{-1} [108, 109]. Nowadays, cell concentrations exceeding 1×10^7 cell.mL^{-1} can be achieved, with specific productivities close to 20–50 pg.cell^{-1}.day^{-1} up to 100 pg. cell^{-1}.day^{-1} and titers reaching approximately 5–20 g.L^{-1} in a fed-batch culture [21, 110]. This 100–500-fold increase in titer responds to several factors, ranging from molecular biology techniques, which led to substantial increases in the expression levels of recombinant cells, to cell culture optimization, which led to continuous improvement of bioreactor culture systems (media development, process control) and their mode of operation. The combination of these achievements has allowed prolonging the production phase duration as well as improving viable cell density and productivity. With regard to culture modes, the suspension culture is the only one providing homogeneous cultures that enable the continuous monitoring and control of the physicochemical environment in which cells are suspended. Besides, it allows direct sampling of the cell suspension for adequate tracking during culture and greatly simplifies the scaling-up procedure.

In the case of anchorage-dependent cells, different systems were addressed. For instance, Vero cells, the first continuous cell line (CCL) to be approved by WHO for the manufacture of viral vaccines for human use, were handled in microcarrier bioreactors and fixed-bed bioreactors to improve the cost-effectiveness of manufacturing and more scalable bioprocesses [111, 112].

Nowadays, fed-batch and perfusion cell cultures are the two current processes for the large-scale industrial production of therapeutic recombinant proteins. From the literature, it is known that the scaling up of the fed-batch process needs larger and more rigid layouts and that perfusion is more appropriate to produce some therapeutic proteins (for example, labile ones) [105].

Additionally, perfusion systems can achieve much higher cell densities compared to fed-batch systems, which leads to increased volumetric productivity and reduced residence time [113]. The steady-state operation of perfusion systems also enables more precise control of process variables, conferring more consistency to culture conditions and product quality attributes. All these properties make it a valuable option for regulatory activities [105].

There is no universal bioreactor suitable for all cell and tissue cultures, but conventional stainless steel stirred bioreactors and disposable bioreactors are the most useful devices for benchtop and industrial applications. The production scale will define the train of bioreactors to be used for scaling up the process. Stainless steel or single-use, disposable stirred tank bioreactors with increasing volumes (1 to 2 l, 10 to 20 l, 100 to 200 l, and 1,000 to 2000 l) can be used for the subsequent steps of the scaling process. Alternatively, other single-use, disposable bioreactors with different designs, such as wave-mixed bioreactors or orbitally shaken bioreactors, can be employed.

Single-use reactor systems demonstrate numerous benefits, including consistent batch-to-batch results, simple operation, money and time savings, and reduced environmental or ecological impacts because of energy and water savings. Very importantly, they also reduce cross-contamination risks [114].

A detailed overview of single-use reactors through historical phases and milestones over time is very well depicted by Jossen and colleagues [115]. Commonly utilized commercial single-use bioreactors, which are currently equipped and operating at scales ranging from cubic centimeters to cubic meters, include stirred systems with a rigid container, stirred systems with a flexible bag, wave-mixed systems with flexible bags (featuring 1D, 2D, and 3D motion), and orbitally shaken systems with flexible bags.

3.1.4 Media Optimization

As has been mentioned before, CHO cells stand to be the most employed therapeutic recombinant protein expression system due to their several properties: they are not susceptible to human virus, they produce posttranslational modifications (PTM) similar to those produced by human cells [116], and they can be used for protein mass production in the biopharmaceutical industry [117]. This last premise is compatible with their ability to grow in suspension using chemically defined serum-free media, thus being suitable for scaling up in bioreactors.

Different culture media are used for growing and maintaining cells and for supporting their functions. To optimize the production process of a desired recombinant protein, the culture media for the producer clone should maximize its growth, improve its metabolic fluxes, and increase its specific productivity, leading to appropriate co- and/or posttranslational modifications, which are needed for successful biological activity of the product. It is important to keep in mind that the composition of the growth medium and culture conditions (pH, O_2, temperature) used for cell maintenance, amplification, and the preparation of the inoculum for seeding a bioreactor may be different from that of the production phase of the recombinant protein.

The components of the various culture medium formulations include a mixture of carbohydrates, amino acids, minerals, vitamins (nutrients), and growth factors in order for cells to behave as expected. The saline concentration of the medium is isotonic (~300 mOsm) and often contains bicarbonate, which acts as a buffer system, together with a carbon dioxide (CO_2) concentration of about 5–10% (V/V) within the environment in which the cells are cultured.

In general, the main carbohydrate source is glucose, but the cells can also metabolize galactose, mannose, and fructose. However, the metabolism of these latter three carbon/energy sources yields the necessary precursors for the pentose phosphate pathway more slowly; therefore, maintaining exponential growth in a culture supplemented exclusively with these energy sources is challenging [118]. Also, in established cell lines, the decoupling action between glycolysis and Krebs cycle due to the lack of some connective enzymes determines that pyruvate does not efficiently enter the cycle and lactic acid increases or accumulates in a batch culture. Nevertheless, for a proper balance between the use and production of cell energy, an active Krebs cycle is required because it is the main contributor to the energetic equivalents (ATP).

Why is glutamine, one of the essential amino acids, ~3–40-fold more required in a cell culture than the other ones? The answer lies in its property to be deaminated by mitochondrial glutaminase releasing ammonium and glutamate. This glutamate enters the Krebs cycle as α-ketoglutarate to be completely oxidated to CO_2, allowing the cells to obtain the energy that is needed for growth and propagation.

Therefore, the decoupling between glycolysis and the Krebs cycle produces inefficient metabolism, characterized by large quantities of glucose consumption and concomitant production of growth-inhibiting by-products such as lactic acid and ammonium. Lactic acid is also responsible for decreasing pH and increasing the osmolality of the culture media, affecting cell viability and product quality, and the ammonium also inhibits the synthesis of some sugar-nucleotides, resulting in the production of glycoproteins with lower branching and lower sialic acid content [119].

Under glucose-limiting conditions, mammalian cells can switch to more energy-efficient metabolism and produce less lactate. This enables higher viable cell densities and therefore product titers, contributing to better performance of the industrial process [120].

It is possible to find a variety of cell culture medium formulations that have been developed to support a wide range of cell types in distinct experimental applications. The first indications of cell cultures using natural media, as indicated in Table 1, date back to 1909, when natural products (e.g., serum, plasma, or other biological fluids or tissue extracts) were the main components of the media [121]. Due to the lack of knowledge about the composition of these fluids for a prolonged period, science

Table 1 Classification of media and description of their composition

Classification	Subclassification	Characteristics/composition
Natural media	None	It may consist of plasma, serum, or other biological fluids such as amniotic, pleural, or cerebrospinal fluid. Also, it may be represented by tissue extracts.
Synthetic media	Serum-containing media	They contain various serum-derived substances, which makes their composition partially known and with component concentrations that change from batch to batch. This sort of media is less reproducible and has contamination risks.
	Serum-free media	They have a defined composition that confers high reproducibility of results and allows the validation of the cultivation process. They may contain discrete proteins or bulk protein fractions (e.g., from animal tissue—Bovine serum albumin, α-globulin, or β-globulin—Or from plant extracts).
	Protein-free media	They do not contain high molecular weight (MW) proteins or protein fractions, but they contain undefined components such as peptide fractions (protein hydrolysates). They make the downstream processing easier, but it is important to emphasize that protein hydrolysates are chemically not defined and may cause problems in reproducibility and the comparability of experiments.
	Animal-free media	They do not contain components of animal or human origin. But there is a subclassification called Xeno free, where animal components are allowed not as supplements but as human-derived components, such as human serum albumin, used as additives.
	Chemically defined media	Crude protein fractions such as hydrolysates and tissue extracts or any other component of unknown composition are utilized not as supplements but as highly purified recombinant proteins, hormones, or growth factors (either from animals or plants), which can be added as appropriate supplements. These media represent the gold standard for producing antibodies and recombinant proteins.

investigated their contents. The focus shifted toward formulating and replacing them with ingredients of a more defined and thus more controllable composition.

In this way, chemical-based synthetic media started replacing them, but bovine serum continued to be an essential component for providing nutrients and unknown growth factors that support cell growth. Despite the fact that serum-containing media are still recommended for difficult-to-grow cultures and used for research, they tend to be completely removed from cell culture technology. The main causes are seasonal and geographical lot-to-lot variation, potential contamination with adventitious agents, high cost, and high content of nonrelated proteins, which increases the efforts and difficulties to purify the protein of interest, mainly in the industrial field. Therefore, in terms of the 3Rs (refinement, reduction, replacement), different strategies to reduce or eliminate the fetal bovine serum have been developed. The past, present, and future of fetal bovine serum were extensively discussed in the third workshop devoted to its replacement in cell and tissue culture methods and reported by van der Valk and colleagues [122].

Nowadays, synthetic media (defined as basal media that have known composition from industrial formulations and contain supplements such as serum, growth factors, and hormones) can be classified, based on the type of added supplements, into serum-containing media, serum-free media, protein-free media, animal-free media, and chemical-defined media [121, 123]. Their features are summarized in Table 1.

The biotechnological industry requires sustaining high-density cultures and high cell productivity to be highly competitive. Regarding this issue, developing and optimizing cell culture media has become critical. The best practices for selecting the proper media for mammalian cell cultures, including having in mind a detailed and rigorous description of the media, components, keys, and clues, can be found in Price (2017) [124]. Cell culture media critically affect cell growth and propagation. To enhance the abovementioned objectives of the biotechnological industry, research went in the direction of efficiently optimizing the medium's composition because it also modulates the critical quality attributes (CQAs). Different approaches have been used to optimize the components for cell culture media. Since this is not the aim of this chapter, a summary of the variety of methods aimed at this purpose can be found in the work of Gangwar and colleagues (2024) [125].

3.2 Downstream Processing

Unlike upstream processing, which refers to the stages leading to the production of the protein of interest from cell cultures, the downstream process refers to those stages of the process that will have a direct impact on defining the final product.

As a general definition, downstream processing (DSP) indicates both the recovery and the purification of a product from a complex mixture. Four stages are vital in DSP as unit operations that enable the isolation of a protein from a complex mixture, which contains not only other proteins but also macromolecules of different physicochemical nature, such as DNA, endotoxins, lipids, and particulate material of various sizes. The following are the four stages:

1. Separation: this is often achieved through filtration, centrifugation, or precipitation.
2. Purification: the elimination of contaminants, other proteins, DNA, viruses, endotoxins, or lipids; this step is routinely carried out through chromatographic steps in batches or in a continuous mode.
3. Concentration: increment of the product concentration; sometimes the same chromatographic step is sufficient to concentrate it on the desired level.
4. Formulation: preparation of the product for its final use; it consists in adjusting the pH and adding stabilizers, which must be perfectly analyzed before being selected as accompanying components of the final product.

The control quality step represents a unit operation that is present during the whole process and ensures the achievement of the desired quality of the therapeutic product.

The DSP approach can be carried out using empirical methods or methods employing algorithmics. The former consists of a set of nonmandatory methodological rules that suggest how to proceed and what problems to avoid, while the latter uses mathematical models and simulations in specialized software [126].

This part of the chapter will focus on the purification stage. The empirical methods consist of a series of general rules that are based on previous knowledge. They can be summarized, from Graselli [127], as follows:

(i) Outline the characteristics of the target protein and key impurities, such as molecular weight (MW), isoelectric point (pI), solubility, molar extinction coefficient, stability, cysteine content, sensitivity to high salt ion concentrations or pH, secondary structure, hydrophilicity, hydrophobicity, sequence similarity to known proteins, susceptibility to oxidation, and potential post-translational modifications, among others.
(ii) Develop analytical assays to detect the protein of interest.
(iii) Minimize risks during the sample handling and its conditioning.
(iv) Minimize the use of additives (as they can affect subsequent purification or interfere with the activity assay).
(v) Eliminate critical impurities.
(vi) Apply methods that use different physicochemical principles for separation at each stage; i.e., use orthogonal properties like molecular size, ionic strength, hydrophobicity, or multimodal assets.
(vii) Minimize the number of steps. A high number of steps critically reduce the global yield and enlarge the time of the process, increasing costs and affecting the stability of the product.
(viii) Combine the stages in a logical manner to optimize the process so that the elution step of one chromatography constitutes the loading of the following one.
(ix) The simpler the process, the better the yield and the quality of the product.

Extracellular or secreted products, such as recombinant proteins from mammalian cells, usually contain impurities such as cell debris, aggregates, nucleic acids, lipids, and polysaccharides. Therefore, it is essential to choose suitable methods to separate the target protein from these coarse impurities in order to prevent its degradation or the loss of biological activity (Table 2). Also, to ensure the process stream is particle-free, a solid-liquid separation procedure, like centrifugation, precipitation, or membrane microfiltration, is required. These methods are gathered as rough fractionation. Ultrafiltration is also routinely carried out to concentrate and prepare samples for the subsequent step before the chromatography-based purification process.

After pretreatments and rough fractionation methods, the sample is ready to continue with a set of fine fractionation procedures to separate the target protein from some proteins of similar size and/or physicochemical properties [128]. Aiming to obtain the highest purity, chromatographic processes are used as fine

Table 2 Rough and fine fractionation methods during downstream processing

Type of fractionation	Method	Type	Characteristics/description
Rough	Precipitation	– Saline (salting out) – Isoelectric (pI) – Solvent fractionation – Nonionic polymer fractionation (PEG)	It is based on the difference of solubility between proteins. By adding different agents, the solubility of the target protein or some contaminants is changed, producing a precipitate. Then the precipitate is removed by centrifugation (contaminants) or dissolved (target protein) to continue with the process.
	Centrifugation	– Differential centrifugation (DCF) – Zonal centrifugation (ZCF)	The centrifugal force is used to move particles radially away from the axis of rotation, separating them by size and shape. DCF is mainly used to separate cells, debris, and precipitates from soluble content without using reagents or media because particulate and nonparticulate components behave differently. Centrifugation remains one of the most commonly used unit operations for effectively harvesting products from mammalian cell culture broth. The disk stack centrifuge used in biotherapeutic bioprocessing constitutes a DCF example [129]. ZCF is used to separate soluble components in a gradient-density medium (CsCl, glycerol, etc.). It is only used for an analytical scale.
	Dialysis		Removal of small molecules in a solution, such as salts, reducing agents, or preservatives, by selective and passive diffusion through a semipermeable membrane—Proteins cannot diffuse due to their size, remaining inside the membrane (the Donnan effect is present) [130].
	Filtration	– Microfiltration (MF) – Nanofiltration (NF) – Ultrafiltration (UF)	Membrane-based separation—Membranes have different porous sizes, allowing or avoiding the movement of certain components of the sample through them. MF and UF are the most used methods for the rough separation of recombinant proteins. MF is used

(continued)

Table 2 (continued)

Type of fractionation	Method	Type	Characteristics/description
			for separating microorganisms, the remaining cells after centrifugation, or protein agglomeration during conservation. UF is mainly used to concentrate proteins and change the saline environment of the protein before initiating a procedure of fine fractionation [128].
Fine	Chromatography	Ion exchange chromatography (IEC)	Separation of ionizable molecules based on differences in charge properties—It is based on the formation of ionic or electrostatic interactions between analyte ions or highly polar molecules and an oppositely charged stationary phase. The sample loading is carried out under a low salt ion concentration and elution with a high-concentration salt ion buffer solution. Also, pH variations can be performed. It is the only chromatographic method highly recommended for any phase of the CIPP process. This feature is due to its particularity of removing various impurities such as target protein variants (isoforms), host cell residual proteins, DNA, culture medium components, endotoxins, and viruses [131, 132].
		Affinity chromatography (AC)	Separation of biomolecules based on their biological function or unique chemical structure—It is an adsorptive technique where a specific interaction occurs between a protein (or a group of similar proteins) and a specific ligand coupled to the chromatographic matrix. It simulates and exploits molecular biorecognition for the purification of a target protein. Proteins and ligands bind through noncovalent interactions, including electrostatic forces, molecular hydrophobicity, van der Waals forces, and hydrogen bonding. Affinity chromatography (AC) is

(continued)

Table 2 (continued)

Type of fractionation	Method	Type	Characteristics/description
			quick, straightforward, and highly efficient. There are numerous classifications of AC. One of them, according to the type of ligands, allows its sorting into AC with biospecific ligands (receptors, antibodies, lectins, vitamins, coenzymes) or AC with pseudobiospecific ligands (triazine dyes, metal ions such as Ni^{2+}, Zn^{2+}, Cu^{2+}, and Co^{2+}). Novel triazine-based affinity resins are still being developed, considering that triazine scaffolds are economical, nontoxic, and chemically defined; they have no cleavable bonds when immobilized; they can be sterilized, and they are resistant to degradation by chemicals or biologics [133]. Besides, there are various affinity tags. They are very useful for tagging proteins and purifying them by using the corresponding ligand. Different sorts of affinity tags have been developed [134].
		Hydrophobic interaction chromatography (HIC)	HIC is a purification technology based on hydrophobicity differences among proteins. Ionic strength controls hydrophobic behavior. When salt concentration increases, hydrophobicity is also augmented. It represents the method of choice following a high ionic strength elution of an IEC step.
		Size-exclusion chromatography (SEC)	Method to separate and purify different proteins based on their molecular size—During the procedure, isocratic conditions are used. Larger molecules elute before smaller ones, which move more slowly because they can enter the mesh of the gel. This method is mainly used as a polishing step (elimination of aggregates of the target protein, removing leachable contaminants such as metals, dyes, protein A,

(continued)

Table 2 (continued)

Type of fractionation	Method	Type	Characteristics/description
			etc., and for buffer exchanging). A great variety of examples can be found in the literature, even the most modern ones that apply machine learning approaches to develop a generic SEC method [135].
		Multimodal chromatography or mix-mode chromatography (MMC)	MMC has emerged as an alternative to the traditional modes, involving the use of more than one mode of separation in the same matrix. The binding modes that are more frequently employed in multimodal ligands comprise ion exchange, hydrogen bonding, and hydrophobic interaction groups [136, 137].

fractionation. Chromatography can be designed as three different stages, known as capture, intermediate, and polishing steps. It is essential to enter the capture stage as fast as possible because the biological activity of the target product could be affected by the surrounding environment and other components. The capture phase should be optimized to provide the highest binding capacity and the highest putative flow rate to minimize biological activity loss. Also, a very important task in this step is to reduce the volume by eliminating the principal molar impurity (water molecules). In conclusion, a concentrated product transferred to an environment that preserves its potency with a significant removal of other contaminants is the main objective of the capture phase. The next step is the intermediate phase, consisting of the heart of the purification process, where the main impurities are removed (other proteins, nucleic acids, lipids, and viruses). Finally, the polishing step configures a product with a high final purity grade, in which the removal of the traces of the components and highly related substances is carried out. The whole chromatographic procedure is known as CIPP (capture, intermediate phase, polishing).

The characteristics of the main chromatographic procedures for downstream application are summarized in Table 3 (obtained from Du and colleagues [128]).

4 Quality Control of Protein-Based Biotherapeutics

The development of biotherapeutics requires ensuring their efficacy, safety, and quality. Efficacy and safety are determined through toxicity studies and in vitro biological assays, but preclinical and clinical trials may also be required depending on whether the drug is in the development phase or is part of the production stage.

Table 3 Comparison of different chromatographic methods. Reproduced with the permission of Frontiers in Bioengineering and Biotechnology from Du M, Hou Z, Liu L, et al. (2022), [1]Progress, applications, challenges and prospects of protein purification technology. Front Bioeng Biotechnol 10:1028691 [128]

Types Characteristics	SEC	IEC	HIC	AC
Separation mechanism	Size	Charge	Hydrophobicity	Specific affinity
Selectivity	Secondary	High	Medium/high	Very high
Load	Low	High	High	High
Purification rate	Medium/low	High	High	High
Biocompatibility	Very nice	Good	Medium/good	Good
Purified protein yield	High	High	Medium/high	High
Capture/concentration phase	+	+++	+++	+++
Intermediate purification phase	+	+++	+++	+++
Polishing	+++	+++	+	++

For a biotherapeutic to be ready for production, it is necessary to fulfill requirements that ensure batch-to-batch consistency through the use of a set of analytical techniques (physicochemical, biological, or immunological) that can verify its purity, biological potency, and stability. These techniques must be consistently applied to generate a robust process. Therefore, during product development, the quality attributes of the process must be defined to meet the requirements regarding the consistency of the production process.

Controls must be carried out at different stages of the process:

(a) Control of starting materials.
(b) Control of the production process itself.
(c) Quality control of the final product.

In the first case (**a**), different information must be considered: the origin of the cell line, precise knowledge of the genotype and phenotype, accuracy of the number of generations in which its production is stable, information about the vector or heterologous gene transfer system, gene sequence, reagents and culture media, the quality and origin of additives if used, and detailed composition and generation of cell banks (master and working cell banks). The cell banks themselves require particular quality control to ensure the suitability of the preserved clones for use in the production of a biotherapeutic.

Nowadays, there are a lot of companies that develop the required cell lines and provide cGMP-compliant cell banks in strict accordance with industry standards. This ensures having access to fully characterized and phenotypically consistent cells.

Regarding **b**, the control of the production process must be carried out at both upstream and downstream levels. At the upstream level, controls should be implemented during the cultivation stage, with criteria leading to rejection or acceptance. Any undesirable variation in the system can lead to modifications, for example, in the content of host cells' proteins, and can decrease the efficacy of the product. Also, a series of analytical orthogonal methods must be carried out downstream to discover if the process affects the desired quality of the biotherapeutic. Additionally, these methods allow controlling the different parameters that are critical during nonchromatographic or chromatographic steps, which endorse the performance of the purification process.

For all these tasks, new concepts must be taken into account:

- Process analytical technology (PAT): the FDA defines the PAT mechanism as designing, analyzing, and controlling pharmaceutical manufacturing processes through the measurement of critical process parameters (CPP), which affect the critical quality attributes (CQA). The aim is to ensure the quality of the final product. In another explanation, PAT enables the real-time monitoring of process parameters and product quality attributes, along with the timely control of process parameters from raw materials to the final drug product, with the goal of ensuring final product quality [138].
- Quality by design (QbD): it is a systematic approach to development that begins with predefined objectives and emphasizes product and process understanding and process control based on sound science and quality risk management. The International Council for Harmonization of Technical Requirements for Pharmaceuticals for Human Use (ICH) guidance notes that the quality by design (QbD) approach for pharmaceutical development involves the precise identification of critical process parameters (CPPs), critical quality attributes (CQAs), and critical control points (CCPs), as well as designing the process with robust control strategies to ensure consistent process performance and validation to achieve the target product profile [138].

The goal is to create a high-quality product that satisfies the customer's demands while decreasing risks for the manufacturer. Thus, the key objective of QbD is to identify, clearly address, and rationalize all variability before the product reaches the market.

These concepts, which biotechnology companies are increasingly using, are greatly described in the work of Rathore [139] with a case study that illustrates the step-by-step development of a purification process using QbD, which aims to produce a biosimilar product: granulocyte colony-stimulating factor (GCSF).

Finally, quality control reaches the objective of the process: the product (**c**). In the field of biotherapeutics, the product can be an innovative product, biosimilar, or biobetter.

A novo biologic drug, biooriginator, innovator, or reference biologic is the first biologic medicine of a specific class or category to be approved, registered, or marketed. When the innovator's patent expires, it opens the door for the commercialization of biosimilars, which are "highly similar" to the original product in terms of structure and function. Although biosimilars may exhibit slight differences in clinically inactive components, they should have no significant differences in terms of safety, purity, and potency compared to the reference product used as the comparator in head-to-head studies. An overview and status of biosimilars as well as the regulatory considerations therefor can be found in Rathore and Bhargava [140].

A biobetter is a new and improved version of an existing biologic but with a similar action. It has been modified to improve manufacturing attributes, pharmacokinetics, efficacy, safety, or stability. Biobetters are also referred to as second-generation biologics. The pharmacokinetic behavior is the most frequently modified aspect to generate biobetters. Well-known examples include ***erythropoietin*** with an extended half-life, known as ***darbepoetin alfa*** or ***Novel Erythropoiesis Stimulating Protein*** (NESP™), which has two additional *N*-glycosylation sites [141]; ***human follicle-stimulating hormone*** (***hFSH***), which is *C*-terminally fused to the *C*-terminal end (CTP) of the ***human chorionic gonadotropin***, resulting in a high density of *O*-glycosylation sites [142]; and the very recently approved Somatotrogon™, a highly *O*-glycosylated CTP-fused ***somatotropin*** [143]. Other examples include proteins covalently attached to polyethylene glycol (PEG), such as ***pegylated granulocyte colony-stimulating factor*** (Pegfilgrastim or Neulasta™) [144, 145]. Even more, biobetters can originate their own biosimilars, for example, ***pegfilgrastim-cbqv***, a biosimilar of ***pegfilgrastim*** [146].

In general, biologically active products need to be extensively characterized throughout product development. The attributes of the product that are essential for assessing safety, efficacy, and potency are identified as critical quality attributes (CQAs). The ICH guidelines define a CQA as "a physical, chemical, biological, or microbiological property or characteristic that should be within an appropriate limit, range, or distribution to ensure the desired product quality" [147]. Physicochemical and functional characterization is based on appropriate and validated analytical methods.

The analysis commonly required at different stages of the manufacturing product and those required to compare biosimilarities are very well and exhaustively depicted in the study by Nupur and colleagues [148]. The work highlights the growing interest in increasing the global harmonization of the regulatory guidelines for biosimilars and reference products, putting emphasis on the design of analytical, nonclinical, and clinical studies. The reading of this review is highly recommended since a comprehensive compilation of assays conducted for various processes and biotherapeutics was carried out, demonstrating their usefulness for the mentioned purposes.

According to Nupur and colleagues [148], orthogonal analytical platforms for different CQAs should consider the primary structure (PS), the high-order structure (HOS), glycosylation (G), product-related variants (PRV_I), and process-related variants (PRV_{II}). All these CQAs are widely used in analytical similarity assessments during the comparison of potential biosimilars with reference products. Many of these analyses could be highly recommended for quality control during production process development and for the final product before formulation.

Nupur and colleagues organized the CQAs in three levels (Fig. 3). The first level is related to the five abovementioned CQAs. The second level is related to features associated with each of the first-level CQAs. In this way, PS includes the mass evaluation of intact proteins or subunit proteins and peptide mapping; HOS comprises the analysis of secondary or tertiary structures and conformation stability; the G CQA involves monosaccharide determination and oligosaccharide patterns; PRV_I

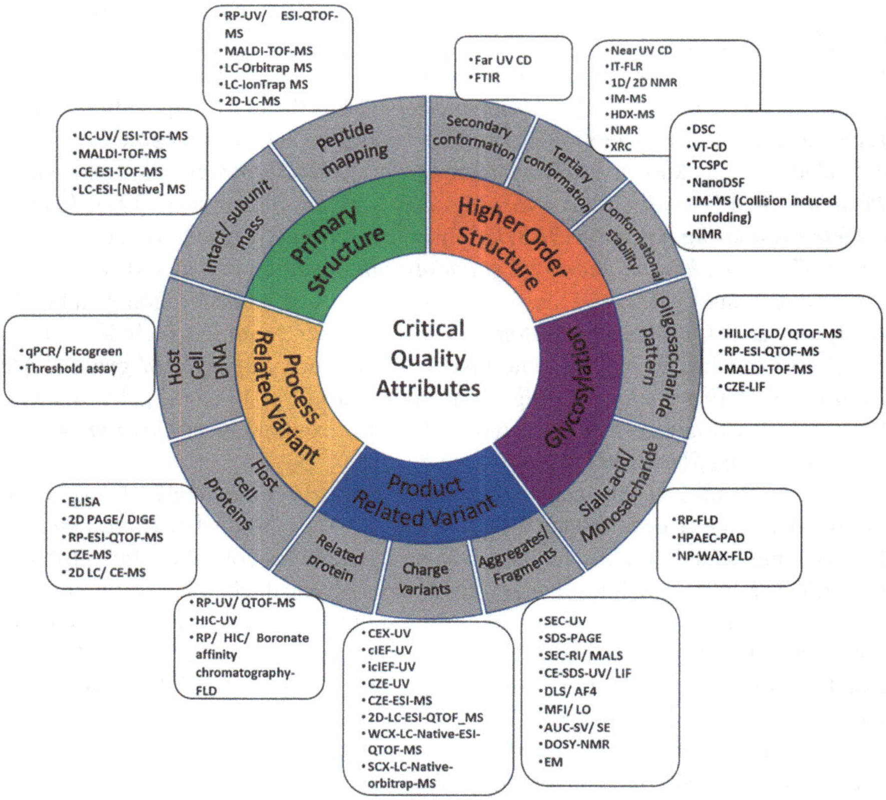

Fig. 3 Orthogonal analytical design for different critical quality attributes (CQAs). Reproduced with permission from Nupur and collaborators: Nupur N, Joshi S, Gulliarme D, et al. (2022), Analytical similarity assessment of bio-similars: Global regulatory landscape, recent studies and major advancements in or-thogonal platforms. Front Bioeng Biotechnol 10:832059 [148]

includes related proteins, charge variants, and fragment/aggregate studies; and, finally, PRV_{II} comprises the evaluation of host cell proteins and host cell DNA. The third level is related to each assay, which is proposed to be carried out with the aim of analyzing each of the CQAs, as can be summarized in Fig. 3.

Potency constitutes another CQA both for the product and for the process control. It can be carried out using in vitro bioassays or by measuring its performance in recognizing its receptor through affinity evaluation methods. This will depend on the sensitivity and detection limits of each method in identifying modifications to product quality. In general, potency assays are used to release batches of medicinal products and to demonstrate the consistency of the production process and the stability of the product.

For a given product, the impact that each CQA might have on its clinical performance must be determined. A risk analysis should be conducted to evaluate which of these CQAs may have the greatest impact on the product itself, establishing a ranking. They are useful to evaluate the performance of the product intended for the clinic. Critical process parameters (CPPs) must also be considered to evaluate how they would affect product quality and link them to the CQAs (e.g., glycosylation, oxidation, deamidation). The control of these parameters can be done during the process or at least at the end, selecting those that show the greatest variability throughout the process. In this sense, the QbD paradigm is set to understand the relationships between CQAs and CPPs [149].

5 Improving Performance and Efficacy: Development of Protein-Based Biobetters

So far, throughout this chapter, the development and production of protein biotherapeutics for the treatment of many diseases have been described. The great progress in the field of biotechnology has increased the number and value of biotherapeutics, which have become an imperative part of the pharmaceutical industry. As has been mentioned above, biotechnological companies focus their attention on the development not only of innovator drugs but also of biosimilars; moreover, the modification of existing biologics to improve their properties has become an increasingly important approach, originating what are called biobetters or second-generation biopharmaceuticals. The most critical drawback of most approved first-generation biotherapeutics lies in their short terminal half-life, which results in rapid degradation and elimination from patients' bloodstream, needing frequent application, thus affecting their quality of life. Thus, the majority of strategies to generate biobetters aim to extend the circulatory half-life of biotherapeutics, allowing the generation of long-lasting biotherapeutics with improved pharmacokinetic properties. Different technologies are available for this purpose, which are mainly based on reducing renal clearance by increasing the molecular size and negative charge. The most important ones include PEGylation (covalent attachment of polyethylene glycol chains to a functional molecule) or other conjugation with other alternative polymers; fusion to albumin or the Fc region of IgG; and introduction of additional *N*- or *O*-glycosylation sites, which constitutes one of the areas of glycoengineering. Many of these techniques have already been used in several FDA-approved drugs and are extensively reviewed in references [150–152].

This section will briefly describe the application of different glycoengineering approaches to a model protein, recombinant ***human alpha interferon-2b*** (***rhIFN-α 2b***).

Interferons[2] (IFNs) are a family of α-helical cytokines, which were first discovered by Isaacs and Lindenman in 1957. In humans, the IFN superfamily constitutes three types: I, II, and III. Type I IFNs exhibit antiviral, antiproliferative, and

[2] See Chapter "Therapeutic Proteins".

immune-modulating activities, these effects being responsible for the production of recombinant IFNs for the treatment of viral and tumor diseases. At present, ***rhIFN-α 2*** is approved for the treatment of follicular lymphoma, hairy cell leukemia, malignant melanoma, AIDS-related Kaposi's sarcoma, and metastatic cell renal carcinoma. Besides, IFN therapy has shown success in controlling infections with hepatitis B and C viruses. Regardless of the administration route, ***rhIFN-α2*** is rapidly inactivated and eliminated from the patient's bloodstream so that high and frequent doses are required to attain the therapeutic window during treatment. As a consequence, toxic effects and adverse symptoms arise, affecting the patient's convenience and quality of life.

With the goal of extending ***rhIFN-α2*** circulation half-life, some glycoengineering strategies have been developed. One of them, named *N*-glycoengineering, consisted in the introduction of four-point mutations into the sequence of ***hIFN-α2*** with the aim of generating potential *N*-glycosylation sites (Asn-Xxx-Ser/Thr, where Xxx can be any amino acid except Pro). The mutated amino acids were selected based on a theoretical analysis that considered both the cytokine structure and the regions involved in receptor interaction. After a screening step, a highly *N*-glycosylated IFN variant with four potential *N*-glycosylation sites (P4N, R23N, K70N, and D77N), called 4 N-IFN, was generated. Importantly, despite the cytokine exhibiting significantly reduced residual in vitro antiproliferative and antiviral activities, it showed a 25-fold increase in the elimination half-life and a 20-fold decrease in the systemic clearance rate compared with the unmodified ***rhIFN-α2*** following subcutaneous administration to rats [153]. Moreover, 4 N-IFN showed a markedly enhanced in vivo antitumor activity in human prostate carcinoma implanted in nude mice [154]. In addition, the hyperglycosylated cytokine showed enhanced stability against different external factors (such as thermal stress, acid pH, and repetitive freeze-thawing cycles) compared to the unmodified IFN [155].

Another glycoengineering strategy consisted in the fusion of a 9-mer peptide containing two potential *N*-glycosylation sites (ANITVNITV) to the *N*-terminus of both native ***rhIFN-α2*** and 4 N-IFN. In this way, through the increment of the MW and the negative charge of both proteins, an important improvement in pharmacokinetic properties was similarly observed in rats [156].

O-glycoengineering, representing the addition of *O*-glycosylation sites to a protein, has also been applied to ***rhIFN-α2***. For this, different approaches have been addressed. One strategy was based on the fusion of the carboxyl-terminal peptide (CTP) of ***human chorionic gonadotropin*** (hCG)β-subunit, which contains four *O*-linked oligosaccharide recognition sites, to each or both *N*- and *C*-terminus of ***hIFN-α2***. Similarly to *N*-glycoengineering, this approach was able to improve the pharmacokinetic properties of the cytokine since the CTP-IFN-CTP variant, with eight additional *O*-glycosylation sites, showed a tenfold longer elimination half-life and a 19-fold decreased plasma apparent clearance compared to the native cytokine. Higher stability against plasma protease inactivation and increased in vitro thermal resistance were also attained through this modification [157]. Another strategy consisted in the design and fusion to the *N*-terminus of ***hIFN-α2*** of two novel peptide tags, named GMOP and mGMOP, which derive from the *N*-terminal region of the human granulocyte and macrophage colony-stimulating factor (hGM-CSF). These peptides not only contain four and six potential *O*-glycosylation sites, respectively, but also comprise a four amino acid epitope (APAR), which is recognized by a monoclonal activity, facilitating the detection, quantification, and purification of the fusion proteins. Both GMOP-IFN and mGMOP-IFN chimeras showed an increased charge and mass due to *O*-glycan addition, which also resulted in an improved pharmacokinetic and stability profile, thus demonstrating the potentiality of both bifunctional peptides for the development of novel long-acting biotherapeutics, facilitating the setting of detection and purification techniques [76].

In conclusion, this case study widely demonstrates the usefulness of glycoengineering strategies to improve protein drugs and develop new biobetters.

Compliance with Ethical Standards This chapter is a review and no humans or animals were involved in the preparation of the chapter.

References

1. Kowey PR, Fisher L, Giardina EG et al (1988) The TPA controversy and the drug approval process: the view of the cardiovascular and renal drugs advisory committee. JAMA J Am Med Assoc 260:2250–2252
2. Walsh G, Walsh E (2022) Biopharmaceutical benchmarks 2022. Nat Biotechnol 40:1722–1760
3. Tripathi NK, Shrivastava A (2019) Recent developments in bioprocessing of recombinant proteins: expression hosts and process development. Front Bioeng Biotechnol 7:420
4. Gupta SK, Shukla P (2018) Glycosylation control technologies for recombinant therapeutic proteins. Appl Microbiol Biotechnol 102:10457–10468
5. Noh SM, Shin S, Min Lee G (2020) Cell line development for therapeutic protein production. In: Min Lee G, HeKildegaard HF (eds) Cell culture engineering: recombinant protein production1st edn. Wiley-VCH Verlag GmbH & Co. KGaA, pp 23–47
6. Gerngross TU (2004) Advances in the production of human therapeutic proteins in yeasts and filamentous fungi. Nat Biotechnol 22:1409–1414
7. Kost TA, Condreay JP, Jarvis DL (2005) Baculovirus as versatile vectors for protein expression in insect and mammalian cells. Nat Biotechnol 23:567–575
8. Park KY, Wi SJ (2016) Potential of plants to produce recombinant protein products. J Plant Biol 59:559–568
9. Rozov SM, Permyakova NV, Deineko EV (2018) Main strategies of plant expression system glycoengineering for producing humanized recombinant pharmaceutical proteins. Biochemistry (Mosc) 83:215–232
10. Szymanski CM, Ruijin Y, Ewing CP et al (1999) Evidence for a system of general protein glycosylation in *Campylobacter jejuni*. Mol Microbiol 32:1022–1030
11. Paliya BS, Sharma VK, Tuohy MG et al (2023) Bacterial glycobiotechnology: a biosynthetic route for the production of biopharmaceutical glycans. Biotechnol Adv 67:108180
12. Eidenberger L, Kogelmann B, Steinkellner H (2023) Plant-based biopharmaceutical engineering. Nat Rev Bioeng 1:426–439
13. Strasser R, Stadlmann J, Schähs M et al (2008) Generation of glyco-engineered Nicotiana benthamiana for the production of monoclonal antibodies with a homogeneous human-like N-glycan structure. Plant Biotechnol J 6:392–402

14. Massoud M, Attal J, Thépot D et al (1996) The deleterious effects of human erythropoietin gene driven by the rabbit whey acidic protein gene promoter in transgenic rabbits. Reprod Nutr Dev 36:555–563
15. Houdebine LM (2009) Production of pharmaceutical proteins by transgenic animals. Comp Immunol Microbiol Infect Dis 32:107–121
16. Owczarek B, Gerszberg A, Hnatuszko-Konka K (2019) A brief reminder of systems of production and chromatography-based recovery of recombinant protein biopharmaceuticals. Biomed Res Int 2019:4216060
17. Geng SL, Zhao XJ, Zhang X et al (2024) Recombinant therapeutic proteins degradation and overcoming strategies in CHO cells. Appl Microbiol Biotechnol 108:182
18. Schütz A, Bernhard F, Berrow N et al (2023) A concise guide to choosing suitable gene expression systems for recombinant protein production. STAR Protoc 4:102572
19. O'Flaherty R, Bergin A, Flampouri E et al (2020) Mammalian cell culture for production of recombinant proteins: a review of the critical steps in their biomanufacturing. Biotechnol Adv 43:107552
20. Gronemeyer P, Ditz R, Strube J (2014) Trends in upstream and downstream process development for antibody manufacturing. Bioengineering (Basel) 1:188–212
21. Xu WJ, Lin Y, Mi CL et al (2023) Progress in fed-batch culture for recombinant protein production in CHO cells. Appl Microbiol Biotechnol 107:1063–1075
22. Hacker DL, Balasubramanian S (2016) Recombinant protein production from stable mammalian cell lines and pools. Curr Opin Struct Biol 38:129–136
23. Lai T, Yang Y, Ng SK (2013) Advances in mammalian cell line development technologies for recombinant protein production. Pharmaceuticals (Basel) 6:579–603
24. Datta P, Linhardt RJ, Sharfstein ST (2013) An 'omics approach towards CHO cell engineering. Biotechnol Bioeng 110:1255–1271
25. Bort JAH, Stern B, Borth N (2010) CHO-K1 host cells adapted to growth in glutamine-free medium by FACS-assisted evolution. Biotechnol J 5:1090–1097
26. Urlaub G, Chasin LA (1980) Isolation of Chinese hamster cell mutants deficient in dihydrofolate reductase activity. Proc Natl Acad Sci USA 77:4216–4220
27. Tian J, He Q, Oliveira C et al (2020) Increased MSX level improves biological productivity and production stability in multiple recombinant GS CHO cell lines. Eng Life Sci 20:112–125
28. Srila W, Baumann M, Riedl M et al (2023) Glutamine synthetase (GS) knockout (KO) using CRISPR/Cpf1 diversely enhances selection efficiency of CHO cells expressing therapeutic antibodies. Sci Rep 13:10473
29. Bandaranayake AD, Almo SC (2014) Recent advances in mammalian protein production. FEBS Lett 588:253–260
30. Ha TK, Lee JS, Lee GM (2019) Platform Technology for Therapeutic Protein Production. In: Lee GML, Kildegaard HF (eds) Cell culture engineering: recombinant protein production1st edn. Wiley-VCH Verlag GmbH & Co. KGaA, pp 1–22
31. Liu L (2015) Antibody glycosylation and its impact on the pharmacokinetics and pharmacodynamics of monoclonal antibodies and Fc-fusion proteins. J Pharm Sci 104:1866–1884
32. Dumont J, Euwart D, Mei B et al (2016) Human cell lines for biopharmaceutical manufacturing: history, status, and future perspectives. Crit Rev Biotechnol 36:1110–1122
33. Lalonde M-E, Durocher Y (2017) Therapeutic glycoprotein production in mammalian cells. J Biotechnol 251:128–140
34. Graham FL, Smiley J, Russell WC et al (1977) Characteristics of a human cell line transformed by DNA from human adenovirus type 5. J Gen Virol 36:59–72
35. Cho MS, Yee H, Chan S (2002) Establishment of a human somatic hybrid cell line for recombinant protein production. J Biomed Sci 9:631–638
36. Moran N (2010) Shire's replacement enzymes validate gene activation. Nat Biotechnol 28: 1139–1140

37. Havenga MJE, Holterman L, Melis I et al (2008) Serum-free transient protein production system based on adenoviral vector and PER.C6 technology: high yield and preserved bioactivity. Biotechnol Bioeng 100:273–283
38. Swiech K, Picanço-Castro V, Covas DT (2012) Human cells: new platform for recombinant therapeutic protein production. Protein Expr Purif 84:147–153
39. Batenchuk C, St-Pierre S, Tepliakova L et al (2011) Chromosomal position effects are linked to sir2-mediated variation in transcriptional burst size. Biophys J 100:L56–L58
40. Martín-Alonso S, Frutos-Beltrán E, Menéndez-Arias L (2021) Reverse transcriptase: from transcriptomics to genome editing. Trends Biotechnol 39:194–210
41. Bollati-Fogolín M, Comini M (2008) Cloning and expression of heterologous proteins in animal cells. In: Castilho LR, Moraes AM, Augusto EFP, Butler M (eds) Animal cell technology: from biopharmaceuticals to gene therapy. Taylor & Francis, pp 39–73
42. Barrett LW, Fletcher S, Wilton SD (2012) Regulation of eukaryotic gene expression by the untranslated gene regions and other non-coding elements. Cell Mol Life Sci 69:3613–3634
43. Furth PA, Hennighausen L, Baker C et al (1991) The variability in activity of the universally expressed human cytomegalovirus immediate early gene 1 enhancer/promoter in transgenic mice. Nucleic Acids Res 19:6205–6208
44. Wenger RH, Moreau H, Nielsen PJ (1994) A comparison of different promoter, enhancer, and cell type combinations in transient transfections. Anal Biochem 221:416–418
45. Osterlehner A, Simmeth S, Göpfert U (2011) Promoter methylation and transgene copy numbers predict unstable protein production in recombinant Chinese hamster ovary cell lines. Biotechnol Bioeng 108:2670–2681
46. Mizushima S, Nagata S (1990) pEF-BOS, a powerful mammalian expression vector. Nucleic Acids Res 18:5322
47. Tossolini I, Gugliotta A, López Díaz F et al (2022) Screening of CHO-K1 endogenous promoters for expressing recombinant proteins in mammalian cell cultures. Plasmid 119-120:102620
48. Nguyen LN, Baumann M, Dhiman H et al (2019) Novel promoters derived from Chinese Hamster Ovary cells via in silico and in vitro analysis. Biotechnol J 14:e1900125
49. Romanova N, Noll T (2018) Engineered and natural promoters and chromatin-modifying elements for recombinant protein expression in CHO cells. Biotechnol J 13:e1700232
50. Das AT, Tenenbaum L, Berkhout B (2016) Tet-on systems or doxycycline-inducible gene expression. Curr Gene Ther 16:156–167
51. Chang MM, Gaidukov L, Jung G et al (2019) Small-molecule control of antibody N-glycosylation in engineered mammalian cells. Nat Chem Biol 15:730–736
52. Haberle V, Stark A (2018) Eukaryotic core promoters and the functional basis of transcription initiation. Nat Rev Mol Cell Biol 19:621–637
53. Cenik C, Derti A, Mellor JC et al (2010) Genome-wide functional analysis of human 5′ untranslated region introns. Genome Biol 11:R29
54. Kruse N, Spencer S, Wirth D (2014) Rational approaches for transgene expression: targeted integration and episomal maintenance. In: Hauser H, Wagner R (eds) Animal cell biotechnology in biologics production. De Gruyter, Berlin, pp 173–215
55. Bode J, Schlake T, Ríos-Ramírez M et al (1995) Scaffold/matrix-attached regions: structural properties creating transcriptionally active loci. Int Rev Cytol 162A:389–454
56. Harraghy N, Buceta M, Regamey A et al (2012) Using matrix attachment regions to improve recombinant protein production. Methods Mol Biol 801:93–110
57. Bhattacharya M, Lyda SF, Lei EP (2024) Chromatin insulator mechanisms ensure accurate gene expression by controlling overall 3D genome organization. Curr Opin Genet Dev 87: 102208
58. Sizer RE, White RJ (2022) Use of ubiquitous chromatin opening elements (UCOE) as tools to maintain transgene expression in biotechnology. Comput Struct Biotechnol J 21:275–283
59. Kim S, Kim YW, Shim SH et al (2012) Chromatin structure of the LCR in the human β-globin locus transcribing the adult δ- and β-globin genes. Int J Biochem Cell Biol 44:505–513

60. Sekhavat A, Sun JM, Davie JR (2007) Competitive inhibition of histone deacetylase activity by trichostatin A and butyrate. Biochem Cell Biol 85:751–758
61. Turan S, Zehe C, Kuehle J et al (2013) Recombinase-mediated cassette exchange (RMCE) - a rapidly-expanding toolbox for targeted genomic modifications. Gene 515:1–27
62. Urnov FD, Rebar EJ, Holmes MC et al (2010) Genome editing with engineered zinc finger nucleases. Nat Rev Genet 11:636–646
63. Cermak T, Doyle EL, Christian M et al (2011) Efficient design and assembly of custom TALEN and other TAL effector-based constructs for DNA targeting. Nucleic Acids Res 39: e82
64. Amiri S, Adibzadeh S, Ghanbari S et al (2023) CRISPR-interceded CHO cell line development approaches. Biotechnol Bioeng 120:865–902
65. Chen JS, Doudna JA (2017) The chemistry of Cas9 and its CRISPR colleagues. Nat Rev Chem 1:0078
66. Kalkan AK, Palaz F, Sofija S et al (2023) Improving recombinant protein production in CHO cells using the CRISPR-Cas system. Biotechnol Adv 64:108115
67. Shin SW, Lee JS (2020) CHO cell line development and engineering via site-specific integration: challenges and opportunities. Biotechnol Bioproc E 25:633–645
68. Nunes NM, Li W, Tian B et al (2010) A functional human Poly(A) site requires only a potent DSE and an A-rich upstream sequence. EMBO J 29:1523–1536
69. Han Z, Moore GA, Mitter R et al (2023) DNA-directed termination of RNA polymerase II transcription. Mol Cell 83:3253–3267.e7
70. West S, Proudfoot NJ (2009) Transcriptional termination enhances protein expression in human cells. Mol Cell 33:354–364
71. López De Silanes I, Paz Quesada M, Esteller M (2007) Aberrant regulation of messenger RNA 3′-untranslated region in human cancer. Cell Oncol 29:1–17
72. Mauro VP (2018) Codon optimization in the production of recombinant biotherapeutics: potential risks and considerations. BioDrugs 32:69–81
73. Stern B, Olsen L, Trösse C et al (2007) Improving mammalian cell factories: the selection of signal peptide has a major impact on recombinant protein synthesis and secretion in mammalian cells. Trends Cell Mol Biol 2:1–17
74. O'Neill P, Mistry RK, Brown AJ et al (2023) Protein-specific signal peptides for mammalian vector engineering. ACS Synth Biol 12:2339–2352
75. Cheung RCF, Wong JH, Ng TB (2012) Immobilized metal ion affinity chromatography: a review on its applications. Appl Microbiol Biotechnol 96:1411–1420
76. Sales MM, Kratje R, Oggero M et al (2021) Bifunctional GM-CSF-derived peptides as tools for O-glycoengineering and protein tagging. J Biotechnol 327:18–27
77. Li F, Vijayasankaran N, Shen A et al (2010) Cell culture processes for monoclonal antibody production. MAbs 2:466–479
78. Kaufman RJ, Sharp PA (1982) Amplification and expression of sequences cotransfected with a modular dihydrofolate reductase complementary dna gene. J Mol Biol 159:601–621
79. Bebbington GR, Renner G, Thomson S et al (1992) High-level expression of a recombinant antibody from myeloma cells using a glutamine synthetase gene as an amplifiable selectable marker. Biotechnology (N Y) 10:169
80. Hauser H (2015) Cell line Development. In: Al-Rubeai M (ed) Animal cell culture. Springer, pp 1–25
81. Sayed N, Allawadhi P, Khurana A et al (2022) Gene therapy: comprehensive overview and therapeutic applications. Life Sci 294:120375
82. Jäger V, Bussow K, Schirrmann T (2015) Transient recombinant protein expression in mammalian cells. In: Al-Rubeai M (ed) Animal cell culture. Springer, pp 27–64
83. Rosazza C, Buntz A, Rieß T et al (2013) Intracellular tracking of single-plasmid DNA particles after delivery by electroporation. Mol Ther 21:2217–2226

84. Grosjean F, Bertschinger M, Hacker DL et al (2006) Multiple glycerol shocks increase the calcium phosphate transfection of non-synchronized CHO cells. Biotechnol Lett 28:1827–1833
85. Walther W, Stein U (2000) Viral vectors for gene transfer: a review of their use in the treatment of human diseases. Drugs 60:249–271
86. Vanaparthy R, Mohan G, Vasireddy D et al (2021) Review of COVID-19 viral vector-based vaccines and COVID-19 variants. Infez Med 29:328–338
87. Shaw A, Cornetta K (2014) Design and potential of non-integrating lentiviral vectors. Biomedicines 2:14–35
88. Priola JJ, Calzadilla N, Baumann M et al (2016) High-throughput screening and selection of mammalian cells for enhanced protein production. Biotechnol J 11:853–865
89. Freshney RI (2010) Culture of animal cells: a manual of basic technique and specialized applications.6th edn. Wiley-Blackwell, Hoboken
90. Alves Rodrigues MT, Moro AM (2008) Regulatory aspects. In: Castilho LR, Moraes AM, Augusto EFP, Butler M (eds) Animal cell technology: from biopharmaceuticals to gene therapy. Walter de Gruyter GmbH & Company KG, Berlin, pp 349–372
91. Ladoux B, Mège RM (2017) Mechanobiology of collective cell behaviours. Nat Rev Mol Cell Biol 18:743–757
92. Horie M, Yamano-Adachi N, Kawabe Y et al (2022) Recent advances in animal cell technologies for industrial and medical applications. J Biosci Bioeng 133:509–514
93. Licata JP, Schwab KH, El H-e Y et al (2023) Bioreactor technologies for enhanced organoid culture. Int J Mol Sci 24:11427
94. Long Q, Liu X, Yang Y et al (2014) The development and application of high throughput cultivation technology in bioprocess development. J Biotechnol 192:323–338
95. Fink M, Cserjan-Puschmann M, Reinisch D et al (2021) High-throughput microbioreactor provides a capable tool for early stage bioprocess development. Sci Rep 11:2056
96. Kusterer A, Krause C, Kaufmann K et al (2008) Fully automated single-use stirred-tank bioreactors for parallel microbial cultivations. Bioprocess Biosyst Eng 31:207–215
97. Porter AJ, Dickson AJ, Racher AJ (2010) Strategies for selecting recombinant CHO cell lines for cGMP manufacturing: realizing the potential in bioreactors. Biotechnol Prog 26:1446–1454
98. Formenti LR, Nørregaard A, Bolic A et al (2014) Challenges in industrial fermentation technology research. Biotechnol J 9:727–738
99. Ambr® 15 Cell culture bioreactor system | Sartorius. https://www.sartorius.com/en/products/fermentation-bioreactors/ambr-multi-parallel-bioreactors/ambr-15-cell-culture. Accessed 29 Jun 2024
100. Micro-matrix Bioreactors (1–10 ml) | AntTeknik.com. https://www.antteknik.com/en/products/?p=micro-matrix-bioreactors-1-10-ml. Accessed 29 June 2024
101. Rish AJ, Drennen JK, Anderson CA (2022) Metabolic trends of Chinese hamster ovary cells in biopharmaceutical production under batch and fed-batch conditions. Biotechnol Prog 38: e3220
102. Ghaffari N, Jardon MA, Krahn N et al (2020) Effects of cysteine, asparagine, or glutamine limitations in Chinese hamster ovary cell batch and fed-batch cultures. Biotechnol Prog 36: e2946
103. Hoskisson PA, Hobbs G (2005) Continuous culture--making a comeback? Microbiology (Reading) 151:3153–3159
104. Vergara M, Torres M, Müller A et al (2018) High glucose and low specific cell growth but not mild hypothermia improve specific r-protein productivity in chemostat culture of CHO cells. PLoS One 13:e0202098
105. MacDonald MA, Nöbel M, Roche Recinos D et al (2022) Perfusion culture of Chinese Hamster Ovary cells for bioprocessing applications. Crit Rev Biotechnol 42:1099–1115

106. Liang K, Luo H, Li Q (2023) Enhancing and stabilizing monoclonal antibody production by Chinese hamster ovary (CHO) cells with optimized perfusion culture strategies. Front Bioeng Biotechnol 11:1112349
107. Karst DJ, Serra E, Villiger TK et al (2016) Characterization and comparison of ATF and TFF in stirred bioreactors for continuous mammalian cell culture processes. Biochem Eng J 110: 17–26
108. Wurm FM (2004) Production of recombinant protein therapeutics in cultivated mammalian cells. Nat Biotechnol 22:1393–1398
109. Yongky A, Xu J, Tian J et al (2019) Process intensification in fed-batch production bioreactors using non-perfusion seed cultures. MAbs 11:1502–1514
110. Masuda K, Kubota M, Nakazawa Y et al (2024) Establishment of a novel cell line, CHO-MK, derived from Chinese hamster ovary tissues for biologics manufacturing. J Biosci Bioeng 137: 471–479
111. Kiesslich S, Vila-Chã Losa JP, Gélinas JF et al (2020) Serum-free production of rVSV-ZEBOV in Vero cells: microcarrier bioreactor versus scale-X™ hydro fixed-bed. J Biotechnol 310:32–39
112. Kiesslich S, Kamen AA (2020) Vero cell upstream bioprocess development for the production of viral vectors and vaccines. Biotechnol Adv 44:107608
113. Angepat S, Gorenflo VM, Piret JM (2005) Accelerating perfusion process optimization by scanning non-steady-state responses. Biotechnol Bioeng 92:472–478
114. Orchard-Webb D, Robu I (2019) Manufacturability of regenerative medicine products for commercial scale. In: Encyclopedia of tissue engineering and regenerative medicine, vol 1–3. Elsevier, pp V3-63–V3-69
115. Jossen V, Eibl R, Eibl D (2019) Single-use bioreactors - an overview. In: Eibl R, Eibl D (eds) Single-use technology in biopharmaceutical manufacture, second edition. Wiley, pp 37–52
116. Krambeck FJ, Bennun SV, Andersen MR et al (2017) Model-based analysis of N-glycosylation in Chinese hamster ovary cells. PLoS One 12:e0175376
117. Sharker SM, Rahman A (2021) A review on the current methods of Chinese hamster ovary (CHO) cells cultivation for the production of therapeutic protein. Curr Drug Discov Technol 18:354–364
118. Torres M, Berrios J, Rigual Y et al (2019) Metabolic flux analysis during galactose and lactate co-consumption reveals enhanced energy metabolism in continuous CHO cell cultures. Chem Eng Sci 205:201–211
119. Savizi ISP, Motamedian E, Lewis NE et al (2021) An integrated modular framework for modeling the effect of ammonium on the sialylation process of monoclonal antibodies produced by CHO cells. Biotechnol J 16:e2100019
120. Nica V, Martínez S, Dietmair S et al (2013) Flux balance analysis of CHO cells before and after a metabolic switch from lactate production to consumption. Biotechnol Bioeng 110:660–666
121. Yao T, Asayama Y (2017) Animal-cell culture media: history, characteristics, and current issues. Reprod Med Biol 16:99–117
122. van der Valk J, Bieback K, Buta C et al (2018) Fetal bovine serum (FBS): past – present – future. ALTEX 35:99–118
123. van der Valk J, Brunner D, De Smet K et al (2010) Optimization of chemically defined cell culture media - replacing fetal bovine serum in mammalian in vitro methods. Toxicol In Vitro 24:1053–1063
124. Price PJ (2017) Best practices for media selection for mammalian cells. In Vitro Cell Dev Biol Anim 53:673–681
125. Gangwar N, Balraj K, Rathore AS (2024) Explainable AI for CHO cell culture media optimization and prediction of critical quality attribute. Appl Microbiol Biotechnol 108:308
126. Amadeo GI (2021) Producción a gran escala de glicoproteínas terapéuticas: Downstream processing. In: Forno AG, Oggero M (eds) Glicoproteínas terapéuticas diseño, expresión en

células de mamífero y análisis de sus glicanos, Primera edición. Ediciones UNL, Santa Fe, pp 185–206
127. Graselli M (2015) Purificar un producto proteico. Para qué y cómo planificarlo. In: Graselli M (ed) Proteínas puras, 2015th ed. Universidad Nacional del Quilmes, Buenos Aires, Bernal, Prov. de Buenos Aires, pp 15–35
128. Du M, Hou Z, Liu L et al (2022) [1]Progress, applications, challenges and prospects of protein purification technology. Front Bioeng Biotechnol 10:1028691
129. Shekhawat LK, Sarkar J, Gupta R et al (2018) Application of CFD in bioprocessing: separation of mammalian cells using disc stack centrifuge during production of biotherapeutics. J Biotechnol 267:1–11
130. Echave J, Fraga-Corral M, Garcia-Perez P et al (2021) Seaweed protein hydrolysates and bioactive peptides: extraction, purification, and applications. Mar Drugs 19:500
131. Adamíková J, Antošová M, Polakovič M (2019) Chromatographic purification of recombinant human erythropoietin. Biotechnol Lett 41:483–493
132. Li Y (2022) Viral removal by column chromatography in downstream processing of monoclonal antibodies. Protein Expr Purif 198:106131
133. Farzi-Khajeh H, Toraby S, Akbarzadeh-Khiavi M et al (2022) Development of biomimetic triazine-based affinity ligands for efficient immunoglobulin G purification from human and rabbit plasma. J Chromatogr A 1684:463559
134. Mishra V (2020) Affinity tags for protein purification. Curr Protein Pept Sci 21:821–830
135. Reinau ME, Åsberg D (2024) Characterization of antibody surface properties and selection of a diverse subset for development of a generic size-exclusion chromatography method. J Chromatogr A 1716:464652
136. Pinto IF, Aires-Barros MR, Azevedo AM (2015) Multimodal chromatography: debottlenecking the downstream processing of monoclonal antibodies. Pharm Bioprocess 3: 263–279
137. Halan V, Maity S, Bhambure R et al (2018) Multimodal chromatography for purification of biotherapeutics – a review. Curr Protein Pept Sci 20:4–13
138. Gillespie C, Wasalathanthri DP, Ritz DB et al (2022) Systematic assessment of process analytical technologies for biologics. Biotechnol Bioeng 119:423–434
139. Rathore AS (2016) Quality by design (QbD)-based process development for purification of a biotherapeutic. Trends Biotechnol 34:358–370
140. Rathore AS, Bhargava A (2020) Biosimilars in developed economies: overview, status, and regulatory considerations. Regul Toxicol Pharmacol 110:104525
141. Egrie JC, Dwyer E, Browne JK et al (2003) Darbepoetin alfa has a longer circulating half-life and greater in vivo potency than recombinant human erythropoietin. Exp Hematol 31:290–299
142. Croxtall JD, McKeage K (2011) Corifollitropin alfa: a review of its use in controlled ovarian stimulation for assisted reproduction. BioDrugs 25:243–254
143. Miller BS, Blair J, Horikawa R et al (2024) Developments in the management of growth hormone deficiency: clinical utility of somapacitan. Drug Des Devel Ther 18:291–306
144. Parker SD, King N, Jacobs TF (2024) Pegfilgrastim. Stat Pearls, Treasure Island
145. Strohl WR (2015) Fusion proteins for half-life extension of biologics as a strategy to make biobetters. BioDrugs 29:215–239
146. Kuehne H, Davis JM, Merewether LA et al (2024) Demonstration of physicochemical and functional similarity of biosimilar Pegfilgrastim-cbqv to Pegfilgrastim. Drugs R D 24:285. https://doi.org/10.1007/s40268-024-00471-9
147. International Conference in Harmonisation. Pharmaceutical development Q8(R2) (2009)
148. Nupur N, Joshi S, Gulliarme D et al (2022) Analytical similarity assessment of biosimilars: global regulatory landscape, recent studies and major advancements in orthogonal platforms. Front Bioeng Biotechnol 10:832059
149. Guerra A, von Stosch M, Glassey J (2019) Toward biotherapeutic product real-time quality monitoring. Crit Rev Biotechnol 39:289–305

150. Torres-Obreque KM, Meneguetti GP, Muso-Cachumba JJ et al (2022) Building better biobetters: from fundamentals to industrial application. Drug Discov Today 27:65–81
151. AlQahtani AD, O'Connor D, Domling A et al (2019) Strategies for the production of long-acting therapeutics and efficient drug delivery for cancer treatment. Biomed Pharmacother 113:108750
152. Kontermann RE (2016) Half-life extended biotherapeutics. Expert Opin Biol Ther 16:903–915
153. Ceaglio N, Etcheverrigaray M, Kratje R et al (2008) Novel long-lasting interferon alpha derivatives designed by glycoengineering. Biochimie 90:437–449
154. Ceaglio N, Etcheverrigaray M, Conradt HS et al (2010) Highly glycosylated human alpha interferon: an insight into a new therapeutic candidate. J Biotechnol 146:74–83
155. Ceaglio N, Etcheverrigaray M, Kratje R et al (2010) Influence of carbohydrates on the stability and structure of a hyperglycosylated human interferon alpha mutein. Biochimie 92:971–978
156. Gugliotta A, Ceaglio N, Kratje R et al (2019) Effect of ANITVNITV peptide fusion on the bioactivity and pharmacokinetics of human IFN-α2b and a hyper-N-glycosylated variant. J Biotechnol 303:46–52
157. Ceaglio N, Gugliotta A, Tardivo MB et al (2016) Improvement of in vitro stability and pharmacokinetics of hIFN-α by fusing the carboxyl-terminal peptide of hCG β-subunit. J Biotechnol 221:13–24

Part III
Biotherapeutics Derived from Nucleic Acids

Top Med Chem
https://doi.org/10.1007/7355_2024_170

Published online: 28 December 2024

Antisense Oligonucleotides (ASOs)

Juan Pablo Tosar

Contents

Abstract Antisense oligonucleotides (ASOs) are short (usually 15–25 nucleotides, nt), single-stranded synthetic oligomers of nitrogenous bases that mimic DNA or RNA molecules. They can recognize specific messenger RNA (mRNA) molecules in a sequence-dependent manner. Their mechanism of action is by either steric blockage, interfering with the binding of other molecules to the target RNA, or by inducing target RNA degradation. The latter is usually the case for DNA-based ASOs (i.e., deoxynucleotide-based), which can form DNA/RNA hybrids. These hybrids are recognized by the ribonuclease RNase H1, which cleaves the target mRNA. Consequently, DNA-based ASOs are useful for downregulating the expression of a specific gene. Alternatively, ASOs can be used to activate gene expression. This is achieved by steric-only RNA-based ASOs, which can alter the mRNA

J. P. Tosar (✉)
Unidad de Bioquímica Analítica, Centro de Investigaciones Nucleares, Facultad de Ciencias, Universidad de la República, Montevideo, Uruguay

Laboratorio de Genómica Funcional, Institut Pasteur de Montevideo, Montevideo, Uruguay

maturation process (i.e., splicing), provoking the inclusion or the exclusion of a specific exon. Since single-stranded nucleic acids are prone to degradation by both extracellular and intracellular nucleases, biotherapeutic ASOs are usually heavily modified to increase their half-life in plasma, tissues, or both. This chapter will introduce the most common types of chemical modifications in currently clinically approved ASOs. Additionally, we will explore the interplay between chemical modifications and the pharmacokinetic and pharmacodynamic properties of different ASOs. Since 2016, the number of FDA-approved therapeutic ASOs has increased substantially. Some ASOs have proven to be extremely effective in clinical trials, while others have shown only limited efficacy. Adverse reactions will also be described, along with solutions derived from basic research on chemical modifications that can improve the therapeutic index of an ASO. The chapter includes several representative case studies, where all these factors are discussed.

Keywords 2'-MOE, Casimersen, Cobomarsen, Eplontersen, Eteplirsen, Fomivirsen, Gapmer, Golodirsen, Gymnosis, Inotersen, LNA, Mipomersen, Morpholino, Nusinersen, Phosphorothioate, Ribose modifications, RNase H, Splice-switching ASOs, Tofersen, Viltolarsen, Volanesorsen

1 Introduction

Oligonucleotides are short, single-stranded or double-stranded DNA or RNA molecules. They can harbor multiple modifications to increase their stability or to affect their recognition by cellular receptors. Many of these modifications are not found in naturally occurring nucleic acids. Based on their structure and mechanism of action, they can be classified into antisense oligonucleotides (ASOs), small interfering RNAs (siRNAs), and aptamers. This chapter will focus mainly on ASOs. Specific chapters on siRNAs and aptamers are found within this book.[1]

Unlike aptamers, which have an antibody-like, specific target-interaction mechanism of action, ASOs and siRNAs regulate gene expression by recognizing messenger RNA (mRNA) molecules by sequence complementarity. Unlike siRNAs, ASOs are single-stranded molecules.

ASOs bind to specific mRNAs and can (a) induce their degradation by recruiting specific ribonucleases (RNases) or (b) sterically interfere with protein complexes like the ribosome or splicing factors, inhibiting translation or inducing alternative splicing, respectively. While the former mechanism of action is always related to mRNA downregulation, splice-switching ASOs can be used to increase the expression of a specific gene.

[1] Chapters "Small Interferring RNAs (siRNAs)" and "Aptamers", respectively.

2 Oligonucleotide Synthesis

The gold standard for oligonucleotide synthesis is based on phosphoramidite chemistry [1], later combined with solid-phased synthesis and automation [2]. Chapter "Nucleic Acids Production" describes solid-phase oligonucleotide synthesis in detail, so this topic will be only briefly introduced in this chapter. In brief, the sequential-cyclic method starts by immobilizing the terminal 3′ nucleoside (based on the desired oligonucleotide sequence) to a solid support (Fig. 1). This nucleoside is protected at the OH of ribose carbon 5 with a DMT (4,4′--dimethoxytrityl) group. Then, the synthesis cycle begins with #1: removal of the DMT protecting group, leaving the free 5'-OH nucleophile. Followed by #2: addition of a 5'-DMT-protected nucleoside, which is added as a phosphoramidite monomer. This nucleoside corresponds to the next base of the desired oligonucleotide, reading in the 3′ → 5′ direction. The 5'-OH group of the first nucleoside attacks the electrophile P atom of the incoming phosphoramidite monomer in a reaction catalyzed by tetrazole in acetonitrile. The coupling reaction attaches both nucleosides by a phosphite triester linkage with a diisopropylamino leaving group. In step #3, this phosphite triester bond is oxidized to the pentavalent phosphate

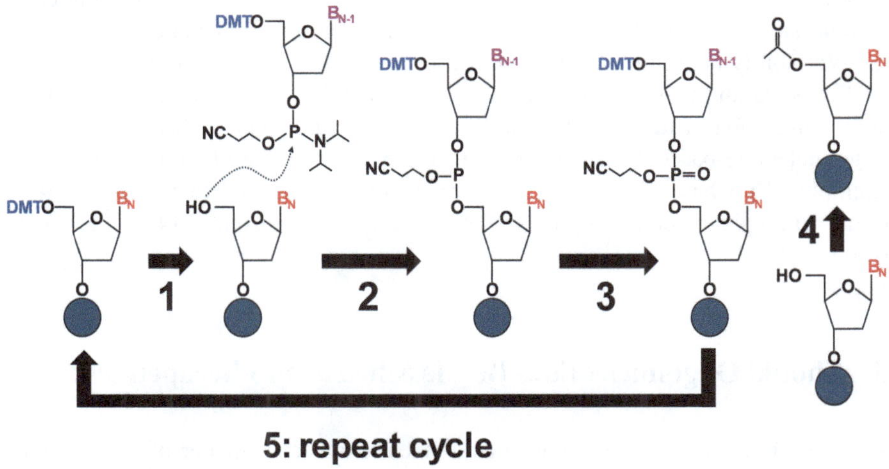

Fig. 1 DNA oligonucleotide synthesis by the phosphoramidite method. Steps are: removal of the DMT protecting group (1), coupling to the next nucleoside in the 3′→ 5′ direction (2), oxidation of the phosphite triester bond (3) and capping of unprotected nucleosides that avoided the coupling reaction (4). The product of step (3) is then used to fuel a new cycle of unprotection, coupling, oxidation and capping (5). Once the desired oligonucleotide length and sequence has been achieved, the 3′ terminal base is released from the solid substrate, leaving a 3′- OH as found in natural nucleic acids. In addition, the solution is heated to remove protecting groups from the phosphate linkages and from the nitrogenous bases (not drawn in the figure for simplicity). All phosphate triesters are converted into phosphodiester bonds, as found in natural nucleic acids. $\mathbf{B_N}$ and $\mathbf{B_{N\text{-}1}}$: thymine, adenine, cytosine or guanine appropriately protected; **DMT**: 4,4′-dimethoxytrityl

triester in a reaction promoted by iodine and catalyzed with lutidine in water. The resulting phosphate triester bond is similar to the standard phosphodiester bond found in DNA or RNA but contains a β-cyanoethyl protecting group that will be removed at the end of the synthetic process. While incoming phosphoramidite monomers in step #2 are added in a large-fold excess compared to the solid phase-attached substrates, it is not possible to achieve 100% coupling efficiency. This means that some unprotected solid phase-attached nucleosides could remain in the reaction. These are acetylated (capped) during stage #4, thus avoiding the possibility that they could work as substrates during the next coupling cycle #5.

It is important to understand that, despite progress made to achieve higher yields and coupling efficiencies across the decades, oligonucleotide synthesis does not allow to obtain a single, chemically homogeneous product, not even after a final step of HPLC purification. For example, even with coupling efficiencies as high as 99.5% and with the capping reaction being almost 100% efficient, this still implies that a small number of molecules in each cycle will not incorporate a new DMT-protected nucleoside and will escape capping. The result is what is called a "deletion mutant": a sequence almost identical to the desired oligonucleotide but lacking one base. Purification of these deletion mutants from the full-length sequence is not straightforward. This problem gets more serious the longer the oligonucleotides are. Yields also sharply decrease with the increase in the oligonucleotide length. A study analyzing deletion mutants in DNA oligonucleotides synthesized by different commercial suppliers found 0.2%–11.7% of the final products contained these errors [3]. Variability between batches from the same supplier was also observed.

The situation is worse for RNA oligonucleotides or DNA oligonucleotides containing modifications. Importantly, modified oligonucleotides comprise all oligonucleotide-based biotherapeutics that are currently used in the clinic. For example, RNA-based biotherapeutics are being administered to humans at 90% purity and typically consist of a complex mixture of stereoisomers [4], as described below.

3 Should Oligonucleotides Be Classified as Biotherapeutics?

The precise definition of a biotherapeutic agent has been a matter of much debate over the years, with even conflicting definitions coexisting[2] based on different interests at stake [5]. These definitions impact, among other things, the regulatory framework that applies to specific medicinal products. With the birth of a "new biotechnology" comprising genetic engineering of whole organisms and cells, it became clear that pharmaceutical products derived from these processes required a different form of regulation compared to traditional small drugs. Whether such a differential status should also apply to other biological macromolecules derived from

[2] See Chapter "Medicinal Chemistry and Biotherapeutics. Overview and Definitions".

live cells or organisms without genetic manipulation was also unclear, with the U.S. and Europe adopting different positions [5].

Unlike small-molecule drugs, biological macromolecules contain orders of magnitude more atoms and involve additional levels of structural complexity. Small molecules are defined by a specific chemical structure, and if two manufacturers produce the same chemical entity, it is possible to prove they are the same (above a given purity threshold) despite differences in the manufacturing process. This is not possible with biotherapeutics, where the process is often considered to be the product. Mainly for this reason, there are no "generic drugs" in biopharma. Instead, once patents expire, other manufacturers can produce and seek approval of "biosimilars." This distinction acknowledges the fact that it is impossible to obtain the same final product when producing biological macromolecules, because even the original or innovative product is chemically heterogeneous and has batch-to-batch variations.

Oligonucleotides (including ASOs, siRNAs, or aptamers) behave similarly to any other biotherapeutic agent in the aforementioned aspects, despite not being produced using any form of biotechnology. Depending on how biotherapeutics are defined, they could be considered either as small-molecule drugs or as biotherapeutics, or both. For instance, the Cambridge Business English Dictionary (published: November 2011) defines biopharmaceuticals as "*medicine and drugs that are produced using biotechnology.*" Under this definition, synthetic oligonucleotides should **not** be considered biotherapeutics and should therefore not be covered in this book. However, the Center for Biologics Evaluation and Research (CBER) of the U.S. Food & Drug Administration (FDA) explains that "biological products" (or "biologics" under FDA terminology) "*include a wide range of products (...) isolated from a variety of natural sources (...) and may be produced by biotechnology methods and other cutting-edge technologies.*" It is important to note the word "may" instead of "must." The same Q&A article [6] further explains the difference between biological products and small-molecule drugs: "*In contrast to most drugs that are chemically synthesized and their structure is known, most biologics are complex mixtures that are not easily identified or characterized. Biological products, including those manufactured by biotechnology, tend to be heat sensitive and susceptible to microbial contamination.*"

Based on the above broad definition of "biologic" by the FDA, it seems clear that therapeutic oligonucleotides should be classified as such. In fact, most scientific publications to date consider ASOs, siRNAs, and aptamers as bona fide biotherapeutics, and the same does the United States Pharmacopeia. However, it should be noted that the FDA still regulates synthetic therapeutic oligonucleotides as small-molecule drugs and not as biologics. For example, all FDA-approved ASOs or siRNAs have been approved based on a New Drug Application (NDA) Approval, like small-molecule drugs and unlike proteins and antibodies that are approved based on a Biologic License Application (BLA) Approval.

In summary, while therapeutic oligonucleotides are complex and of high molecular weight, biologically inspired molecules are not synthesized using living biological systems. Therefore, neither the FDA nor the European Medicines Agency

(EMA) classifies them as "biologics" (FDA) or "biological medicinal products" (EMA). Nevertheless, they share many of the complexities that characterize biotherapeutics and that distinguish them from conventional small-molecule drugs.

4 Chemical Modifications of Therapeutic ASOs and Structure-Function Relationships

Because most therapeutic ASOs are administered as naked oligonucleotides (i.e., without forming a new entity with a protecting agent), a key issue is their stability against extracellular and intracellular nucleases (i.e., DNases and RNases).

DNases are enzymes that degrade DNA molecules, while RNases catalyze the hydrolysis of RNA. Both types of enzymes are abundant and highly active in human biofluids such as the blood [7, 8]. Consequently, most naked RNAs spiked in either blood serum, urine, or cerebrospinal fluid are degraded in seconds at 37 °C [9].

Both DNases and RNases catalyze similar reactions where a water molecule in the active site is deprotonated, generating a hydroxide that can perform a nucleophilic attack on the phosphate connector. The final product is the breakage of the phosphodiester bond, leaving a 5′ fragment containing a 3′-phosphate group and a 3′ fragment containing a free 5'-OH. In the case of RNA, the catalytic cycle involves the deprotonation of the ribose's 2'-OH, and the consequent nucleophilic attack on the phosphate group to generate a 2′,3′ cyclic phosphate intermediate. This intermediate is then attacked by the deprotonated water molecule in the active site, to resolve the 2′,3′ cyclic phosphate into an acyclic 3′-phosphate group.

DNases and RNases can also be classified into exonucleases and endonucleases. The former cleave nucleic acids in a processive way, starting from one of the extremes and advancing in either the 5′ → 3′ or the 3′ → 5′ direction. Endonucleases, in contrast, cleave internal phosphodiester bonds and can have preferences for specific nt or short sequence motifs. Two well-known examples are restriction enzymes (double-stranded DNA endonucleases codified in many bacterial genomes) and RNase A family members (single-stranded RNA endonucleases, abundantly present in human blood plasma).

Based on the catalytic mechanisms of these enzymes explained above, it should be easy to understand that modifications in either the phosphate group (for both DNA and RNA) or the ribose's 2'-OH group (for RNA) can render oligonucleotides resistant to enzymatic degradation, a desirable feature for therapeutic nucleic acids.

Thus, nuclease protection is essential to achieve meaningful oligonucleotide concentrations in the extracellular space and hence measurable biological effects once the oligonucleotides enter their target cells. For naked nucleic acids, this means introducing chemical modifications in their backbone (i.e., at the level of the phosphodiester bonds that link nucleosides together), in the ribose, or both.

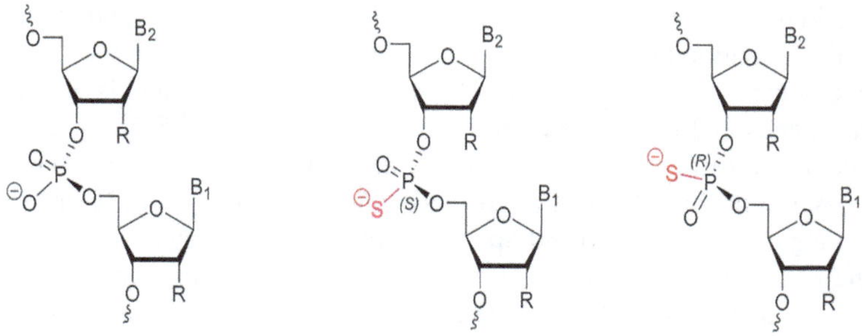

Fig. 2 Natural phosphodiester linkage (left) or the two possible configurations of the PS modification: *S*p (middle) and *R*p (right) diasteromers. R: -H for DNA, -OH for RNA, or -methoxyethyloxy in the case of 2'-MOE ASOs. Extracted from Jahns et al. Nucleic Acids Res. (2022) [11]. ©: the authors, under CC BY 4.0

4.1 First-Generation ASOs: The Phosphorothioate (PS) Modification

Fomivirsen (Vitravene®) was the first ASO approved for human use by the FDA in 1998. Developed by Isis Pharmaceuticals (now Ionis Pharmaceuticals, Inc.) and licensed to Novartis, it was indicated as an antiviral capable of inhibiting cytomegalovirus (CMV) eye infections in patients with acquired immunodeficiency syndrome (AIDS). Despite being effective, it was voluntarily withdrawn from the market in 2002 in Europe and in 2006 in the United States of America due to the decline of AIDS-related CMV infections (due to the availability of the highly active antiretroviral therapy for HIV).

From a chemical point of view, ***fomivirsen*** is a 21-mer[3] modified DNA oligonucleotide. Its sequence is complementary to the CMV mRNA, and its administration inhibits viral early protein synthesis. It was administered by intravitreal injection in the eye and reached peak concentrations in the retina two days after injection. Work in non-human primates showed an elimination half-life of 78 h in the monkey retina [10].

How could a short single-stranded DNA oligonucleotide achieve such high stability, avoiding degradation by both extracellular and intracellular nucleases? The answer is evident when looking at the chemical structure: all phosphodiester bonds connecting adjacent nucleosides were substituted with phosphorothioates (PS). We will refer to oligonucleotides containing this modification as PS ASOs.

A phosphorothioate is a functional group in which one of the oxygens not involved in the phosphodiester is replaced by a sulfur (Fig. 2). This substitution does not strongly affect nucleic acid structure and, hence, does not interfere with

[3]Denotes the length of the oligonucleotide (from Greek meros: "part"). The way to annotate is by first indicating the number of nt, followed by a hyphen, and finally the "mer" particle. For example, an oligonucleotide of 20 nt would usually be called a "20-mer."

nucleic acid hybridization. However, it can strongly inhibit endo- and exonuclease-mediated degradation [12].

Unlike phosphate groups, the phosphoros atom in the PS group is bound to four different ligands and is therefore chiral. This means that the PS group can exist in two structurally different configurations, called the *R*p and the *S*p diasteromers (Fig. 2). While having identical atoms and identical bonds, the different tridimensional arrangements of the *R*p and *S*p diasteromers confer distinct susceptibilities to both endo- and exonucleases [13]. Due to differences in protein binding and nuclease resistance, *R*p and *S*p diasteromers can impact the pharmacokinetic and pharmacodynamic properties of PS-modified therapeutic nucleic acids [11].

PS groups can be incorporated into synthetic nucleic acids by enzymatic reactions involving DNA or RNA polymerases, using PS-modified (deoxy)nucleoside 5'-O-(1-thiotriphosphates). These enzymes accept only the *S*p diasteromer [12] and therefore yield stereoselective products. However, as mentioned earlier, oligonucleotide biotherapeutics are generally manufactured by solid-phase synthesis using the phosphoramidite method (Fig. 1). Here, PS groups are introduced by substituting the oxidation reaction (step #3 in Fig. 1) with a sulfurization reaction, employing different sulfur transfer agents.

An advantage of introducing PS modifications during solid-phase oligonucleotide synthesis is that PS bonds can be placed at any desirable site on the oligonucleotide. For example, all phosphates can be substituted with PS groups (as in ***fomivirsen***), or only those in certain positions. However, an important disadvantage is that the process is not stereoselective, yielding a racemic mixture of *R*p/*S*p diasteromers. For an oligonucleotide with *n* PS groups, this means that 2^n distinct stereoisomers are obtained per reaction. And this even without considering truncated products and deletion mutants.

Methods for stereoselective oligonucleotide synthesis have shown that the *R*p and *S*p diasteromers behave differently in their hybridization strength and nuclease sensitivity [14, 15]. Thus, novel methods for stereoselective ASO design, alternating *R*p and *S*p diasteromers at defined positions, could theoretically achieve a better balance between activity and stability. However, attempts to do so did not find significant benefits compared to stereo-random PS ASOs that could justify the added complexity and costs [16].

4.2 2'-MOE Ribose Modifications and "Gapmer" Design for Efficient RNase H1-Mediated mRNA Degradation

Mipomersen (Kynamro®) was the second ASO approved by the FDA, this time in 2013; fifteen years after the approval of ***fomivirsen***. The EMA rejected it twice, in 2012 and 2013, due to important side effects. The FDA approved its indication for the treatment of homozygous familial hypercholesterolemia as an adjunct to lipid-lowering medications and diet. However, it imposed constraints on prescription and stricter pharmacovigilance due to the side effects observed during clinical trials, related to a serious risk of liver damage. The ASO was complementary to the apolipoprotein B-100 (ApoB) mRNA, and induced its degradation. The ApoB-100 protein is involved in cholesterol transport from the liver to other tissues, being an integral component of very low-density, low-density, and intermediate-density lipoproteins (VLDL, LDL, and IDL, respectively); collectively referred to as "bad cholesterol." By lowering ApoB-100 protein levels in the liver, ***mipomersen*** reduced LDL cholesterol levels in the circulation of patients with homozygous [17] as well as heterozygous familial hypercholesterolemia [18].

Mipomersen was developed by Ionis Pharmaceuticals and licensed to Genzyme Corporation. The company requested that the FDA withdraw its approval in 2019, as they were retiring the product from the market.

As mentioned earlier, ***mipomersen***'s mechanism of action involves targeted degradation of the apolipoprotein B-100 mRNA. This enzymatic degradation is mediated mainly by members of the ribonuclease H (RNase H) family.

RNase H enzymes cleave RNA in the context of RNA/DNA hybrids [19]. These hybrids are formed during DNA replication in the nucleus and mitochondria and are called R-loops. RNase H enzymes are involved in R-loop processing, but also in ribonucleotide excision repair, a mechanism preventing misincorporation of RNA nucleosides in DNA molecules [20]. The existence of RNase H enzymes in human cells can therefore be exploited for targeted mRNA degradation using ASOs with a DNA backbone. The DNA backbone is here essential because RNase H enzymes do not cleave double-stranded RNA.

The human genome codifies for two RNase H enzymes, termed RNase H1 and RNase H2. While RNase H2 is mostly nuclear, RNase H1 is expressed both in the nucleus and the cytoplasm, and is thought to be the main enzyme involved in ASO-mediated mRNA degradation [21]. Due to their localization in both the nucleus and the cytoplasm, RNase H1-dependent ASOs can therefore target either unprocessed mRNAs or non-coding RNAs in the nucleus, or processed mRNAs in the cytoplasm [22].

Importantly, the PS modification does not interfere with the formation of DNA/RNA hybrids and the RNA cleavage by RNase H1. In other words, from an RNase H1 perspective, PS-modified ASOs behave just like DNA.

A problem with first-generation PS ASOs like ***fomivirsen*** is that their nuclease stability is not optimal. As explained in the previous sub-section of this chapter, nearly half of the PS bonds in each molecule are in a chiral conformation that is susceptible to either exonuclease or endonuclease-mediated degradation. This was probably not a problem for ***fomivirsen*** because it was injected directly into the eye. However, ASOs intended for systemic delivery, such as ***mipomersen*** (that needed to target the liver), required further optimization.

Mipomersen was a second-generation ASO that followed a now-popular "gapmer" design. It contained 20 nt complementary to the coding region of the ApoB mRNA, where all the 19 phosphodiester bonds were substituted for phosphorothioates. In addition, the first five and the last five residues had a

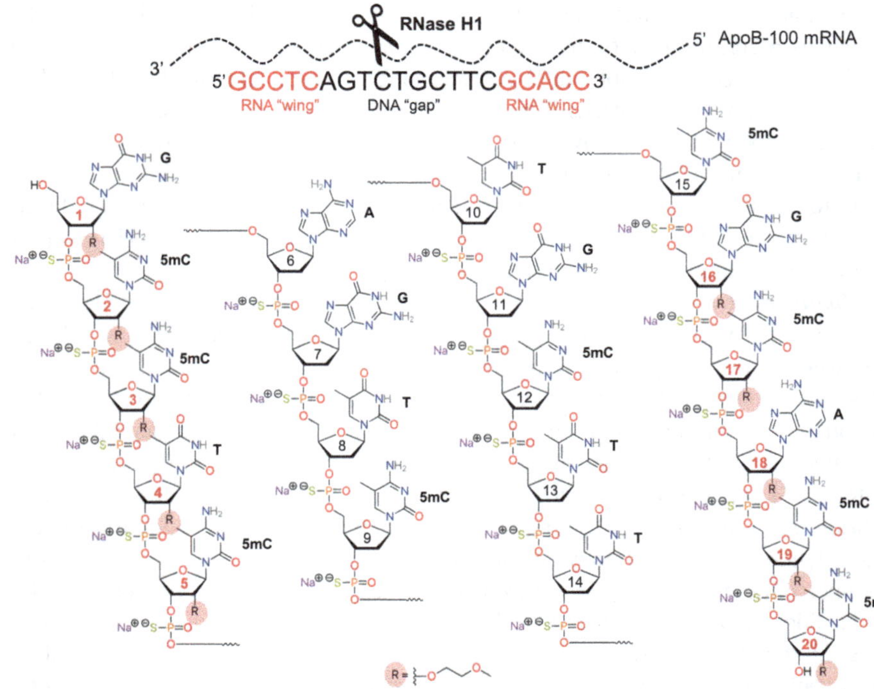

Fig. 3 ***Mipomersen*** (Kynamro™), a characteristic second-generation PS ASO with gapmer design. Top: schematic representation of the gapmer, with 2'-MOE residues (RNA "wing") represented in red, and 2′-deoxy residues (DNA "gap") represented in black. Bottom: structural formula of ***mipomersen*** sodium salt. Note the sulfur atoms represented in yellow, characteristic of the PS groups that substitute every phosphodiester bond in this molecule. R: methoxyethyloxy (highlighted in red circle). Also, note that all C residues are changed to 5mC. U residues are also methylated, but 5 mU is actually T. The image in the bottom panel was modified from a loyalty-free image deposited in Wikimedia Commons

2′-methoxyethyl (MOE)-modification in the ribose. These 5-mer "wings" flanked an internal region of 10 DNA (i.e., 2′-deoxy) nucleosides (termed the "gap") (Fig. 3).

X-ray crystallography has shown that oligonucleotides containing a 2'--MOE-modification in the ribose are structurally similar to RNA [23]. The modification even improves the stability of the hybrids that are formed upon the recognition of complementary RNA sequences. In addition, it vastly improves nuclease stability compared to PS-DNA oligonucleotides or to unmodified RNA molecules.

Oligonucleotides containing a 2'-MOE modification are no longer capable of inducing RNase H1-mediated target RNA degradation. This makes sense considering RNase H cleaves the RNA strand of a RNA/DNA hybrid, and 2'-MOE oligonucleotides tend to form standard A-form dsRNA complexes [23]. For this reason, ***mipomersen*** adopted a gapmer design where the 2'-MOE 5-mer wings conferred increased exonuclease protection and improved target recognition, while the 10-mer internal "gap" enabled RNase H1-mediated cleavage of ApoB-100 mRNA.

Besides PS bonds and 2'-MOE-modifications in the ribose of the most external nucleosides, ***mipomersen*** also contained modifications in the nitrogenous bases. In concrete, cytosine residues were all substituted for 5-methyl cytosine [24], presumably as a way to avoid recognition of innate immune sensors like TLR9 [25].

Other second-generation gapmers also relying on RNase H1-mediated mRNA degradation are ***Inotersen*** (Tegsedi®), ***Volanesorsen*** (Waylivra®), and ***Tofersen*** (Qalsody®). ***Eplontersen*** (Wainua®) is a derivative of ***inotersen*** that will be described in the pharmacokinetics section of this chapter.

Inotersen is indicated for the treatment of early stages of nerve damage in patients with polyneuropathy of hereditary transthyretin-mediated amyloidosis (hATTR). Mutations in the gene encoding transthyretin (*TTR*) can provoke the misfolding of the *TTR* protein, which can then form amyloid deposits that build up in the heart, nerves, and other organs. ***Inotersen*** inhibits hepatic production of transthyretin through RNase H1-mediated *TTR* mRNA degradation. Weekly subcutaneous administration of the drug improved the course of neurological disease and quality of life for patients with hATTR, but serious side effects were observed during clinical trials [26]. Nevertheless, both the FDA and EMA approved the ASO for clinical use in 2018, due to the unmet medical need for these neglected patients.

Volanesorsen was designed for the treatment of familial chylomicronemia syndrome, a rare genetic disease characterized by the loss of lipoprotein lipase activity, and the consequent buildup of chylomicrons (large triglyceride-rich lipoproteins produced in the intestine). The target gene in this case is apolipoprotein C3 (*APOC3*), which encodes a protein present in chylomicrons and other lipoproteins and that inhibits lipoprotein lipase. Phase 3 trials showed a reduction of both plasma apolipoprotein C3 and triglyceride levels in patients that received the drug versus participants in the placebo group [27]. It was approved for medical use by EMA in May, 2019. However, the FDA rejected its marketing authorization in the United States of America due to concerns on safety issues.

Tofersen targets the superoxide dismutase 1 (*SOD1*) mRNA and is indicated in patients with amyotrophic lateral sclerosis (ALS) associated with mutations in *SOD1*, which accounts for approximately 20% of the cases of familial ALS [28]. During phase 3 trials, adults with SOD1 ALS received either eight doses of ***tofersen*** or placebo for 28 weeks by intrathecal administration. Despite lower levels of SOD1 in cerebrospinal fluid or neurofilament light chain in plasma (a surrogate marker of neuronal death) observed in the treatment group, clinical endpoints were not met [29]. However, an open-label extension of the trial, where all participants received ***tofersen***, showed improvements in participants who started ***tofersen*** at trial entry. Based on the unmet medical need for ALS treatment, ***tofersen*** was approved by the FDA under the accelerated approval pathway on April 25, 2023. EMA approved ***tofersen***, under exceptional circumstances, on February 22, 2024.

4.3 ASO Biotherapeutics and Pharmaceutical Industry

Different pharmaceutical companies have helped develop and market first and second-generation PS ASOs. However, groundbreaking work and initial clinical phases have been sponsored in all cases by Ionis Pharmaceuticals, Inc. The company, based in Carlsbad (California), was founded in 1989 as Isis Pharmaceuticals and later changed its name to Ionis. It was established by Stanley T. Crook with the mission of establishing a new medicinal chemistry based on oligonucleotides [30]. The clinical success of the oligonucleotide synthesis, design, and evaluation platform developed at Ionis is remarkable. ***Fomivirsen*** was the first FDA-approved product of the company and later licensed to Novartis. ***Mipomersen*** was developed by Ionis into phase 2 and later licensed to Genzyme Corporation. ***Inotersen*** and ***volanesorsen*** are commercialized by Akcea Therapeutics, a spinout of Ionis that the company purchased back in 2020. ***Tofersen*** was also developed by Ionis and licensed to co-developer Biogen, headquartered in Cambridge (Massachusetts). As will be discussed next, ***nusinersen*** (Spinraza™), a splice-switching fully modified second-generation ASO, was also developed by Ionis (in collaboration with Yimin Hua and Adrian Krainer from the Cold Spring Harbor Laboratory) and co-developed by Biogen, who markets the product. Anecdotally, Biogen was co-founded by Phillip Sharp in 1978, who shared the 1993 Nobel Prize in Physiology or Medicine for the discovery of mRNA splicing. ***Nusinersen*** is, precisely, a drug targeting and interfering with this important mRNA maturation process. It will be further discussed in the section on splice-switching ASOs.

4.4 Locked Nucleic Acid (LNA)-Containing Gapmers

The quest for alternative chemistries to the 2'-MOE substitution that could also confer nuclease resistance led to the employment of locked nucleic acids (LNA) in combination with the PS substitution. In LNAs, the 2′ oxygen and the 4′ carbon of the nucleic acid ribose are linked together, forming a new furan ring and generating a constrained nt (Fig. 4). The consequence is a modified nucleic acid with reduced

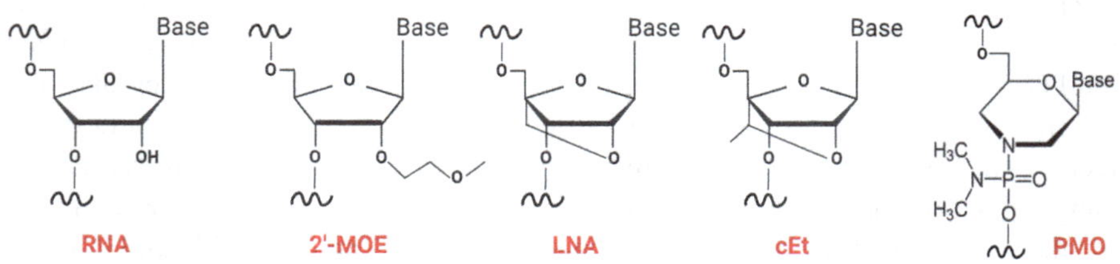

Fig. 4 Representative ribose modifications in biotherapeutic oligonucleotides. Left: unmodified RNA (2'-OH). 2'-MOE: 2-Methoxyethyl RNA. LNA: locked nucleic acid. cEt: constrained ethyl nucleic acid. PMO: phosphorodiamidate morpholino oligomer. In the case of PMOs, both the ribose and the nucleic acid backbone are substantially modified

conformational flexibility and a consequent increase in its binding affinity. The increase is remarkable: 4 °C to 8 °C in the T_m of an ASO/RNA hybrid per base modified with LNA [31]. Methylated versions of LNA produced by methylfuran ring production, known as "constrained ethyl" (cEt), are also being tested.

LNA-based gapmers showed increased affinity for their targets. Unfortunately, this comes with a cost. Several reports have shown increased incidence of hepatotoxicity for highly active ASOs containing LNA and cEt modifications [32]. Apparently, this toxicity is due to increased protein binding and the consequent delocalization of nuclear, cytoplasmic, and/or mitochondrial proteins. Fortunately, a single 2-O-MOE modification at position 2 of the gap was shown to substantially reduce toxicity with minimal impairment on antisense activity [32].

4.4.1 Other LNA-Based ASOs

Due to their high binding affinity, LNA-based ASOs are also useful to interfere with the function of noncoding RNA molecules by steric blockage. For example, ***cobomarsen*** is a LNA-based ASO that targets microRNA miR-155. MicroRNAs are small RNAs (20–22 nt) that do not code for proteins.[4] They bind to the 3′ untranslated region of mRNAs, inhibiting their translation. Because some microRNAs bind and repress the translation of tumor suppressor mRNAs, several microRNAs have been found to play important roles in cancer. For example, increased miR-155 expression is crucial for the initiation and progression of B-cell lymphoma [33].

4.5 Splice-Switching ASOs

4.5.1 Fully Modified 2'-MOE ASOs for Exon Inclusion in Spinal Muscular Atrophy

Nusinersen (Spinraza®) is a first-of-its-kind ASO that, instead of engaging RNase H1 to degrade specific mRNAs, alters mRNA maturation (splicing) in the nucleus to induce the expression of a gene and produce higher levels of its encoded protein. Thus, it has a completely different mechanism of action compared to ***fomivirsen***, ***mipomersen***, ***inotersen***, ***volanesorsen,*** and ***tofersen***, discussed above.

When activating the expression of a gene is desired, it is critical to avoid RNase H1-mediated mRNA degradation. For this reason, ***nusinersen*** moves away from the gapmer design. In contrast, it is 100% modified with PS bonds and 2'-MOE groups in the ribose. In other words, it is a stabilized RNA molecule that acts based on the steric block of its target mRNA.

---

[4] See Chapter "Small Interferring RNAs (siRNAs)".

Nusinersen is indicated for the treatment of spinal muscular atrophy (SMA), a rare genetic disease that affects the gene encoding the survival of motor neuron 1 protein (SMN1). Disruptions in this gene induce global splicing defects, especially in spinal motor neurons. As a result, motor neurons die, leading to muscular atrophy and ultimately death in the most severe forms of the disease. It is diagnosed in infancy or early childhood and, until approval of the drug, was the most common cause of infant death due to a genetic disease.

SMN1 is encoded in an unstable genomic region susceptible to duplications, and, as a result, the human genome contains several copies of *SMN1* and *SMN2*, an almost identical gene. *SMN1* and *SMN2* are nearly identical in sequence, except for a C to T transition in exon 7 of *SMN2* [34]. Despite being present in a coding sequence, this mutation is actually silent. In other words, the SMN2 protein should theoretically have exactly the same amino acid composition as SMN1. However, this silent mutation overlaps a splicing enhancer. As a result, during mRNA maturation (i.e., splicing), exon 7 of *SMN2* is spliced out. The *SMN2* gene therefore results in an mRNA lacking exon 7, producing a shorter protein product that is unstable and has markedly reduced functionality [35]. Because *SMN2* is mainly nonfunctional, SMA results from biallelic mutations in the *SMN1* gene. The severity of the disease mostly depends on copy number variations in *SMN2* between patients. While the capacity of the *SMN2* gene to compensate for the loss of the SMN1 protein is very limited, having several copies of *SMN2* in the genome is beneficial. For example, patients with infantile-onset SMA (also called SMA type 1, accounting for roughly 60% of the cases) typically have two copies of *SMN2*, while patients with late-onset SMA (types 3 and 4) might have between 3 and 8 copies of the gene [36]. Clinical manifestations of each SMA type vary greatly. Children with type 1 SMA, for example, are unable to sit without support and have trouble breathing, feeding, and swallowing. They usually die before age 2. Children with type 3 SMA, in contrast, might be able to walk and have an almost normal life expectancy.

Nusinersen is a fully modified second-generation RNA ASO, targeting an intron between exon 7 and exon 8 of the *SMN2* gene. It contains PS bonds and 2'-MOE ribose modifications in all possible positions. Consequently, it does not induce RNase H1-mediated mRNA degradation. As in other second-generation ASOs, all cytosines are converted to 5mC to reduce immune responses. This ASO was designed based on the work of the Uruguayan-American biochemist Dr. Adrian Krainer (from the Cold Spring Harbor Laboratory, USA) and Dr. Frank Bennett (one of the founding members of Ionis Pharmaceuticals, formerly known as Isis). Their teams showed that, in mice, targeting an intronic sequence of the *SMN2* pre-mRNA with an ASO-induced exon 7 inclusion and the consequent production of a functional SMN2 protein. This change compensated for the loss of the SMN1 protein in neurons and improved clinical symptoms in mouse models of SMA [37–39]. Based on this groundbreaking work, both researchers won the 2019 Breakthrough Prize in Life Sciences "for the development of an effective antisense oligonucleotide therapy for children with the neurodegenerative disease spinal muscular atrophy."

The stability of PS- and 2'-MOE-modified ASOs in the central nervous system is remarkable. Following 7 days of daily intracerebroventricular administration of the

ASO in mice, it still promoted ~90% SMN2 exon 7 inclusion in the spinal cord half a year after completing the 7-day treatment – compared to ~10% in saline or mismatch ASO-treated mice [39].

Nusinersen was a first-of-its-kind drug when it was approved by the FDA on December 23, 2016 (EMA: 30 May, 2017). It was the first approved treatment for SMA patients, at least in the U.S. and Europe, and it was almost the first nucleic acid-based drug that altered mRNA splicing to increase the expression of a gene. A few months earlier, the FDA approved ***eteplirsen***, another splice-switching ASO (see below). However, ***nusinersen*** was the first drug to induce exon inclusion to rescue the function of a different, mutated gene. It is intrathecally injected by lumbar puncture. Following four initial loading doses administered 14–30 days apart, the drug needs to be injected every four months. The treatment is probably lifelong.

FDA approval was based on results of ENDEAR, a randomized, double-blind, sham-controlled, phase 3 efficacy and safety trial of ***nusinersen*** in infants with SMA with two copies of the *SMN2* gene [40]. As mentioned earlier, infant-onset SMA (i.e., type 1 SMA) is a fatal disease, with patients usually dying before two years of age. Primary endpoints were motor milestone response and event-free survival (time to death or the use of permanent assisted ventilation). During a pre-specified interim analysis, 41% of the trial participants on ***nusinersen*** had a motor milestone response, versus 0% in the sham control group. Based on this observation, the trial was terminated, and the drug was approved. During a final analysis that occurred later, the likelihood of event-free survival and overall survival were both found to be significantly higher in the ***nusinersen*** group.

The NURTURE study is an open-label study of ***nusinersen*** given pre-symptomatically in 15 children with two *SMN2* copies (type 1 SMA) and 10 children with three *SMN2* copies. Results after five years of trial initiation were published in 2023 [41]. All 15 children with two *SMN2* copies achieved sitting without support. Of them, 14 could walk with assistance, and 13 could walk alone. Children with three *SMN2* copies achieved all World Health Organization (WHO) motor milestones, mainly within normal developmental timeframes.

While the drug is overwhelmingly effective, it is also highly expensive. As of 2019, treatment with Spinraza® could cost USD 750,000 the first year and then USD 375,000 each year (based on a list price of USD 125,000 per dose). A five-year treatment can therefore surpass two million U.S. dollars.

4.6 Phosphorodiamidate Morpholino Oligomers (PMOs) and their Use for Exon Inclusion in Duchene Muscular Dystrophy

PMO-based ASOs were developed in the early 90's at AntiVirals Inc. (currently Sarepta Therapeutics), a company founded in 1980 to move antisense therapy into the clinic [1]. By that time, several DNA analogues were being tested, looking for better pharmacological properties and nuclease resistance. PMOs are oligomers containing conventional nitrogenous bases (i.e., A, C, G, T, and their derivatives), but, unlike DNA or RNA, they do not contain a ribose ring. In contrast, nitrogenous bases are attached to a backbone of methylenemorpholine rings (Fig. 4) linked through phosphorodiamidate groups.

An important chemical property of PMOs is that the anionic phosphodiester group of DNA or RNA is substituted for a neutral phosphorodiamidate group. Consequently, PMOs are uncharged at the physiological pH, and their binding to both extracellular and intracellular proteins is dramatically different and lower than the wild entities. A second important consequence is that PMOs are insensitive to the action of nucleases, as they are not nucleic acids (but can bind to them in a sequence-specific manner). This can be viewed both as an advantage and as a disadvantage. It is a disadvantage because PMO-based ASOs cannot be used to induce mRNA degradation based on RNase H1. However, when only steric hindrance is intended, like in the case of splice-switching ASOs, the stability conferred by the PMO modification is beneficial.

The fact that PMOs (or chemically similar molecules, porting morpholino-like moieties, collectively referred to as "morpholinos," Fig. 4) do not engage with RNase H1 does not mean that they cannot be used to downregulate gene expression in a sequence-dependent manner. As steric blockers, they can still inhibit translation of specific mRNAs, possibly by blocking the formation of a translation initiation complex. In fact, morpholinos have been used for decades in developmental biology to silence specific genes at the translation level. The company Gene Tools, LLC, was spun off from AntiVirals, Inc. to commercialize morpholinos for the research community [42].

The widespread acceptance of morpholinos as gene silencing tools by the developmental biology research community is understandable considering that, in those applications, morpholinos can simply be microinjected into a cell's cytoplasm. In contrast, medicinal chemistry applications require an efficient in vivo delivery route. As will be discussed in a later section, most ASOs on the market rely on their spontaneous uptake by cells, a process that tends to be highly inefficient. This has not prevented Sarepta Therapeutics from moving several PMO-based products into the clinic, based on intravenous injections and spontaneous uptake. The efficacy of these FDA-approved drugs is, however, still controversial.

Eteplirsen (Exondys 51®) is a 30-mer PMO-based ASO indicated for the treatment of some types of Duchene muscular dystrophy (DMD), a severe disease affecting predominantly boys and provoking progressive muscle loss and paralysis. As of 2024, there is still no cure for the disease, with available treatments only targeting symptoms. ***Eteplirsen*** targets the cause of the disease, but clinical benefits are still unclear.

DMD is caused by mutations in the *DMD* gene, coding for dystrophin, a protein that is important for the connection between the actin cytoskeleton of muscle cells and the extracellular matrix. Loss of dystrophin leads to problems with calcium signaling in muscular tissue, affects the muscle's contractile activity, and ultimately leads to the death of muscular cells and their replacement by fatty and fibrotic tissue

[43]. In turn, the disease is fatal, with a median life expectancy of 22 years. The *DMD* gene is codified in the X chromosome, and while mutations in this gene are recessive, boys inheriting the mutation do not have an extra copy of the gene and develop the disease. The prevalence of DMD is estimated at about 10 in 100,000 live male births. The *DMD* gene is extremely long (2.4 megabases). It has 79 exons flanked by long introns. Not surprisingly, several different mutations can lead to DMD. However, mutations in DMD can also lead to Becker muscular dystrophy (BMD), which is a milder disease with a slower progression and a later onset. When looking at *DMD* mutations in patients with DMD and BMD, those in DMD patients tend to change the reading frame, while those in BMD do not (i.e., they are typically deletions involving a number of nucleotides that is divisible by three). In other words, deletions and insertions in DMD patients can be shorter but change the way the ribosome reads the mRNA, ultimately leading to the emergence of a premature termination codon (PTC). PTCs are problematic in several ways: (i) they can induce mRNA degradation by a process termed nonsense-mediated decay, (ii) they can interfere with transcription by a still not well understood epigenetic mechanism [44], and (iii) if a protein with a PTC is produced, it will lack its C-terminal protein domains. In the case of dystrophin, the protein has several internal spectrin-like repeats, but its C-terminal domains are important for function. Dystrophin proteins produced by BMD patients tend to have a lower number of spectrin-like repeats but are still functional.

Eteplirsen's mechanism of action is to sterically block exon 51 of the *DMD* gene, leading to the exclusion of this exon from the mature *DMD* mRNA. For this reason, it only makes sense to treat DMD patients with ***eteplirsen*** when they have frameshifting deletions or insertions in exon 51. These mutations are found in approximately 14% of DMD patients [43]. The beauty of this mechanism of action is that DMD is converted, by the action of the ASO, into an induced form of BMD, which is known to be a milder disease. The resultant protein is shorter (it lacks all the amino acids encoded by exon 51), but it is still produced at higher levels. Because the reading frame is rescued, the PTC disappears, and the protein recovers partial functionality. At least in theory.

Eteplirsen received accelerated approval by the FDA in 2016 based on results from a phase 2, placebo-controlled clinical trial and its open-label extension [45]. The trial involved 12 patients, but two patients randomized to the ***eteplirsen*** arm had to be excluded due to a rapid progression of the disease. After several back-and-forth between the company and the FDA on how best to analyze efficacy data derived from these studies, an analysis of the data by the FDA showed dystrophin levels in treated patients reaching between 0.25% and 2.5% of the levels present in normal subjects. The increase in the percentage of dystrophin-positive muscle fibers was more pronounced. The company also provided data based on the natural history of the disease, justifying clinical benefits in treated patients. Initially, an FDA advisory committee voted against approval of the drug, considering efficacy was not demonstrated. However, the FDA still granted accelerated approval in September 2016. Two members of the FDA review team resigned. The journals Nature Biotechnology and Nature Medicine [46] published a critical editorial stating that ***eteplirsen***'s approval "sends a signal that evidence from poorly designed clinical trials – if presented in the context of vocal patient advocacy – can be sufficient to obtain accelerated approval of a rare-disease drug." The EMA rejected approval of the drug in 2018.

Results from an open-label, phase 3 trial were published in 2021 [47]. While the levels of dystrophin protein were elevated seven-fold versus baseline after 96 weeks of weekly administration of the ASO (30 mg/kg), mean dystrophin levels were only increased from 0.24% to 0.63% of the protein levels found in normal subjects.

Since the controversial approval of ***eteplirsen*** in 2016, the FDA has approved several other PMO-based ASOs targeting other exons of the *DMD* gene, also mutated in a relevant fraction of DMD patients. These drugs include: ***golodirsen*** (Vyondys 53®) and ***viltolarsen*** (Viltepso®), both inducing exon 53 skipping, and ***casimersen*** (Amondys 45®), which targets exon 45. Like ***eteplirsen***, ***golodirsen*** and ***casimersen*** were developed and are marketed by Sarepta Therapeutics, which also has an FDA-approved gene therapy for DMD. Vitolarsen, in contrast, was developed by the Japanese pharmaceutical company Nippon Shinyaku.

5 Pharmacokinetics and Toxicology of Therapeutic ASOs

ASOs containing PS linkages bind tightly to albumin and other plasma proteins. This interaction is mostly dependent on the PS bonds, with the nucleobases or other chemical modifications playing a minor role [48]. The capacity of binding plasma proteins avoids the rapid renal clearance that is observed with ASOs containing natural phosphodiester bonds (i.e., PO groups). The neutral PMO ASOs do not bind to plasma proteins and are also rapidly eliminated by glomerular filtration [49].

After systemic delivery by intravenous or subcutaneous injection, PS ASOs are absorbed and distributed broadly to peripheral tissues [49]. Tissue elimination is due to nuclease digestion (typically by endonucleases followed by exonucleases). The shorter fragments no longer bind tightly to plasma proteins and are eliminated in urine. Because tissue elimination half lives depend on nuclease stability, substitutions that increase nuclease resistance, such as 2'-MOE modifications in the ribose or the use of LNAs, can have a strong impact. For example, while PS ASOs show tissue elimination half-lives of 48 h, 2'-MOE or LNA-containing gapmers extend the tissue elimination half-life to two to four weeks [49].

PS ASOs do not cross the blood-brain barrier. When they are injected intrathecally, such as in the case of ***nusinersen***, the half-life in cerebrospinal fluid (CSF) is as high as four to six months after initial clearance [50]. This is probably due to the fact that ***nusinersen*** is fully modified with 2'-MOE groups and due to the lower extracellular RNase concentration of CSF compared to plasma or serum. This extraordinary long half-life permits that maintenance doses of ***nusinersen*** are applied once every four months after the initial dosing regimen.

PMO ASOs are totally resistant to nucleases and have tissue elimination half-lives of 7–14 days as intact molecules in urine [49].

5.1 Uptake by Cells

Eplontersen (Wainua®) is a gapmer ASO developed by Ionis Pharmaceuticals and AstraZeneca, indicated for the treatment of hATTR in adults. It was approved by the FDA in December 2023. Like ***inotersen***, it directs RNase H1 to degrade the *TTR* mRNA in the liver, helping to avoid toxic amyloid deposits of the mutant TTR protein in the peripheral nervous system (and in other tissues such as the heart).

Eplontersen shares the same nucleotide sequence as ***inotersen***, with a similar gapmer design containing wings with 2'-MOE groups in the ribose. It has a lower number of PS bonds in these wings than *inotersen*, possibly as a way to attenuate sequence-dependent proinflammatory effects [51]. However, the main difference between both ASOs is the conjugation of ***eplontersen*** to a triantennary *N*-acetylgalactosamine (GalNAc$_3$) moiety in the 5′-position.[5] This modification facilitates delivery of the ASO to hepatocytes, through asialoglycoprotein receptor (ASGPR)-mediated uptake. This is important because hepatocytes are the cells that produce the TTR protein and release it into the circulation. While PS ASOs tend to accumulate in the liver, they are mostly internalized by endothelial cells and, to a lesser extent, Kupffer cells [52].

ASGPR is a lectin that acts as the receptor of glycoproteins that have lost sialic acid and therefore expose galactose residues. This receptor is expressed at very high levels (around 500,000 copies per cell) and almost exclusively in hepatocytes. Conjugation of nucleic acids to GalNAc$_3$ improves the potency of the ASOs up to ten-fold when hepatocytes are the targeted cell type [53]. Indeed, ***eplontersen*** demonstrated a 50-fold increase in potency compared to the unconjugated ***inotersen*** in human hepatocyte cell culture, and a 30-fold increase in potency in mouse models of the disease [54]. This increased potency enables lower dosing regimens and, consequently, fewer side effects. For example, the recommended dose of ***inotersen*** is 284 mg by subcutaneous injection every week. ***Eplontersen***, in contrast, is administered by weekly subcutaneous injections of only 45 mg.

During a phase 3 clinical trial of ***inotersen***, several adverse effects were registered, such as a fatal case of intracranial hemorrhage with severe thrombocytopenia [26]. In contrast, all thrombocytopenia events registered in a phase 3 clinical trial of ***eplontersen*** were rated as mild [55]. Subcutaneous administration of PS ASOs often leads to mild injection site reactions [49]. However, injection site reactions were similarly frequent in the ***eplontersen*** group compared to placebo during phase 3 clinical trials [55].

Coupling oligonucleotides to GalNAc$_3$ for efficient hepatocyte delivery is a strategy that is being followed in several other clinical trials involving ASOs, as well as small interfering RNAs (siRNAs). For example, the FDA-approved ***givosiran*** (Givlaari®) also uses this delivery route.[6] Attempts to find other ligand/

---

[5] The 3x GalNAc conjugation strategy is described with more detail in Chapter "Small Interferring RNAs (siRNAs)", as most siRNAs currently on the market use this strategy.

[6] See Chapter "Small Interferring RNAs (siRNAs)".

receptor interactions that could be exploited for targeted delivery of oligonucleotides to other cell types or tissues have proven extremely difficult [56].

5.1.1 Spontaneous Uptake of Naked Oligonucleotides

Most ASOs discussed in this chapter are unconjugated. Their pharmacokinetics depends largely on their capacity to bind plasma proteins and their stability against nucleases. Another critical aspect on oligonucleotide-based therapeutics is their capacity to enter cells spontaneously, even if unconjugated. Except for PMOs that are neutral, most ASOs are negatively charged at the physiological pH. ASOs are therefore incapable of passively diffusing across the plasma membrane. How, then, are they internalized?

Early work suggested that PS ASOs are predominantly taken up, in vivo, by endothelial cells in the liver through the interaction with type A-I/A-II scavenger receptors [52]. These receptors show low substrate specificity and tend to bind polyanionic molecules, inducing their internalization using different endocytic pathways. Type A-I scavenger receptors were also shown to be involved in the uptake of PMO ASOs by muscle cells [57]. However, later studies in knockout mice lacking the scavenger receptor SR-A did not show changes in the biodistribution or pharmacokinetic properties of therapeutic PS ASOs [58]. Stabilin-1 and Stabilin-2, belonging to the class H of scavenger receptors, have also been suggested as specific receptors for PS ASOs in the liver [59]. It is possible that several receptors on the cell surface contribute to ASO uptake with varying efficiencies, depending on the ASO's chemical structure and the differential concentration of the receptors on the surface of different cell types.

Cell internalization of the ASOs is dependent on endocytosis, usually following ASO/receptor interactions (i.e., receptor-mediated endocytosis). However, cell uptake is not the end of the story. Most ASOs (> 99%) internalized by endocytosis will end up being degraded in lysosomes and never reach the cytosol or the nucleus [56]. This is also the case for ligand-conjugated ASOs such as ***eplontersen***. In other words, most events of ASO uptake are not productive from a functional perspective. Some ASOs manage to escape from endosomal compartments into the cytosol, but the mechanism of endosomal escape is still not well understood. As it will be discussed in later chapters describing siRNAs and mRNA-based therapeutics,[7] endosomal escape is currently considered the rate limiting step of RNA biotherapeutics [60].

Some authors have coined the term "gymnosis" (for the Greek word gymnos, meaning naked) to explain the functional uptake of naked RNA molecules, including ASOs [61]. An important aspect of this term is that it involves functionality and not simply uptake (i.e., it includes endosomal escape for naked oligonucleotides being internalized by endocytosis). Importantly, gymnotic uptake seems to be dependent

[7] See Chapter "Small Interferring RNAs (siRNAs)".

on the ASO chemical structure, with LNA-containing gapmers being active and unmodified PS ASOs being in general not active [61]. It has been hypothesized that the increased hydrophobicity of the phosphorothioate bond (compared to the standard phosphodiester linkage) might explain why PS ASOs are better at escaping from endosomal compartments than natural nucleic acids [56]. However, unpublished results from our research group show that extracellular nuclease stability is what mainly dictates gymnotic uptake [62]. When co-incubating different RNA molecules with RNase inhibitors, the efficiency of gymnotic uptake increased dramatically irrespectively of the RNA type. These results suggest that the high stability of LNA- or 2'-MOE-containing PS ASOs is the key determinant for gymnotic RNA uptake, at least in cell culture conditions. If this proves to be the case, new chemistries also conferring nuclease resistance will likely function by gymnotic uptake.

Compliance with Ethical Standards This chapter is a review and no humans or animals were involved in the preparation of the chapter.

References

1. Beaucage SL, Caruthers MH (1981) Deoxynucleoside phosphoramidites-a new class of key intermediates for deoxypolynucleotide synthesis. Tetrahedron Lett 22:1859–1862
2. Caruthers MH (1985) Gene synthesis machines: DNA chemistry and its uses. Science 80(230): 281–285
3. Filges S, Mouhanna P, Ståhlberg A (2021) Digital quantification of chemical oligonucleotide synthesis errors. Clin Chem 67:1384–1394
4. Obexer R, Nassir M, Moody ER et al (2024) Modern approaches to therapeutic oligonucleotide manufacturing. Science 384:eadl4015
5. Rader RA (2008) (Re)defining biopharmaceutical. Nat Biotechnol 26:743–751
6. FDA (2018) What are "biologics" questions and answers. https://www.fda.gov/about-fda/center-biologics-evaluation-and-research-cber/what-are-biologics-questions-and-answers
7. Barra GB, Santa Rita TH, de Vasques JA et al (2015) EDTA-mediated inhibition of DNases protects circulating cell-free DNA from ex vivo degradation in blood samples. Clin Biochem 48:976–981
8. Blank A, Dekker CA (1981) Ribonucleases of human serum, urine, cerebrospinal fluid, and leukocytes. Activity staining following electrophoresis in sodium dodecyl sulfate-polyacrylamide gels. Biochemistry 20:2261–2267
9. Costa B, Li Calzi M, Castellano M et al (2023) Nicked tRNAs are stable reservoirs of tRNA halves in cells and biofluids. Proc Natl Acad Sci USA 120:e2216330120
10. Perry CM, Balfour JAB (1999) Fomivirsen. Drugs 57:375–380
11. Eckstein F (2014) Phosphorothioates, essential components of therapeutic oligonucleotides. Nucleic Acid Ther 24:374–387
12. Eckstein F (1985) Nucleoside phosphorothioates. Annu Rev Biochem 54:367–402
13. Jahns H, Taneja N, Willoughby JLS et al (2022) Chirality matters: stereo-defined phosphorothioate linkages at the termini of small interfering RNAs improve pharmacology in vivo. Nucleic Acids Res 50:1221–1240
14. Koziolkiewicz M, Krakowiak A, Kwinkowski M et al (1995) Stereodifferentiation-the effect of P chirality of oligo (nucleoside phosphorothioates) on the activity of bacterial RNase H. Nucleic Acids Res 23:5000–5005

15. KoziołKiewicz M, Wójcik M, Kobylańska A et al (1997) Stability of stereoregular oligo (nucleoside phosphorothioate)s in human plasma: Diastereoselectivity of plasma 3′--exonuclease. Antisense Nucleic Acid Drug Dev 7:43–48
16. Wan WB, Migawa MT, Vasquez G et al (2014) Synthesis, biophysical properties and biological activity of second generation antisense oligonucleotides containing chiral phosphorothioate linkages. Nucleic Acids Res 42:13456–13468
17. Raal FJ, Santos RD, Blom DJ et al (2010) Mipomersen, an apolipoprotein B synthesis inhibitor, for lowering of LDL cholesterol concentrations in patients with homozygous familial hypercholesterolaemia: a randomised, double-blind, placebo-controlled trial. Lancet 375:998–1006
18. Stein EA, Dufour R, Gagne C et al (2012) Apolipoprotein B synthesis inhibition with Mipomersen in heterozygous familial hypercholesterolemia: results of a randomized, double-blind, placebo-controlled trial to assess efficacy and safety as add-on therapy in patients with coronary artery disease. Circulation 126:2283–2292
19. Cerritelli SM, Crouch RJ (2009) Ribonuclease H: the enzymes in eukaryotes. FEBS J 276: 1494–1505
20. Lockhart A, Pires VB, Bento F et al (2019) RNase H1 and H2 are differentially regulated to process RNA-DNA hybrids. Cell Rep 29:2890–2900.e5
21. Wu H, Lima WF, Zhang H et al (2004) Determination of the role of the human RNase H1 in the pharmacology of DNA-like antisense drugs. J Biol Chem 279:17181–17189
22. Liang XH, Sun H, Nichols JG et al (2017) RNase H1-dependent antisense oligonucleotides are robustly active in directing RNA cleavage in both the cytoplasm and the nucleus. Mol Ther 25: 2075–2092
23. Teplova M, Minasov G, Tereshko V et al (1999) Crystal structure and improved antisense properties of 2'-O-(2- methoxyethyl)-RNA. Nat Struct Biol 6:535–539
24. Geary RS, Baker BF, Crooke ST (2015) Clinical and preclinical pharmacokinetics and pharmacodynamics of Mipomersen (Kynamro®): a second-generation antisense oligonucleotide inhibitor of apolipoprotein B. Clin Pharmacokinet 54:133–146
25. Pollak AJ, Zhao L, Vickers TA et al (2022) Insights into innate immune activation via PS-ASO-protein-TLR9 interactions. Nucleic Acids Res 50:8107–8126
26. Benson MD, Waddington-Cruz M, Berk JL et al (2018) Inotersen treatment for patients with hereditary transthyretin amyloidosis. N Engl J Med 379:22–31
27. Witztum JL, Gaudet D, Freedman SD et al (2019) Volanesorsen and triglyceride levels in familial Chylomicronemia syndrome. N Engl J Med 381:531–542
28. Li HF, Wu ZY (2016) Genotype-phenotype correlations of amyotrophic lateral sclerosis. Transl Neurodegener 5:3
29. Miller TM, Cudkowicz ME, Genge A et al (2022) Trial of antisense oligonucleotide Tofersen for SOD1 ALS. N Engl J Med 387:1099–1110
30. Crooke ST (2022) Establishing an environment in which rigorous scientific inquiry is practiced: a personal journey. Nucleic Acids Res 50:7216–7223
31. Khvorova A, Watts JK (2017) The chemical evolution of oligonucleotide therapies of clinical utility. Nat Biotechnol 35:238–248
32. Shen W, De Hoyos CL, Migawa MT et al (2019) Chemical modification of PS-ASO therapeutics reduces cellular protein-binding and improves the therapeutic index. Nat Biotechnol 37: 640–650
33. Anastasiadou E, Seto AG, Beatty X et al (2021) Cobomarsen, an oligonucleotide inhibitor of miR-155, slows DLBCL tumor cell growth in vitro and in Vivo. Clin Cancer Res 27:1139–1149
34. Chen X, Harting J, Farrow E et al (2023) Comprehensive SMN1 and SMN2 profiling for spinal muscular atrophy analysis using long-read PacBio HiFi sequencing. Am J Hum Genet 110:240–250
35. Lorson CL, Hahnen E, Androphy EJ et al (1999) A single nucleotide in the SMN gene regulates splicing and is responsible for spinal muscular atrophy. Proc Natl Acad Sci USA 96:6307–6311

36. Butchbach MER (2016) Copy number variations in the survival motor neuron genes: implications for spinal muscular atrophy and other neurodegenerative diseases. Front Mol Biosci 3:7
37. Hua Y, Vickers TA, Okunola HL et al (2008) Antisense masking of an hnRNP A1/A2 Intronic splicing silencer corrects SMN2 splicing in transgenic mice. Am J Hum Genet 82:834–848
38. Hua Y, Vickers TA, Baker BF et al (2007) Enhancement of SMN2 exon 7 inclusion by antisense oligonucleotides targeting the exon. PLoS Biol 5:729–744
39. Hua Y, Sahashi K, Hung G et al (2010) Antisense correction of SMN2 splicing in the CNS rescues necrosis in a type III SMA mouse model. Genes Dev 24:1634–1644
40. Finkel RS, Mercuri E, Darras BT et al (2017) Nusinersen versus sham control in infantile-onset spinal muscular atrophy. N Engl J Med 377:1723–1732
41. Crawford TO, Swoboda KJ, De Vivo DC et al (2023) Continued benefit of nusinersen initiated in the presymptomatic stage of spinal muscular atrophy: 5-year update of the NURTURE study. Muscle Nerve 68:157–170
42. Summerton JE (2017) Invention and early history of morpholinos: from pipe dream to practical products. Methods Mol Biol 1565:1–15
43. Duan D, Goemans N, Takeda S et al (2021) Duchenne muscular dystrophy. Nat Rev Dis Prim:7
44. García-Rodríguez R, Hiller M, Jiménez-Gracia L et al (2020) Premature termination codons in the DMD gene cause reduced local mRNA synthesis. Proc Natl Acad Sci U S A 117:15664–16464
45. Mendell JR, Rodino-Klapac LR, Sahenk Z et al (2013) Eteplirsen for the treatment of Duchenne muscular dystrophy. Ann Neurol 74:637–647
46. (2016) Railroading at the FDA. Nat Med 22: 1193
47. McDonald CM, Shieh PB, Abdel-Hamid HZ et al (2021) Open-label evaluation of Eteplirsen in patients with Duchenne muscular dystrophy amenable to exon 51 skipping: PROMOVI trial. J Neuromuscul Dis 8:989–1001
48. Gaus HJ, Gupta R, Chappell AE et al (2019) Characterization of the interactions of chemically-modified therapeutic nucleic acids with plasma proteins using a fluorescence polarization assay. Nucleic Acids Res 47:1110–1122
49. Crooke ST, Baker BF, Crooke RM et al (2021) Antisense technology: an overview and prospectus. Nat Rev Drug Discov 20:427–453
50. Chiriboga CA, Swoboda KJ, Darras BT et al (2016) Results from a phase 1 study of nusinersen (ISIS-SMN Rx) in children with spinal muscular atrophy. Neurology 86:890–897
51. Diep JK, Yu RZ, Viney NJ et al (2022) Population pharmacokinetic/pharmacodynamic modelling of eplontersen, an antisense oligonucleotide in development for transthyretin amyloidosis. Br J Clin Pharmacol 88:5389–5398
52. Bijsterbosch MK, Manoharan M, Rump ET et al (1997) In vivo fate of phosphorothioate antisense oligodeoxynucleotides: predominant uptake by scavenger receptors on endothelial liver cells. Nucleic Acids Res 25:3290–3296
53. Prakash TP, Graham MJ, Yu J et al (2014) Targeted delivery of antisense oligonucleotides to hepatocytes using triantennary N-acetyl galactosamine improves potency 10-fold in mice. Nucleic Acids Res 42:8796–8807
54. Viney NJ, Guo S, Tai LJ et al (2021) Ligand conjugated antisense oligonucleotide for the treatment of transthyretin amyloidosis: preclinical and phase 1 data. ESC Hear Fail 8:652–661
55. Coelho T, Marques W, Dasgupta NR et al (2023) Eplontersen for hereditary transthyretin amyloidosis with polyneuropathy. JAMA 330:1448–1458
56. Dowdy SF (2017) Overcoming cellular barriers for RNA therapeutics. Nat Biotechnol 35:222–229
57. Miyatake S, Mizobe Y, Tsoumpra MK (2019) Scavenger receptor class A1 mediates uptake of Morpholino antisense oligonucleotide into dystrophic skeletal muscle. Mol Ther Nucleic Acids 14:520–535
58. Butler M, Crooke RM, Graham MJ et al (2000) Phosphorothioate oligodeoxynucleotides distribute similarly in class a scavenger receptor knockout and wild-type mice. J Pharmacol Exp Ther 292:489–496

59. Miller CM, Donner AJ, Blank EE et al (2016) Stabilin-1 and Stabilin-2 are specific receptors for the cellular internalization of phosphorothioate-modified antisense oligonucleotides (ASOs) in the liver. Nucleic Acids Res 44:2782–2794
60. Dowdy SF (2023) Endosomal escape of RNA therapeutics: how do we solve this rate-limiting problem? RNA 29:396–401
61. Stein CA, Hansen JB, Lai J et al (2009) Efficient gene silencing by delivery of locked nucleic acid antisense oligonucleotides, unassisted by transfection reagents. Nucleic Acids Res 38:e3
62. Castellano M, Blanco V, Li Calzi M et al (2024) Ribonuclease activity undermines immune sensing of naked extracellular RNA. bioRxiv. 2024.04.23.590771

Top Med Chem
https://doi.org/10.1007/7355_2024_171

Published online: 28 December 2024

Small Interfering RNAs (siRNAs)

Juan Pablo Tosar

Contents

Abstract Small interfering RNAs (siRNAs) are short (~21 nucleotides), double-stranded RNA molecules that can be administered to patients to degrade specific messenger RNAs (mRNAs). They do so by a mechanism called RNA interference (RNAi), which cells use in the context of their antiviral defenses. Briefly, one of the two strands of a siRNA is selected by a host protein, and this strand is used to recognize (and cleave) mRNAs with perfect sequence complementarity. Thus, RNAi provides a potent and predictable way to silence the expression of theoretically any desired gene, acting at the posttranscriptional level. However, it took 20 years for the first siRNA biotherapeutic to reach the market. This chapter will explain the many

J. P. Tosar (✉)
Unidad de Bioquímica Analítica, Centro de Investigaciones Nucleares, Facultad de Ciencias, Universidad de la República, Montevideo, Uruguay

Laboratorio de Genómica Funcional, Institut Pasteur de Montevideo, Montevideo, Uruguay

challenges that basic and applied researchers had to face to move RNAi into the clinic, with delivery into cells being one of the major struggles. This problem was solved by the use of either lipid nanoparticles (***patisiran***) or chemically modified siRNAs conjugated to the ligand of an abundant cell surface receptor (***givosiran***, ***lumasiran***, ***inclisiran***, ***vutrisiran***, ***nedosiran***). Despite the latter strategy having proven useful, with several products now in the market, its use is currently restricted to the liver. Nevertheless, several clinical trials using siRNAs directed to other cell types are ongoing. The RNAi revolution might be just starting.

Keywords Amvuttra, Argonaute2, Dicer, Gene silencing, Givlaari, Leqvio, Lipid nanoparticle, MicroRNA, Onpattro, Oxlumo, Rivfloza, RNA interference, RNAi, RNA-induced silencing complex

1 Introduction

In 1998, the molecular basis for RNA interference (RNAi) was untangled by Nobel Prize winners Andrew Fire and Craig Mello [1]. Briefly, RNAi can be triggered by the introduction of double-stranded RNA (dsRNA) into a cell. Long, dsRNA molecules are processed into short RNA species termed small interfering RNAs (siRNAs). If one of the strands of a siRNA is complementary to a host cell's messenger RNA (mRNA), the siRNA directs a protein complex into this target mRNA and induces its degradation. Thus, RNAi enables the sequence-specific degradation of any target mRNA by simply adding a dsRNA molecule into a cell. Silencing genes at mRNA levels could allow targeting even targets considered "undruggable."

However, despite the promise and the initial enthusiasm surrounding RNAi, it took 20 years until the US Food and Drug Administration (FDA) approved the first siRNA-based drugs in 2018. This chapter will present the quest to move siRNAs into the clinic and the unexpected challenges that had to be overcome during this process. Many of these challenges were related to the initially incomplete understanding of cells' innate immune response to exogenous dsRNA molecules and the delivery problem common to most RNA-based therapeutics. Finally, it will present recent approvals of new siRNAs and future perspectives for this biotherapeutic family.

1.1 The Discovery of RNAi

RNAi started based on an unexpected observation by a group of researchers working for an American plant biotechnology company. The authors wanted to test whether the enzyme chalcone synthase (CHS) was the rate-limiting step in the biosynthesis of anthocyanins by petunias [2]. Considering that anthocyanins are responsible for the colors of plant flowers, the generation of transgenic plants overexpressing chalcone synthase could potentially give place to flowers with more vivid colorings. Surprisingly, when the authors introduced a transgenic copy of the gene into violet petunias, the flowers either were fully white or presented patterns alternating between white and violet. No flowers were visibly darker than the parental genotype, which was the expected outcome of the experiment. In contrast, the introduction of a transgenic copy of CHS inhibited the synthesis of the pigments by strongly downregulating the expression of CHS itself (a 50-fold reduction in its mRNA levels). The authors termed this phenomenon "cosuppression." Simultaneously, an independent group from the Netherlands arrived at almost identical results and suggested that this mechanism of gene silencing could operate in naturally occurring regulatory circuits [3]. However, the mechanism responsible for this transgene-induced suppression of the endogenous target gene remained elusive.

A few years later, a group of Italian researchers observed a similar phenomenon when working with *Neurospora crassa*, a mold that is widely used as a model for genetic studies. Cultures of this mold are orange due to the presence of carotenoids. The authors ectopically expressed some of the enzymes involved in carotenoid biosynthesis, and to their surprise, the transformants were white [4]. They called this phenomenon "quelling" while acknowledging evident similarities with the "cosuppression" mechanism earlier described in petunias. For example, both phenomena were reversible. In *N. crassa*, white transformants reverted to orange (wild type) while still having copies of the transgenes in their genomes. The authors also noticed that the levels of both the endogenous and the exogenous genes were substantially affected at the mRNA level in transformed molds. Interestingly, mRNA levels rose to levels similar to those found in wild-type cultures in revertants.

Several hypotheses were formulated to explain the mechanism of cosuppression in plants or quelling in fungi. For example, the Dutch team reporting one of the petunia papers reasoned that the suppression of CHS mRNA induced by the ectopic expression of extra copies of this gene [3] resembled their earlier work expressing a transcript that was antisense to the CHS mRNA [5]. While antisense-induced gene silencing was easy to explain, how could the sense transcripts give rise to a similar phenotype? The authors speculated that sense genes may involve "interference of RNA" strands with the transcription process itself, potentially forming a triple helix between sense mRNAs and the duplex DNA [3]. Although this explanation was still far from the actual RNAi mechanism, it did pose a direct action of mRNA as the trans factor capable of inducing the silencing of a homologous gene. In contrast, the authors of the *N. crassa* paper speculated on a DNA-centric mechanism involving DNA methylation [4], citing earlier evidence in plants [6].

The breakthrough came in 1998 with the publication of the article "Potent and Specific Genetic Interference by Double-Stranded RNA in *Caenorhabditis elegans*" [1]. Andrew Fire, the first author, had earlier published a study showing that antisense RNA could be used to silence specific genes in the nematode *C. elegans* and described this phenomenon as "antisense interference" [7]. In their 1998 paper, Fire and colleagues compared the silencing potential of sense RNA, antisense RNA, and a dsRNA comprising both sense and antisense sequences [1]. Surprisingly, dsRNA was at least two orders of magnitude more effective at silencing a specific mRNA than either strand alone. The effect was both specific and potent: only a few molecules of injected dsRNA were required per affected cell. This suggested a catalytic effect and argued against the stoichiometric interference earlier suggested for antisense RNAs. In addition, despite the fact that the exact mechanism was still not understood, a direct observation of dsRNA-induced mRNA degradation (i.e., "RNA interference" (RNAi)) suggested a posttranscriptional mechanism of gene silencing. Finally, RNAi was transmittable in *C. elegans*. DsRNAs could be injected into one organ, and gene silencing could be observed not only in distant organs but even in the worm's progeny.

The observation that RNAi could be transported between cells strongly suggested the existence of a dsRNA-derived molecule that could mediate these systemic effects. In 1999, Hamilton and Baulcombe found that a small antisense RNA of approximately 25 nucleotides (nt) accumulated every time posttranscriptional gene silencing was experimentally induced in plants [8]. One year later, Gregory Hannon and colleagues identified a nuclease activity in extracts of the fruit fly *Drosophila melanogaster*, which was responsible for RNAi, and termed this nuclease RISC (RNA-induced silencing complex) [9]. They chromatographically purified RISC and found that it coeluted with small RNAs of approximately 25 nt comprising both sense and antisense sequences, suggesting that RISC could bind double-stranded small RNAs produced from longer dsRNA precursors. Almost simultaneously, Zamore, Tuschl, Sharp, and Bartel used a cell-free *Drosophila* system and found that long dsRNAs are indeed processed into a family of small RNAs of 21–23 nt, in a reaction that is independent of the target mRNA [10]. They also showed that these small dsRNAs are bound by a protein complex (i.e., RISC) that hydrolyzes adenosine triphosphate (ATP) to unwind the double-stranded substrate and retain only one of its two strands. They proposed that this protein/small RNA complex was then directed to the complementary mRNAs, which would be cleaved by RISC in a sequence-specific manner.

The mysterious RNAi phenomenon was being rapidly solved, and a coherent and precise molecular pathway was emerging. Further work by the group of Thomas Tuschl using the *Drosophila* in vitro system added the missing pieces to the puzzle [11]. They showed that the nuclease processing long dsRNA precursors into small double-stranded species belonged to the RNase III family. This family of enzymes makes two staggered cuts into a dsRNA, leaving a 3′ overhang of 2 nt in each strand. In agreement with this finding, chemically synthesized 21- or 22-nt RNA duplexes with 3′ overhangs of 2 nt were more efficient in reducing target RNA expression than blunt-ended dsRNAs. They called these species small interfering RNAs (siRNAs). Finally, they showed that target mRNA cleavage occurred at a precise location, 11 or 12 nt downstream of the first base of the siRNA strand that remains bound by RISC. Based on this finding, the authors predicted that it would be possible "to design a pair of 21 or 22-nt RNAs to cleave a target RNA at almost any given position."

In 2001, the group of Gregory Hannon identified the RNAse III enzyme responsible for the initial processing of dsRNAs into siRNAs and called it Dicer

[12]. Shortly after, several groups showed that Dicer was actually at the crossroads between RNAi and the microRNA (miRNA) pathway [13, 14], a mechanistically related pathway involved in "self" gene expression regulation (see below). Thus, any narrative describing the discovery of RNAi starts merging with the discovery of microRNAs from 2001 onward. Today, we understand that both are evolutionarily related pathways that can be considered two sides of the same coin.

The discovery of Dicer only cracked the first half of the RNAi pathway. Dicer processed long dsRNA precursors into short 21–22 nt siRNAs, but how could these siRNAs induce target mRNA degradation? Presumably, this involved a different protein within RISC with the capacity to cleave RNA in the context of a siRNA: mRNA hybrid. However, the identity of such a protein remained elusive.

In 2002, Thomas Tuschl's group used biotinylated siRNAs to pull down RISC from human cell extracts and identified AGO1 and AGO2, two proteins belonging to the Argonaute protein family [15]. Of note, Argonaute2 was already identified as an integral component of RISC in *Drosophila* cells [16]. Furthermore, Argonaute protein family members were already known to be involved in RNAi in the nematode *C. elegans* [17], in the fungus *Neurospora crassa* [18], and in plants [19]. By that time, it was evident that RNAi was an ancestral eukaryotic pathway, and the molecular mechanism involving Dicer and Argonaute proteins was conserved in at least fungi, plants, and animals.

Figure 1 depicts the RNAi pathway as we conceive it today. Readers interested in how the story beyond the discovery of RNAi unfolded might appreciate some of the excellent review articles that have been written on the topic [20]. Figure 1, right, depicts the functionally and mechanistically related microRNA pathway, also involving Dicer and Argonaute proteins.

1.2 The MicroRNA Pathway

MicroRNAs (miRNAs) were independently and simultaneously codiscovered in 1993 in the nematode *C. elegans* by the groups of Victor Ambros [21] and Gary Ruvkun [22]. In that organism, the product of the *lin-4* gene was known to be capable of silencing LIN-14, a transcription factor important for the temporal control of larval development. Surprisingly, Ambros and colleagues [21] showed that the product of the *lin-4* gene was not a protein. In contrast, the gene gave rise to two RNA transcripts of 61 nt and 22 nt, both of which contained a sequence that was complementary to a repeated sequence element in the 3′ untranslated region (3' UTR) of the *lin-14* mRNA.

Among several models proposed by the authors, one included the possibility of the 61 nt *lin-4* transcript being posttranscriptionally processed into 22 nt small RNAs [21], capable of inhibiting the translation of the *lin-14* mRNA by binding to its 3' UTR [21, 22]. As can be concluded from an analysis of Fig. 1, this model follows the logic of what we now understand as the microRNA pathway. However, by that time, this seemed to be just a particular feature of *C. elegans*. In addition, there was still no

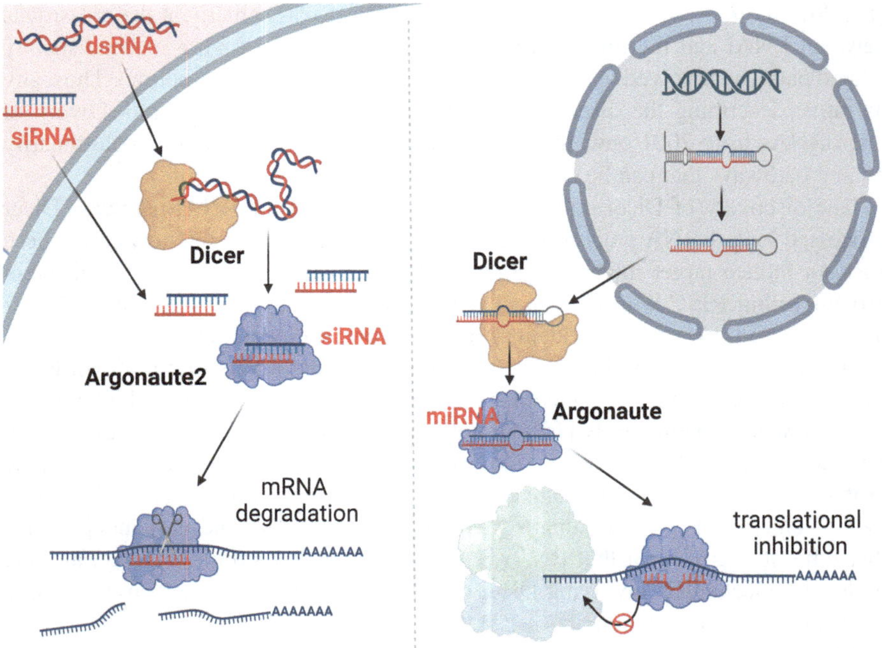

Fig. 1 Schematic representation of the RNA interference (RNAi) pathway (left) and the microRNA (miRNA) pathway (right). The RNAi pathway is fed by dsRNAs that can be internalized from the extracellular space or generated intracellularly (e.g., by viral replication). DsRNA molecules are recognized by the RNase III enzyme Dicer, which produces 21–22 nt small interfering RNAs (siRNAs) bearing a 3′ overhang of usually two single-stranded nt. In the biotherapeutic setup, siRNAs are delivered directly into cells. These siRNAs are recognized by the Argonaute2 protein, which selects one of the siRNA strands and degrades the other. The retained siRNA strand (termed the "antisense" strand, shown in red) directs Argonaute2 to a fully complementary mRNA. Argonaute2's slicer activity degrades the mRNA by direct endonucleolytic cleavage. In the miRNA pathway (right), miRNA genes are transcribed into primary miRNA transcripts that have a characteristic stem-loop structure. These primary miRNAs are cleaved by a nuclear RNase III enzyme (Drosha) to produce pre-miRNAs. Pre-miRNAs are then exported into the cytoplasm and are recognized and cleaved by Dicer to produce mature miRNAs. Mature miRNAs are structurally similar to siRNAs, but the two strands are not fully complementary. Like siRNAs, mature miRNAs are recognized by one of the four Argonaute proteins that are coded in the human genome (including Argonaute2 itself). These proteins select one of the two miRNA strands (the "antisense" strand, shown in red) and use it to recognize complementary mRNAs. Unlike siRNAs, miRNA/mRNA pairing is imperfect and restricted to a "seed" region comprising 2–8 nt of the miRNA. For this reason, even Argonaute proteins with slicer activity like Argonaute2 are unable to cleave the target mRNA. Nevertheless, they recruit other proteins that usually inhibit mRNA translation or that degrade the target mRNA by indirect mechanisms. Created with BioRender.com

conceptual connection between this observation and the RNAi pathway, which would be discovered in the same organism 5 years later [1].

An entertaining and insightful history of the early days of microRNA discovery can be read in the pen of one of its discoverers [23]. In addition, readers can also consult an updated and detailed review of metazoan microRNAs, in case they would like to learn more about this small RNA family, which is important for the regulation of human gene expression [24].

Unlike siRNAs, human miRNAs only pair with their target mRNAs using short-sequence motifs, usually restricted to positions 2–8 of their sequence (known as the "seed" motif). Although additional sequence matches in the 3′ end of the miRNA usually reinforce this "seed" match, miRNA/mRNA pairing is mostly imperfect. This might sound counterintuitive, but it allows individual miRNAs to recognize and target several mRNAs. In addition, the imperfect pairing between miRNAs and mRNAs usually avoids Argonaute-dependent mRNA degradation. In contrast, miRNAs usually act cooperatively by inhibiting mRNA translation or inducing mRNA destabilization, with the individual effect of a single miRNA sitting on an mRNA 3' UTR being rather small. In summary, miRNAs and mRNAs form complex interaction networks, thanks to the promiscuity of miRNA/mRNA recognition and the need for several miRNAs to act cooperatively in order to inhibit a specific mRNA.

1.3 Advantages of siRNA vs miRNAs as Drugs

Evolution-shaped animal miRNA/mRNA pairing rules to efficiently regulate gene expression, not to provide an attractive mechanism for the biopharmaceutical industry. As briefly mentioned in the chapter on antisense oligonucleotides,[1] there are investigational oligonucleotide-based drugs that target and inhibit specific miRNAs. However, as a tool for mRNA recognition and degradation, siRNAs are more convenient than miRNAs because a) the perfect pairing between siRNAs and mRNAs increases substrate specificity and therefore diminishes toxicity, b) siRNAs induce site-specific Argonaute2-mediated mRNA degradation, and c) siRNAs are potent (i.e., a single, well-designed siRNA can induce the degradation of several mRNA molecules).

2 Moving siRNAs into the Clinic: The Long and Winding Road

2.1 Earliest Clinical Trials

RNAi was (re)discovered in 1998 [1], and by 2001, it was clear that siRNAs could be used to silence specific genes *à la carte* in human cells [25]. By 2004, several clinical trials had already started. Most of these seminal clinical studies were based on

[1] See Chapter "Antisense Oligonucleotides (ASOs)".

unconjugated (i.e., naked) siRNA administration in the eye, in the context of age-related macular degeneration (AMD) and other related ophthalmological diseases. There was a clear rationale on why to choose the eye and AMD for initial efforts to move siRNAs into the clinic. First, AMD is a major cause of vision loss or blindness in adults, so there was a clear unmet medical need for better treatments at that time. In addition, earlier research has shown that wet forms of AMD are caused by excess angiogenesis in the eye, mediated by the vascular endothelial growth factor (VEGF). Thus, using siRNAs against VEGF could be an alternative to the use of anti-VEGF monoclonal antibodies[2] such as ***ranibizumab*** (Lucentis®) and ***bevacizumab*** (Avastin®) or to anti-VEGF aptamers[3] such as ***pegaptanib*** (Macugen®). Last, the aptamer ***pegaptanib*** and the antisense oligonucleotide ***fomivirsen*** had shown that naked nucleic acids could be safely administered in the eye by intravitreal injection and that this delivery route was effective in reaching retinal cells.

Clinical trial NCT00363714 was a phase 1 trial that started in 2004 and was cosponsored by Allergan and Sirna Therapeutics. The latter company was founded in 2003 after a $53 million cash injection to Ribozyime Pharmaceuticals, aiming to shift from one RNA-based approach (ribozymes) to the new RNAi field. The company was later acquired by Merck for $1.1 billion in 2006, shortly after the announcement of the study's completion [26]. Results of the trial were published in 2010 and showed that the treatment was well tolerated [27].

OPKO Health also started phase 1 clinical trials in 2004 (NCT00722384). Their experimental product, ***bevasiranib***, was also a siRNA targeting VEGF administered by intravitreal injection in the eyes of patients with wet AMD. The company started a phase 2 study in 2005 and a phase 3 study in 2007 (COBALT, NCT00499590). This study evaluated the safety and effectiveness of ***bevasiranib***, administered after an initial treatment with the anti-VEGF monoclonal antibody Lucentis®, compared to a control group treated with Lucentis® alone. Despite the initial enthusiasm, the company announced in 2009 the early termination of the study due to disappointing preliminary results. Based on a press release by the company (March 6, 2009), there was evidence for the activity of the siRNA. However, the trial, as structured, seemed unlikely to meet its primary endpoint.

2.2 Concerns on the Potential Immunogenicity of siRNAs

OPKO Health's disappointing phase 3 trial still showed, according to the company's press release, some evidence of effectiveness. However, basic science studies showed that surface-bound toll-like receptor 3 (TLR3), a double-stranded RNA sensor expressed by several human cells, was able to recognize any siRNA with a

[2] See Chapter "Monoclonal Antibodies as Biotherapeutics".

[3] See Chapter "Aptamers".

length of 21 nt or higher and trigger an interferon response. This interferon response also abrogated angiogenesis but through a sequence-independent mechanism that was unrelated to RNAi [28, 29]. The authors showed that naked siRNAs were unable to enter human or mouse endothelial or retinal epithelial cells, which is a prerequisite for efficient RNAi. Thus, the apparent proof of effectiveness was probably linked to an unspecific, potentially harmful, inflammatory response.

RNAi was discovered in *C. elegans*, an organism that uses the RNAi pathway in a way quite different from that of human beings and other mammals. In the nematode, RNAi serves key functions related to endogenous gene expression regulation, including protection against transposable elements, while also being a major pathway to protect cells against viral infection. The nematode's genome also codes for an RNA-dependent RNA polymerase that can produce secondary siRNAs after initial mRNA cleavage events. In addition, *C. elegans* exhibits systemic [30] and environmental [31] RNAi. As a consequence, the worms can be fed on bacteria expressing a dsRNA, and the ingested dsRNA molecules can trigger RNAi even in cells distant from the intestine.

Systemic and environmental RNAi do not work in human cells. We humans mostly rely on the miRNA and the Piwi-interacting RNA (piRNA) pathways for gene expression regulation and transposon silencing, respectively, and on the interferon pathway for antiviral responses. Consequently, human cells do not favor the internalization of exogenous dsRNA molecules. On the contrary, human cells actively sense exogenous RNA molecules based on a complex set of single-stranded and double-stranded RNA sensors (including TLR3), which, upon exogenous RNA recognition, trigger the production of interferons and other inflammatory mediators.

Once RNAi was demonstrated in human cells [25], the pharmaceutic industry rushed to push siRNAs into the clinic. However, pretending that RNAi would work in humans as well as it worked in nematodes and other organisms proved to be quite naïve. Initial failures in clinical trials showed that more basic science was still needed to understand how human cells sense and defend themselves against exogenous dsRNA molecules. Once the innate immune response against dsRNA was better understood, more rationale approaches were designed that circumvented these issues.

2.3 The Delivery Problem

In addition to the problem of synthetic siRNAs inducing an interferon-mediated inflammatory response, human and other mammalian cells do not effectively internalize long dsRNAs or siRNAs. As seen in Chapter "Antisense Oligonucleotides (ASOs)", antisense oligonucleotides (ASOs) do enter cells and reach the cytosol even when injected in naked form. However, siRNAs are double-stranded RNA molecules, and hence, their net negative charge is too high for efficient spontaneous internalization through the plasma membrane and the membranes of endocytic vesicles [32–34].

2.4 A Complicated Patent Landscape

The early history of RNAi research has already been briefly presented. After the seminal paper from Fire and Melo, which won them the Nobel Prize [1], a major breakthrough was the development of a cell-free system derived from *Drosophila* embryos that recapitulated many of the features of RNAi [35]. Using this cell-free system, the basic molecular mechanisms beyond RNAi were later uncovered. More specifically, a collaboration between Philip Zamore (University of Massachusetts Medical School (UMass)), Thomas Tuschl (Max Planck Institute for Biophysical Chemistry (Max Planck)), Phillip Sharp (Massachusetts Institute of Technology (MIT)), and David Bartel (Whitehead Institute for Biomedical Research and MIT) showed that dsRNAs are enzymatically processed into short 21–23 nt siRNAs, directing sequence-specific mRNA cleavage [10]. Based on this discovery, the authors wrote a patent that was assigned to UMass, Max Planck, MIT, and the Whitehead Institute. This patent is colloquially referred to as "Tuschl I."

To investigate the therapeutic potential of siRNAs, Tuschl, Zamore, Sharp, and Bartel, together with Paul Schimmel, founded Alnylam Pharmaceuticals in 2002. As will be seen later in this chapter, this company has had a major role in moving siRNAs into the clinic.

Thomas Tuschl's laboratory also showed that 21-nucleotide siRNA duplexes could be used to silence desired genes in human cells [25]. This research resulted in a second patent teaching methods to make siRNAs with 3′ overhangs for targeted mRNA degradation. This patent is referred to as "Tuschl II." According to the former CEO of Alnylam Pharmaceuticals, Max Planck intended to license both patents exclusively to Alnylam, but UMass refused. Therefore, while Alnylam secured exclusive rights on Tuschl II, Tuschl I was licensed nonexclusively to several companies, including Alnylam and its competitor, Sirna Therapeutics [36]. In 2006, Merck bought Sirna Therapeutics for $1.1 billion.

Then the patent wars began. Tuschl II claims were unable to be granted based on the interpretation that these claims were already part of Tuschl I. In 2009, Max Planck and Alnylam sued UMass, Whitehead, and MIT on issues regarding licensing agreements related to both patents. The parties arrived at an agreement 2 years later, and Tuschl II claims were finally granted by several patent offices worldwide. Alnylam emerged from this trial with a stronger IP position.

2.5 2010–2015: Big Pharma Moves Away from RNAi

The four issues described above – (i) unsuccessful clinical trials, (ii) concerns about the immunogenicity of siRNAs, (iii) the problem of siRNA delivery into cells, and (iv) the unclear patent scenario – made big pharmaceutical companies worry about the readiness of RNAi. Between 2010 and 2015, companies such as Merck, Roche, Pfizer, and Novartis started selling, shrinking, or shutting down their earlier investments in RNAi [37, 38]. In 2014, Alnylam bought Sirna Therapeutics from Merck for $175 million [36]. This is roughly 15% of what Merck paid when buying Sirna Therapeutics in 2006. With this acquisition, Alnylam consolidated exclusive rights on both Tuschl I and Tuschl II patents and became the dominant player in the RNAi ecosystem [36].

2.6 siRNA Delivery Using Lipid Nanoparticles and the Discovery of Onpattro®

Efforts at Alnylam to achieve siRNA delivery into cells made them explore the use of chemical modifications and the conjugation of siRNAs to lipid molecules. In 2004, they showed that a siRNA targeting ApoB had biological activity when conjugated to cholesterol but not when used in naked form [39]. One year later, researchers at the rival Sirna Therapeutics showed effectiveness when siRNAs were incorporated into a specialized liposome to form a stable nucleic-acid-lipid particle [40]. These siRNA/lipid particles are conceptual precursors of current lipid nanoparticles (LNPs), comprising an ionizable lipid, a phospholipid, cholesterol, and a PEGylated lipid [41]. A similar strategy was used by researchers at Alnylam in 2006, who showed that siRNAs encapsulated inside lipid nanoparticles could reduce ApoB mRNA levels when injected systemically in nonhuman primates [42]. This study was a collaboration with researchers at Protiva Biotherapeutics Inc., a Canadian company developing liposomal delivery technologies (currently Arbutus Biopharma).

A key aspect of RNA-containing LNPs is the presence of an ionizable lipid. These lipids are positively charged at acidic pH, favoring complexation with RNA molecules. They become neutral at the physiological pH of 7.4, which minimizes their toxicity. However, upon endocytosis, endosomes engulfing these LNPs are protonated, and hence the LNPs become positively charged again. This attracts and presumably breaks apart the negatively charged endosomal membrane, forming a transient pore that enables RNA to escape into the cytosol [41] (Fig. 2).

The 2006 Alnylam-Protiva collaboration already used an ionizable lipid for the LNP formulation (DLinDMA) [42]. However, the therapeutic index of these LNPs was poor [36]. Attempts to improve the pharmacological properties of siRNA-containing LNPs [43] led to the discovery of DLin-MC3-DMA (MC3) [44]. DlinDMA and DLin-MC3-DMA were compared indirectly in two clinical trials using identical siRNAs targeting the transthyretin (*TTR*) mRNA [36]. This gene, if mutated, codes for a toxic protein that produces transthyretin (TTR)-mediated amyloidosis (ATTR). Surprisingly, the robust and durable suppression of serum TTR levels was only achieved when using the new-generation MC3 lipids [45, 46].

The MC3-containing LNP/anti-*TTR* siRNA formulation led to ***patisiran*** (Onpattro®), the first approved siRNA-based therapy. ***Patisiran*** was approved by

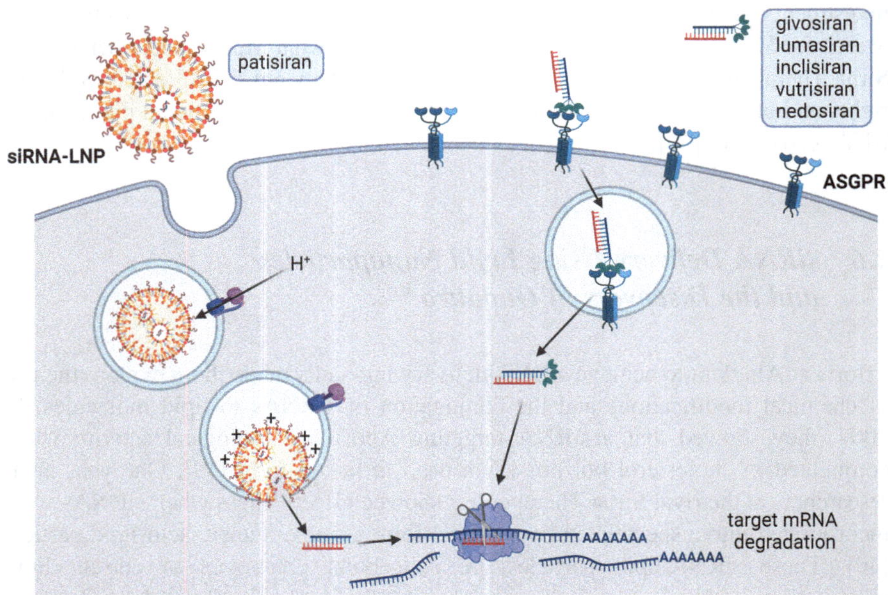

Fig. 2 Strategies for siRNA delivery into hepatocytes. Patisiran (left) is a siRNA encapsulated inside a lipid nanoparticle (LNP). When injected intravenously, LNPs interact with plasmatic lipoproteins and get loaded with the apolipoprotein E (ApoE), which facilitates endocytosis in hepatocytes mediated by the low-density lipoprotein receptor (LDLR) [45, 53]. Once inside endosomes, LNPs containing ionizable lipids get positively charged on their surface due to the acidification of the endosomal lumen. Consequently, the LNPs fuse with the endosomal membrane, and the siRNA escapes into the cytosol. Givosiran and all other FDA-approved siRNAs (right) are conjugated to a trivalent GalNAc molecule in the 3′ end of the sense strand. GalNAc molecules are recognized by the asialoglycoprotein receptor (ASGPR), which is expressed at high levels in the membrane of hepatocytes. This recognition induces an efficient receptor-dependent endocytosis of the siRNAs, which later escape from the endosomal compartments by yet unknown mechanisms. Created with BioRender.com

the FDA for the treatment of hereditary ATTR on August 10, 2018, 20 years after the discovery of RNAi [1].

2.7 The Onpattro® Story and Connections with mRNA Therapies and Vaccines

Patisiran is a two-in-one biotherapeutic agent combining RNAi and LNP technology. As explained above, the biotherapeutic was codeveloped by a collaboration between scientists at US-based Alnylam and at Canada-based Protiva Biotherapeutics. The latter merged in 2008 with Inex (its parent company) to form Tekmira Pharmaceuticals. Tekmira, a result of this merger, focused on developing RNAi drugs, but in 2009, some of its scientists founded AlCana Technologies (a fusion between Alnylam and Canada [36]) to find new ionizable lipids that, like MC3 [44], could improve LNP pharmacological properties. AlCana later changed its name to Acuitas Therapeutics and currently focalizes on the development of LNPs for messenger RNA (mRNA) therapy and vaccines.

This complex scenario led again to several patent-related litigations. In 2011, Tekmira sued Alnylam for patent infringement. Alnylam sued Tekmira shortly after. These legal battles were settled in 2012, with Alnylam paying Tekmira $65 million for a license of their LNP technology, clearing the way for the development of ***patisiran***.

While the Onpattro® story reached a happy ending with the first-ever FDA-approved siRNA drug, the legal struggles behind the property of the LNP technology made Alnylam move away from LNPs [36]. However, the LNP saga continued in the mRNA therapeutics field.

Messenger RNA (mRNA) therapy consists in delivering the mRNA that codes for a protein that a cell is unable to make. If this mRNA codes for a surface protein from a pathogen (e.g., SARS-CoV-2's spike protein), the same technology serves to make mRNA-based vaccines. Like siRNAs, mRNA therapeutics need to face the fact that mRNA is not efficiently internalized by cells in naked form. It also needs to circumvent the problem of innate immune sensors and the inflammatory response that is elicited by unmodified extracellular mRNAs. While the dsRNA-specific TLR3 is involved in the response against siRNAs [28], mRNAs are recognized by the endosomal Toll-like receptor 7 (TLR7), which senses uridine in the context of single-stranded RNA [47]. The use of pseudouridine or methylated pseudouridine makes mRNAs invisible to TLR7 and other RNA sensors [48]. This discovery was a major breakthrough in the development of mRNA therapies and mRNA-based vaccines. For this reason, Katalin Karikó and Drew Weissman were awarded the Nobel Prize in 2023. However, the other key component of these therapies and vaccines is the LNP used for mRNA delivery.

The clinical and commercial success of COVID-19 mRNA vaccines has been accompanied by several lawsuits from Tekmira (currently Arbutus Biopharma), Acuitas Therapeutics, and Alnylam, suing Moderna, Pfizer, BioNTech, and/or CureVac (manufacturers and/or distributors of mRNA vaccines and mRNA therapeutics) for the infringement of LNP patents. The scenario that will emerge from these lawsuits will undoubtedly mark the future of the mRNA therapeutics field. It should be noted that while COVID-19 vaccines are the only marketed mRNA products currently available, Moderna, BioNTech, and CureVac have several mRNA biotherapeutics in their pipeline, with many already navigating phase 3 trials.

2.8 The Successful 3x GalNAc Technology for Delivery into Hepatocytes

Building on early positive data on siRNA-cholesterol conjugates [39] and moving away from the complex patent landscape involving LNP technologies, Alnylam's scientists looked for other siRNA conjugates that could solve the delivery problem [36]. In 2014, they showed that the conjugation of a siRNA to multivalent (3x) *N*-acetylgalactosamine (GalNAc) moieties enabled the efficient internalization of the siRNA by hepatocytes, even after subcutaneous injections in mice [49].

GalNAc conjugation exploits the asialoglycoprotein receptor (ASGPR), which is highly expressed on the surface of hepatocytes. This receptor recognizes glycoproteins with exposed terminal galactose moieties and clears them from circulation through clathrin-mediated endocytosis. GalNAc-conjugated siRNAs still need to escape from endosomes in order to be functional. As we mentioned, this is a rate-limiting step for all RNA-based therapeutics, with most endocytosed RNA ending up in degradation compartments, such as the lysosome [32, 33, 50]. However, the increased endocytosis rate conferred by the GalNAc/ASGPR interaction greatly enhances the efficiency of the overall process.

A major limitation of the GalNAc conjugation technology is that it only allows for efficient delivery into hepatocytes. However, several genetic or metabolic diseases involve the liver, so this feature could even be considered advantageous.

During the second decade of the twenty-first century, Alnylam sponsored several clinical trials for different GalNAc-conjugated siRNAs or collaborated with other companies to do so. Not all of these trials were successful. For example, increased mortality was demonstrated in the case of ***revusiran***, a first-generation GalNAc conjugate targeting *TTR* for the treatment of hereditary transthyretin-mediated amyloidosis (hATTR) [51]. Because ***revusiran*** was the GalNAc version of the FDA-approved ***patisiran***, it could be assumed that the GalNAc platform could be responsible for the observed adverse events. However, this was disproved by other clinical trials employing the GalNAc technology in other disease models. The success in these trials led to a succession of FDA approvals for several GalNAc-conjugated siRNAs from 2018 onward: ***givosiran*** (2019), ***lumasiran*** (2020), ***inclisiran*** (2021), ***vutrisiran*** (2022), and ***nedosiran*** (2023).

Givosiran (Givlaari®) was approved by the FDA on November 20, 2019 (European Medicines Agency (EMA): 2020), for the treatment of adults with acute hepatic porphyria (AHP). It is the second FDA-approved siRNA and the first to employ the GalNAc conjugation technology. AHP is a family of rare genetic diseases characterized by defects in heme biosynthesis. Consequently, heme biosynthesis intermediates accumulate and build up in the liver, causing toxicity in the liver or in other tissues. One of the most frequent forms of AHP involves mutations that induce the expression of the hepatic delta-aminolevulinic acid (ALA) synthase 1 (*ALAS1*). This provokes the accumulation of ALA and other metabolites, producing neurotoxicity. ***Givosiran*** is a subcutaneously administered siRNA targeting *ALAS1* mRNA, conjugated to a trivalent GalNAc ligand. Results from a double-blind, placebo-controlled, phase 3 trial showed clinical benefits (74% lower rates of porphyria attacks) and a reduction in urine ALA levels [52].

2.9 siRNAs Currently in the Market

Patisiran (Onpattro®) and ***givosiran*** (Givlaari®) have already been described due to their historical importance as the first two FDA-approved siRNAs. As mentioned earlier, they both target a gene expressed predominantly in the liver (*TTR* and *ALAS1*, respectively) in the context of rare, monogenetic diseases (hATTR and AHP, respectively). They were both developed by Alnylam Pharmaceuticals and have annual treatment costs per patient in the range of 0.4–0.6 million USD. Despite these similarities, they employ radically different strategies for delivery into the liver, with ***patisiran*** employing LNPs and ***givosiran*** using the GalNAc conjugation technology for ASGPR-mediated uptake into hepatocytes (Fig. 2).

Lumasiran (Oxlumo®) was also developed by Alnylam and received FDA and EMA approval in 2020. It is indicated for the treatment of primary hyperoxaluria type 1 (PH1), a rare genetic disease characterized by recurrent kidney and bladder stones. The disease is caused by a deficiency in the liver-specific alanine–glyoxylate aminotransferase (AGT), an enzyme involved in the conversion of glyoxylate to glycine. In the absence of this enzyme, glyoxylate is oxidized and converted to oxalate. The increase in oxalate plasma levels induces the formation of calcium oxalate stones in the kidney and/or the bladder. If left untreated, plasma oxalate levels can increase up to levels sufficient to cause end-stage kidney disease or systemic oxalosis (oxalate deposition in other tissues such as bones). ***Lumasiran***'s mechanism of action is the degradation of the mRNA encoding glycolate oxidase, an enzyme acting upstream of AGT (and producing glyoxylate, which is the precursor of the toxic oxalate).

In a phase 3, randomized, placebo-controlled clinical trial recruiting patients 6+ years of age, the monthly subcutaneous injection of ***lumasiran*** achieved a reduction in oxalate levels in the urine of treated patients, with effects seen as early as 1 month after treatment [54]. Adverse effects were mainly mild and transient injection-site reactions.

Despite the fact that PH1 is mostly a kidney disease, this third FDA-approved siRNA is similar to ***patisiran*** and ***givosiran*** in the sense that it also targets the liver, where most endogenous oxalate is produced. It follows the same delivery strategy as ***givosiran*** (3x GalNAc). Also like ***givosiran***, it is administered by subcutaneous injection, with doses 1–3 months apart depending on the treatment stage. Because ***lumasiran*** is indicated for both children and adults and the dose is proportional to body weight measurements, annual treatment costs vary significantly between the pediatric and adult populations. In adults, annual treatment costs can surpass $one million [55].

Inclisiran (Leqvio®) is the fourth FDA-approved siRNA, and like the others, it was also developed by Alnylam Pharmaceuticals. It was approved by EMA in 2020 in the UK and the FDA in 2021 and in China in 2023. Its approval marks a turning point in the field of oligonucleotides (including both ASOs and siRNAs) because, for the first time, the treatment is not restricted to a rare genetic disease [56]. On the contrary, ***inclisiran*** was approved as a cholesterol-lowering treatment for patients with atherosclerotic cardiovascular disease. This condition affects more than 500 million individuals worldwide and accounts for 19 million annual deaths, being among the leading causes of global mortality [57]. In 2020, global rights to develop, manufacture, and commercialize the biotherapeutic were conferred from Alnylam to Novartis. As expected, based on a much broader population of potential patients benefiting from the drug, ***inclisiran***'s list price is roughly an order of magnitude lower than its siRNA predecessors, despite being chemically similar. Nevertheless, the drug still has the potential to become a sales blockbuster in the near future.

Inclisiran's mechanism of action is the silencing of proprotein convertase subtilisin/kexin type 9 (PCSK9) in hepatocytes, where it is internalized via ASGPR-mediated uptake (i.e., it uses the 3x GalNAc strategy) [58]. PCSK9 binds the low-density lipoprotein (LDL) receptor inside endosomes and targets the receptor for lysosomal degradation. In the absence of PCSK9, the LDL receptor is recycled to the plasma membrane, where it can participate in the internalization of more LDL particles. As a consequence, plasmatic LDL cholesterol levels go down, which is known to be beneficial in atherosclerosis.

Phase 3 trials enrolled patients with atherosclerosis cardiovascular disease who had elevated LDL cholesterol levels despite receiving statin therapy at the maximum tolerated dose. At day 510 posttreatment (284 mg of subcutaneously injected ***inclisiran***, on day 1, day 90, and every six months, or placebo), LDL cholesterol levels were roughly reduced to half the levels at baseline (vs 1–4% in the placebo group). Mild injection-site adverse events were observed in the treated group [59]. A significant reduction in LDL cholesterol levels was also observed in another phase 3 clinical trial enrolling patients with heterozygous familial hypercholesterolemia [60]. However, the treatment was unsuccessful in patients with homozygous familial hypercholesterolemia despite the robust silencing of the *PCSK9* gene [61].

In the European Union, ***inclisiran*** was approved for use in adults with primary hypercholesterolaemia (heterozygous familial and nonfamilial) or mixed dyslipidaemia as an adjunct to diet. In addition, the biotherapeutic is to be used in combination with statins in cases where statins alone are ineffective but can be used alone in statin-intolerant patients [62]. In the United States, ***inclisiran*** was initially approved as an adjunct to diet and statin therapy in atherosclerotic cardiovascular disease and heterozygous familial hypercholesterolemia. However, the FDA expanded these indications in 2023, enabling broader use in patients with primary hyperlipidemia (high LDL cholesterol) who are at an increased risk of heart disease.

Vutrisiran (Amvuttra®) is indicated for the treatment of the polyneuropathy of hereditary transthyretin-mediated (hATTR) amyloidosis. It has already been commented how clinical trials on ***revusiran***, the GalNAc version of ***patisiran***, failed due to increased mortality issues among the treatment group [51]. ***Vutrisiran*** has an identical sequence to the ill-fated ***revusiran*** but has a stabilized siRNA backbone [36]. This stabilized siRNA backbone, containing additional phosphorothioate linkages at the 5′ ends of both strands and fewer 2′-fluoro substitutions (see subsection below on siRNA modifications), improved the molecule's metabolic stability. ***Vutrisiran*** was compared against ***patisiran*** and an external placebo group in a phase 3 trial, improving clinical outcomes and with an acceptable safety profile

[63]. Compared to ***patisiran***, also developed by Alnylam, it has the advantage of subcutaneous administration every 3 months instead of an intravenous injection every 3 weeks. It was approved by the FDA and by EMA in 2022.

Nedosiran (Rivfloza®) is a siRNA that, like Alnylam's ***lumasiran***, is indicated for the treatment of PH1. The main difference is that instead of targeting the mRNA coding for glycolate oxidase, it targets hepatic lactate dehydrogenase (LDHA). This enzyme catalyzes the conversion of glyoxylate to oxalate, the last step of oxalate formation in the liver [64]. It also uses the GalNAc conjugation strategy to target hepatocytes. However, in this case, the biotherapeutic was developed by Dicerna Pharmaceuticals, a company that was later acquired by Novo Nordisk. In 2020, Alnylam and Dicerna signed a cross-licensing agreement to market their competing products ***lumasiran*** and ***nedosiran***, both indicated for the same disease but using a different molecular target. Based on data from a phase 2 clinical trial and an interim analysis of a phase 3 trial [65], the FDA approved ***nedosiran*** for adults and children ≥9 years old in October 2023.

3 Chemical Modifications in Therapeutic siRNAs

The chapter on ASOs[4] explains that a key aspect governing the pharmacokinetics and pharmacodynamics of therapeutic oligonucleotides are backbone and ribose modifications. This is not different in the case of therapeutic siRNAs, although there are different structural constraints.

To be effective, siRNAs need to be recognized by the protein Argonaute2 (AGO2). This protein first selects one of the two siRNA strands and degrades the other. The siRNA strand that remains bound to AGO2 is called the "antisense strand" because it is antisense to the mRNA that is going to be recognized and cleaved. Although the human genome codes for eight Argonaute proteins (four from the Ago clade and four from the Piwi clade), only AGO2 is active in the RNAi pathway and possesses *slicer* activity (i.e., the capacity to cleave a specific mRNA using siRNA-guided recognition).

A first dilemma regarding therapeutic siRNAs is determining which of the two strands is going to be used as the antisense strand. If the wrong strand is loaded and retained in AGO2, this could give place to toxic effects caused by off-target silencing (i.e., degradation of a nondesired mRNA). Endogenous siRNAs are symmetric molecules containing 3′ overhangs at both ends. However, AGO2 has intrinsic RNA strand selection capability, dictated by specific duplex parameters, such as thermodynamics, 5′ nucleotide identity, and structure [66]. The presence of chemical modifications in either the sense of the antisense strand of a synthetic siRNA duplex can also greatly bias the AGO2 strand selection process [67].

---

[4] See Chapter "Antisense Oligonucleotides (ASOs)".

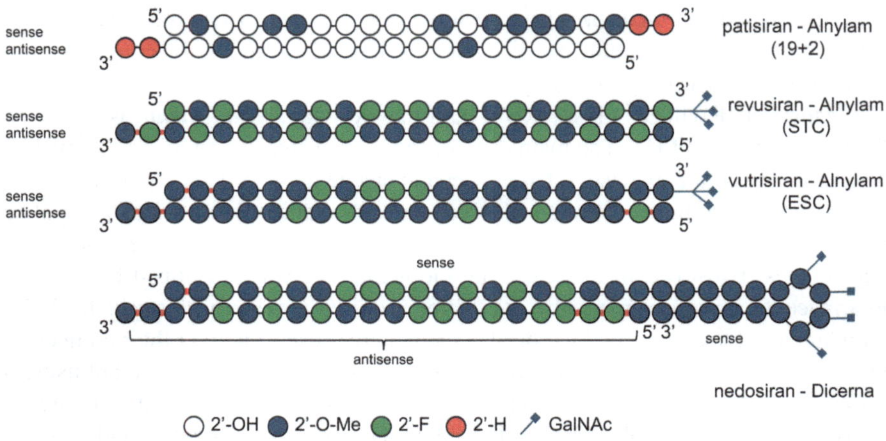

Fig. 3 Schematic representation of four siRNAs that made their way into, at least, phase 3 clinical trials. ***Patisiran*** (Onpattro®) was the first FDA-approved siRNA. It is encapsulated inside a lipid nanoparticle and has a 19 + 2 configuration, with 3′ overhangs of two DNA nt (dTdT) at both ends. It also bears just a few modified RNA bases (2'-O-Me, where a methyl group is added to the 2'-OH group of the ribose). ***Revusiran*** is a siRNA containing a trivalent GalNAc linkage that uses Alnylam's "standard template chemistry" (STC). Despite being highly modified, it was still metabolically labile, and the required high doses were toxic for several human subjects. ***Vutrisiran*** (Amvuttra®) has an identical nucleotide sequence but uses Alnylam's more advanced "enhanced stabilization chemistry" (ESC). Other FDA-approved siRNAs, like ***givosiran*** and ***inclisiran***, also follow the ESC strategy. Last, ***nedosiran*** (Rivloza®) uses a discontinuous hairpin structure, with four GalNAc moieties attached to the bases of the sense strand forming the loop

First-generation siRNAs, like ***patisiran***, follow a 19 + 2 design, comprising 19 RNA bases forming dsRNA interactions, and 3′ overhangs containing two consecutive deoxythymidine (dT) moieties. Strand selectiveness in this case is conferred by the differential thermodynamic stability of both ends and the presence of chemical modifications (Fig. 3). However, strand selectiveness can also be improved using an asymmetric siRNA design, where only one of the siRNA ends contain 3′ overhangs [68, 69]. From ***givosiran*** onward, all FDA-approved siRNAs have an asymmetric design. For example, Dicerna, the manufacturer of ***nedosiran***, uses a nick dsRNA with an overall stem-loop structure.

Targeting ligands like trivalent or tetravalent GalNAc-containing chemistries also introduce asymmetry. Because human Argonaute2 has binding pockets for both the 5′ and 3′ ends of the siRNA guide strand (i.e., the antisense strand) [70], targeting ligands are always introduced in either the 5′ or the 3′ ends of the passenger strand (i.e., the sense strand).

Chemical modifications, in addition to influencing the strand selection process, also affect the pharmacokinetics and pharmacodynamic properties of a siRNA. In particular, they can render the siRNA more resistant to endo- and exonucleases or alter its recognition by surface-bound or intracellular RNA sensors. Chemical modifications present in therapeutic siRNAs comprise different combinations of backbone and ribose modifications [71]. However, unlike ASOs, where virtually all phosphodiester bonds are changed to phosphorothioates (PS), siRNAs need to preserve the structural characteristics of a natural dsRNA molecule. Otherwise, they would not be recognized or would not bind Argonaute proteins efficiently. This adds constraints on the extent of modifications, which a siRNA molecule can accept before losing activity. Where to add each modification is mostly empiric, but several designs containing PS linkages have been proven useful and are frequent in several marketed products (Fig. 3).

Ribose modifications are also frequent in therapeutic siRNAs. ***Patisiran*** contains mostly unmodified RNA bases, but some positions (especially in the sense strand) contain 2'-OMe base modifications. More recently approved siRNAs, based on the 3x GalNAc conjugation technology, no longer contain unmodified riboses. These siRNAs are more exposed to nucleases, and therefore chemical modifications become critical. All positions of ***givosiran***, ***inclisiran***, ***lumasiran***, or ***vutrisiran*** contain either 2'-OMe or 2'-F ribose modifications (Fig. 3). In addition, because 5′ exonucleases have been found to be the most prevalent nuclease activity in endolysosomal compartments [72], these siRNAs also contain PS bonds in their extremes.

An example of the importance of chemical modifications and siRNA metabolic stabilization is the story of ***revusiran*** and ***vutrisiran***. Both siRNAs have essentially the same sequence, and they are both conjugated to GalNAc and employ the same administration route. However, ***revusiran*** was associated with increased mortality in clinical trials and was therefore discontinued. This was shown to be caused by the poor metabolic stability of the siRNA, requiring the use of relatively high doses in clinical trials. Further research showed that the inclusion of two additional PS linkages at the 5′ ends of both strands (for a total of six PS bonds), and a higher proportion of 2'-OMe vs 2'-F ribose modifications, increased the metabolic stability of the siRNAs. As a result, siRNAs could be used at much lower doses and keep their effectiveness while reducing toxicity [36]. For example, phase 3 clinical trials on ***revusiran*** used daily doses of 500 mg, while phase 3 clinical trials on ***vutrisiran*** (same sequence but more modified) were based on 25 mg doses injected subcutaneously every 3 months [51]. Only the latter reached the market.

4 Future Challenges

So far, siRNAs available in the market are restricted to the treatment of diseases where the liver is the primary organ that needs to be targeted. ***Patisiran*** is encapsulated in LNPs containing ionizable lipids, but these LNPs accumulate in the liver when injected intravenously [45, 53]. Alternatively, GalNAc-conjugated siRNAs can be administered subcutaneously, but these siRNAs are only useful when targeting hepatocytes due to the presence of a high number of ASGPR molecules in the cell's surface ($>10^6$/cell) [32].

To make reality the promise of "drugging even the undruggable targets," new strategies for siRNA delivery into nonhepatic cells need to be developed. Preclinical studies on Huntington's disease (HD) have shown that siRNAs can be used to target mutated alleles, but not wild-type alleles, in neurons [73, 74]. Efforts to design conjugated and modified siRNAs that can cross the blood-brain barrier also showed promising results for the treatment of Alzheimer's disease (AD) [75]. In addition, clinical trials for the treatment of AD using intrathecally administered siRNAs against the amyloid precursor protein are ongoing [76]. Other clinical trials targeting the liver, eyes, skin, lungs, and kidneys have been studied [77].

Another frontier is to improve the pharmacological properties of therapeutic siRNAs to make them more tolerable and effective or to make them more stable and extend the time between dosages. This has been in part already achieved, with some siRNAs retaining therapeutic effects even when administered quarterly or even twice yearly [36, 78].

Lastly, as is often the case when innovative drugs for orphan or rare genetic diseases get into the market, the cost-effectiveness of the treatments becomes an important issue that can influence widespread accessibility to life-saving drugs.

Compliance with Ethical Standards This chapter is a review and no humans or animals were involved in the preparation of the chapter.

References

1. Fire A, Xu S, Montgomery MK et al (1998) Potent and specific genetic interference by double-stranded RNA in caenorhabditis elegans. Nature 391:806–811
2. Napoli C, Lemieux C, Jorgensen R (1990) Introduction of a chimeric chalcone synthase gene into petunia results in reversible co-suppression of homologous genes in trans. Plant Cell 2:279–289
3. Van Der Krol AR, Mur LA, Beld M et al (1990) Flavonoid genes in petunia: addition of a limited number of gene copies may lead to a suppression of gene expression. Plant Cell 2:291–299
4. Romano N, Macino G (1992) Quelling: transient inactivation of gene expression in Neurospora crassa by transformation with homologous sequences. Mol Microbiol 6:3343–3353
5. van der Krol AR, Mur LA, de Lange P et al (1990) Antisense chalcone synthase genes in petunia: visualization of variable transgene expression. MGG Mol Gen Genet 220:204–212
6. Matzke MA, Primig M, Trnovsky J et al (1989) Reversible methylation and inactivation of marker genes in sequentially transformed tobacco plants. EMBO J 8:643–649
7. Fire A, Albertson D, Harrison SW et al (1991) Production of antisense RNA leads to effective and specific inhibition of gene expression in *C. elegans* muscle. Development 113:503–514
8. Hamilton AJ, Baulcombe DC (1999) A species of small antisense RNA in posttranscriptional gene silencing in plants. Science 286:950–952
9. Hammond SM, Bernstein E, Beach D et al (2000) An RNA-directed nuclease mediates post-transcriptional gene silencing in Drosophila cells. Nature 404:293–296
10. Zamore PD, Tuschl T, Sharp PA et al (2000) RNAi: double-stranded RNA directs the ATP-dependent cleavage of mRNA at 21 to 23 nucleotide intervals. Cell 101:25–33
11. Elbashir SM, Lendeckel W, Tuschl T (2001) RNA interference is mediated by 21- and 22-nucleotide RNAs. Genes Dev 15:188–200
12. Bernstein E, Caudy AA, Hammond SM et al (2001) Role for a bidentate ribonuclease in the initiation step of RNA interference. Nature 409:363–366

13. Hutvágner G, McLachlan J, Pasquinelli AE et al (2001) A cellular function for the RNA-interference enzyme dicer in the maturation of the let-7 small temporal RNA. Science 293:834–838
14. Grishok A, Pasquinelli AE, Conte D et al (2001) Genes and mechanisms related to RNA interference regulate expression of the small temporal RNAs that control *C. elegans* developmental timing. Cell 106:23–34
15. Martinez J, Patkaniowska A, Urlaub H et al (2002) Single-stranded antisense siRNAs guide target RNA cleavage in RNAi. Cell 110:563–574
16. Hammond SM, Boettcher S, Caudy AA et al (2001) Argonaute2, a link between genetic and biochemical analyses of RNAi. Science 293:1146–1150
17. Tabara H, Sarkissian M, Kelly WG et al (1999) The rde-1 gene, RNA interference, and transposon silencing in *C. elegans*. Cell 99:123–132
18. Catalanotto C, Azzalin G, Macino G et al (2000) Gene silencing in worms and fungi. Nature 404:245
19. Fagard M, Boutet S, Morel JB et al (2000) AGO1, QDE-2, and RDE-1 are related proteins required for post-transcriptional gene silencing in plants, quelling in fungi, and RNA interference in animals. Proc Natl Acad Sci U S A 97:11650–11654
20. Sen GL, Blau HM (2006) A brief history of RNAi: the silence of the genes. FASEB J 20:1293–1299
21. Lee RC, Feinbaum RL, Ambros V (1993) The *C. elegans* heterochronic gene lin-4 encodes small RNAs with antisense complementarity to lin-14. Cell 75:843–854
22. Wightman B, Ha I, Ruvkun G (1993) Posttranscriptional regulation of the heterochronic gene lin-14 by lin-4 mediates temporal pattern formation in *C. elegans*. Cell 75:855–862
23. Ambros V (2008) The evolution of our thinking about microRNAs. Nat Med 14:1036–1040
24. Bartel DP (2018) Metazoan microRNAs. Cell 173:20–51
25. Elbashir SM, Harborth J, Lendeckel W et al (2001) Duplexes of 21-nucleotide RNAs mediate RNA interference in cultured mammalian cells. Nature 411:494–498
26. (2006) A billion dollar punt. Nat Biotechnol 24:1453
27. Kaiser PK, Symons RCA, Shah SM et al (2010) RNAi-based treatment for neovascular age-related macular degeneration by Sirna-027. Am J Ophthalmol 150:33–39
28. Kleinman ME, Yamada K, Takeda A et al (2008) Sequence- and target-independent angiogenesis suppression by siRNA via TLR3. Nature 452:591–597
29. Cho WG, Albuquerque RJC, Kleinman ME et al (2009) Small interfering RNA-induced TLR3 activation inhibits blood and lymphatic vessel growth. Proc Natl Acad Sci U S A 106:7137–7142
30. Winston WM, Molodowitch C, Hunter CP (2002) Systemic RNAi in *C. elegans* requires the putative transmembrane protein SID-1. Science 295:2456–2459
31. Winston WM, Sutherlin M, Wright AJ et al (2007) Caenorhabditis elegans SID-2 is required for environmental RNA interference. Proc Natl Acad Sci U S A 104:10565–10570
32. Dowdy SF (2017) Overcoming cellular barriers for RNA therapeutics. Nat Biotechnol 35:222–229
33. Dowdy SF (2023) Endosomal escape of RNA therapeutics: how do we solve this rate-limiting problem? RNA 29:396–401
34. Dowdy SF, Setten RL, Cui XS et al (2022) Delivery of RNA therapeutics: the great endosomal escape! Nucleic Acid Ther 32:361–368
35. Tuschl T, Zamore PD, Lehmann R et al (1999) Targeted mRNA degradation by double-stranded RNA in vitro. Genes Dev 13:3191–3197
36. Maraganore J (2022) Reflections on Alnylam. Nat Biotechnol 40:641–650
37. Krieg AM (2011) Is RNAi dead? Mol Ther 19:1001–1002
38. Conde J, Artzi N (2015) Are RNAi and miRNA therapeutics truly dead? Trends Biotechnol 33: 141–144
39. Soutschek J, Akinc A, Bramlage B et al (2004) Therapeutic silencing of an endogenous gene by systemic administration of modified siRNAs. Nature 432:173–178

40. Morrissey DV, Lockridge JA, Shaw L et al (2005) Potent and persistent in vivo anti-HBV activity of chemically modified siRNAs. Nat Biotechnol 23:1002–1007
41. Han X, Zhang H, Butowska K et al (2021) An ionizable lipid toolbox for RNA delivery. Nat Commun 12:7233
42. Zimmermann TS, Lee ACH, Akinc A et al (2006) RNAi-mediated gene silencing in non-human primates. Nature 441:111–114
43. Semple SC, Akinc A, Chen J et al (2010) Rational design of cationic lipids for siRNA delivery. Nat Biotechnol 28:172–176
44. Jayaraman M, Ansell SM, Mui BL et al (2012) Maximizing the potency of siRNA lipid nanoparticles for hepatic gene silencing in vivo. Angew Chemie - Int Ed 51:8529–8533
45. Akinc A, Maier MA, Manoharan M et al (2019) The Onpattro story and the clinical translation of nanomedicines containing nucleic acid-based drugs. Nat Nanotechnol 14:1084–1087
46. Coelho T, Adams D, Silva A et al (2013) Safety and efficacy of RNAi therapy for transthyretin amyloidosis. N Engl J Med 369:819–829
47. Diebold SS, Kaisho T, Hemmi H et al (2004) Innate antiviral responses by means of TLR7-mediated recognition of single-stranded RNA. Science 303:1529–1531
48. Karikó K, Buckstein M, Ni H et al (2005) Suppression of RNA recognition by toll-like receptors: the impact of nucleoside modification and the evolutionary origin of RNA. Immunity 23:165–175
49. Nair JK, Willoughby JLS, Chan A et al (2014) Multivalent N-acetylgalactosamine-conjugated siRNA localizes in hepatocytes and elicits robust RNAi-mediated gene silencing. J Am Chem Soc 136:16958–16961
50. Du Rietz H, Hedlund H, Wilhelmson S et al (2020) Imaging small molecule-induced endosomal escape of siRNA. Nat Commun 11:1809
51. Judge DP, Kristen AV, Grogan M et al (2020) Phase 3 multicenter study of Revusiran in patients with hereditary transthyretin-mediated (hATTR) amyloidosis with cardiomyopathy (ENDEAVOUR). Cardiovasc Drugs Ther 34:357–370
52. Balwani M, Sardh E, Ventura P et al (2020) Phase 3 trial of RNAi therapeutic Givosiran for acute intermittent porphyria. N Engl J Med 382:2289–2301
53. Akinc A, Querbes W, De S et al (2010) Targeted delivery of RNAi therapeutics with endogenous and exogenous ligand-based mechanisms. Mol Ther 18:1357–1364
54. Garrelfs SF, Frishberg Y, Hulton SA et al (2021) Lumasiran, an RNAi therapeutic for primary hyperoxaluria Type 1. N Engl J Med 384:1216–1226
55. Gang X, Liu F, Mao J (2023) Lumasiran for primary hyperoxaluria type 1: what we have learned? Front Pediatr 10:1052625
56. Obexer R, Nassir M, Moody ER et al (2024) Modern approaches to therapeutic oligonucleotide manufacturing. Science 384:eadl4015
57. Vasan RS, Enserro DM, Xanthakis V et al (2022) Temporal trends in the remaining lifetime risk of cardiovascular disease among middle-aged adults across 6 decades: the Framingham study. Circulation 145:1324–1338
58. Migliorati JM, Jin J, Zhong XB (2022) siRNA drug Leqvio (inclisiran) to lower cholesterol. Trends Pharmacol Sci 43:455–456
59. Ray KK, Wright RS, Kallend D et al (2020) Two phase 3 trials of Inclisiran in patients with elevated LDL cholesterol. N Engl J Med 382:1507–1519
60. Raal FJ, Kallend D, Ray KK et al (2020) Inclisiran for the treatment of heterozygous familial hypercholesterolemia. N Engl J Med 382:1520–1530
61. Raal F, Durst R, Bi R et al (2024) Efficacy, safety, and tolerability of Inclisiran in patients with homozygous familial hypercholesterolemia: results from the ORION-5 randomized clinical trial. Circulation 149:354–362
62. Lamb YN (2021) Inclisiran: first approval. Drugs 81:389–395
63. Adams D, Tournev IL, Taylor MS et al (2023) Efficacy and safety of vutrisiran for patients with hereditary transthyretin-mediated amyloidosis with polyneuropathy: a randomized clinical trial. Amyloid 30:1–9

64. Syed YY (2023) Nedosiran: First approval. Drugs 83:1729–1733
65. Groothoff J, Sellier-Leclerc AL, Deesker L et al (2024) Nedosiran safety and efficacy in PH1: interim analysis of PHYOX3. Kidney Int Reports 9:1387–1396
66. Noland CL, Doudna JA (2013) Multiple sensors ensure guide strand selection in human RNAi pathways. RNA 19:639–648
67. Varley AJ, Desaulniers JP (2021) Chemical strategies for strand selection in short-interfering RNAs. RSC Adv 11:2415–2426
68. Park JH, Hong SW, Yun S et al (2014) Effect of siRNA with an asymmetric RNA/dTdT overhang on RNA interference activity. Nucleic Acid Ther 24:364–371
69. Chang CI, Yoo JW, Hong SW et al (2009) Asymmetric shorter-duplex siRNA structures trigger efficient gene silencing with reduced nonspecific effects. Mol Ther 17:725–732
70. Schirle NT, MacRae IJ (2012) The crystal structure of human argonaute2. Science 336:1037–1040
71. Friedrich M, Aigner A (2022) Therapeutic siRNA: state-of-the-art and future perspectives. BioDrugs 36:549–571
72. Nair JK, Attarwala H, Sehgal A et al (2017) Impact of enhanced metabolic stability on pharmacokinetics and pharmacodynamics of GalNAc-siRNA conjugates. Nucleic Acids Res 45:10969–10977
73. DiFiglia M, Sena-Esteves M, Chase K et al (2007) Therapeutic silencing of mutant huntingtin with siRNA attenuates striatal and cortical neuropathology and behavioral deficits. Proc Natl Acad Sci U S A 104:17204–17209
74. Conroy F, Miller R, Alterman JF et al (2022) Chemical engineering of therapeutic siRNAs for allele-specific gene silencing in Huntington's disease models. Nat Commun 13:5802
75. Zhou Y, Zhu F, Liu Y et al (2020) Blood-brain barrier-penetrating siRNA nanomedicine for Alzheimer's disease therapy. Sci Adv 6:eabc7031
76. Cohen S, Ducharme S, Brosch JR et al (2023) Interim phase 1 Part A results for ALN-APP, the first investigational RNAi therapeutic in development for Alzheimer's disease. Alzheimers Dement 19:e082650
77. Ahn I, Kang CS, Han J (2023) Where should siRNAs go: applicable organs for siRNA drugs. Exp Mol Med 55:1283–1292
78. Hu B, Zhong L, Weng Y et al (2020) Therapeutic siRNA: state of the art. Signal Transduct Target Ther 5:101

Top Med Chem
https://doi.org/10.1007/7355_2024_172

Published online: 24 December 2024

Aptamers

Victoria Calzada

Contents

Abstract There is a new class of structured functional oligonucleotides called aptamers with a wide range of applications and a promising future. Aptamers are small oligonucleotides with structural motifs that bind with high affinity and specificity to their target. In the last decades, aptamer has emerged as a new opportunity in the biotherapeutic field. This section summarizes the most important characteristics of aptamers and their applications in the biopharmaceutical field.

Keywords Aptamers, Biopharmaceutical, Drug delivery, Oligonucleotides

---

V. Calzada (✉)
Área de Radiofarmacia-Centro de Investigaciones Nucleares, Facultad de Ciencias, Universidad de la República, Montevideo, Uruguay
e-mail: vcalzada@fcien.edu.uy

1 Definition and Characteristics

Aptamers are small oligonucleotides with very particular structural motifs that bind to a target of interest with high affinity and specificity. They were first named by Ellington and Szostak in the 1990s as a word composed by the Latin term "aptus," meaning to fit, and the Greek term "meros," which means the part [1]. Based on their chemical nature [2], these short 20- to 100-nucleotide sequences can fold into unique three-dimensional structures under specific conditions (Fig. 1), acquiring intrinsic stability. They can form hairpins, bulges, pseudoknots, as well as G-quadruplex structures [3].

The aptamer–target binding is a consequence of conformational and weak interactions, including hydrophobic and hydrogen bonding, electrostatic interactions, π–π stacking, van der Waals forces, or their combinations [4, 5]. Thus, aptamers can bind to many types of small molecules, such as metal ions [6], hormones [7], proteins [8], sugars [9], and also viral particles [10] and tumoral cells [11], among others [12], allowing a variety of applications.

Aptamers can be synthesized entirely in vitro, which is desirable because it is a cell and animal free methodology. It also ensures batch-to-batch reproducibility in their production. Therefore, aptamers can be produced in large quantities and at low cost.

In addition, aptamers exhibit diverse characteristics that make them very valuable for in vivo applications. These characteristics include their relatively small size, potentially enabling more effective penetration on tissue, even across the blood-brain barrier. Typical aptamers are approximately ten-fold smaller than antibodies [13, 14]. Besides, aptamers are non-immunogenic and nontoxic [15].

Because of their chemical nature, aptamers are characterized by thermal stability, as well as the possibility of being stable to a wide variety of environmental conditions (e.g., pH value, chemical composition, etc.). Aptamers are aqueous-soluble molecules, which facilitates their chemical modifications to be

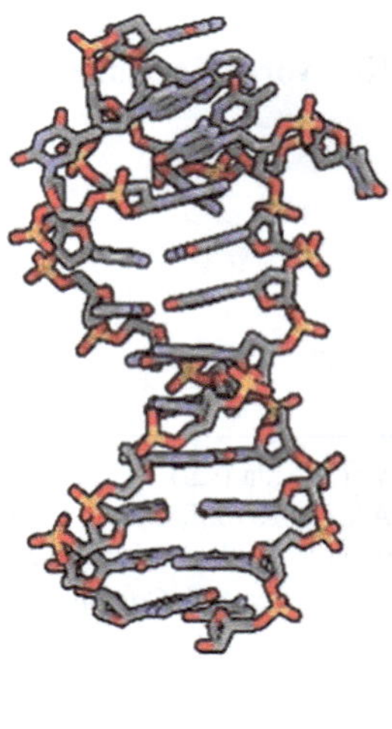

Fig. 1 General representation of an aptamer. We thank Dr. Pablo Dans for the design and drawing of this figure

functionalized to create a variety of applications [16]. For example, the chemical modification of the 2′-ribose significantly enhances the stability of RNA aptamers [17]. Additionally, conjugation of aptamers to other substances such as polyethylene glycol (PEG),[1] to increase blood circulation [18], as well as iRNA [19], drugs [20], and diverse strategies have been explored as biotherapeutics.

2 The History of Aptamers

The Polymerase Chain Reaction (PCR) has revolutionized the field of molecular biology with immediate impact. Thanks to this, the discovery of aptamers is attributable to several groups simultaneously. In 1990, Ellington and Szostak, who named aptamers, reported RNA molecules that bind to a small organic dye [1]. Independently, the group of Larry Gold described for the first time the in vitro process of selecting aptamers: the Systematic Evolution of Ligand by Exponential Enrichment (SELEX) [21]. Also, the application of in vitro selection of oligonucleotides was also reported by Robert and Joyce that same year [22]. Years later, evidence of aptamers in nature was confirmed with the discovery of functional noncoding RNA molecules [23] and riboswitches, which display key roles as regulatory elements in controlling gene expression [24].

The large number of oligonucleotide sequences resulted from the selection process necessitated extensive sequencing methodologies. The introduction of high-throughput sequencing methods in 2007 accelerated this process to identify aptamers [25, 26]. Strategies to sequence DNA and RNA in a rapid and cost-effective manner were also called deep sequencing, next-generation sequencing (NGS), high-throughput sequencing, or massively parallel sequencing [25].

Currently, artificial intelligence is advancing knowledge in the field of aptamers [27].

3 Systematic Evolution of Ligand by Exponential Enrichment (SELEX)

Aptamers are selected from libraries of oligonucleotide sequences of 20- to 100-nucleotides long. Typically, the multiple-round process called SELEX is performed, which involves several steps. Initially, the round of selection starts when the random oligonucleotide library (10^4–10^6 unique sequences) is incubated with the target. Target-bound sequences are then partitioned from the rest of the unbound library pool and PCR-amplified. Now, an enriched pool is ready for a novel selection cycle [17].

[1] See Chapter “Encapsulation and Structural Modifications of Biotherapeutics”.

Conceptually, in SELEX aptamers are developed by an iterative process, whereby as selection progresses, enrichment sequences capable of binding to the target dominate the pool [28]. However, many critical factors influence aptamer selection, and the rational design of SELEX procedures can lead to higher success rates in identifying aptamers [29]. In fact, many SELEX strategies have been developed, such as cell-SELEX [28] and in vivo SELEX [30], among others [29]. While most aptamers are attributable to the SELEX method, several groups report the non-SELEX selection of aptamers [31, 32].

Once selection is performed, sequence identification and bioinformatic analysis are necessary for further experimental evaluation. Finally, assessing the affinity and specificity of the individual sequences for the target is essential to identifying the candidate(s).

4 Aptamers as Biotherapeutics

Years after the first SELEX report, in December 2004, the Food and Drug Administration of the United States of America (FDA) approved the first therapeutic aptamer for clinical application in the treatment of neovascular age-related macular degeneration [33, 34]. Over the past few decades, other aptamers have shown potential for clinical use, and a new aptamer was recently approved (Table 1).

4.1 Avacincaptad Pegol

Avacincaptad pegol (IZERVAY™; formerly Zimura®) was approved by the FDA in August 2023 for the treatment of geographic atrophy secondary to age-related macular degeneration [43]. Geographic atrophy is the advanced form of age-related macular degeneration [44] that can lead to irreversible vision loss and a high burden of disease. Its incidence is estimated to affect up to five million individuals globally [45]. The pathogenesis of geographic atrophy is considered multifactorial and not fully understood [35]. However, chronic inflammation and overactivation of the complement system are thought to contribute to the disease [46, 47].

Avacincaptad pegol is a 39-nucleotide RNA aptamer attached to PEG [35], with the specific capability to inhibit complement protein C5. C5 plays a central role in both innate and adaptive immune responses, and its inhibition by the aptamer prevents the formation of key terminal fragments of the complement system, C5a and C5b, potentially reducing inflammation and cell death [48]. These findings suggest that ***avacincaptad pegol*** may slow disease progression and potentially alter the trajectory of disease for patients with geographic atrophy [35, 48].

Table 1 Aptamers used in clinic or involved in clinical trials as ***biotherapeutics***

Name	Type	Target	Clinical trial	Indications	Reference
Pegaptanib sodium (Macugen®)	RNAmod, pegyl	VEGF165	FDA approved	Neovascular age-related macular degeneration (discontinued)	[34]
Avacincaptad pegol (Izervay®)	RNAmod, pegyl	C5	FDA approved	Geographic atrophy	[35]
Pegpleranib (Fovista®)	DNA/ RNA, pegyl	PDGF-B	Phase III	Neovascular age-related macular degeneration	[36]
ApTOLL	DNA	TLR4	Phase IIb	Stroke	[37]
Olaptesed pegol (***NOX-A12***)	L-RNA, pegyl	CXCL12	Phase IIa	Relapsed/refractory chronic lymphocytic leukemia	[38]
AS1411	DNA	Nucleolin	Phase IIa	Metastatic refractory renal cell carcinoma	[39]
Emapticap pegol (NOX-E36)	L-RNA, pegyl	CCL2	Phase IIa	Type 2 Diabetes Mellitus	[40]
Lexaptepid pegol (NOX-H94)	L-RNA, pegyl	Hepcidin	Phase IIa	Anemia	[41]
Rondaptivon pegol	RNA, pegyl	VWF	Phase IIa	Severe and nonsevere hemophilia A	[42]

4.2 ApTOLL

Toll-like receptors are transmembrane receptors that represent a key molecular link between innate and adaptive responses, mediating tissue injury, and inflammation. Particularly, accumulating evidence highlights the crucial role of Toll-like receptor 4 (TLR4) in inflammatory responses. TLR4 recognizes a variety of ligands, which are released in response to ischemic injuries [49, 50] resulting in the activation of the TLR4 inflammatory signaling pathway and the acute-inflammatory phase. Timely control of the inflammatory response is crucial to prevent the extension of the injury.

ApTOLL is an unmodified DNA aptamer designed to antagonize TLR4 and, therefore, to block the inflammatory response produced after injuries, such as acute ischemic stroke [49] or acute myocardial infarction [50]. ***ApTOLL*** has shown a proven neuroprotective effect in preclinical models of stroke with an excellent safety profile and adequate pharmacokinetic profile in healthy volunteers [51]. These preliminary findings from clinical trials in patients with acute ischemic stroke (Phase Ib/IIa trial) showed a reduction in mortality from 18% to 5% [37, 51, 52]. ***ApTOLL*** displays a high therapeutic potential in this area.

4.3 NOX-A12

An alternative way of making therapeutic aptamers was developed by Furste and colleagues. To extend the stability of aptamers, they developed a methodology to select L-oligonucleotide aptamers named "Spiegelmers" (from the German "spiegel," for mirror) [53, 54]. L-oligonucleotide aptamers are perfect mirror images of their D-aptamer counterparts of the same sequence. In these aptamers, all chirality centers located in the (deoxy)ribose sugars of the nucleotide are mirror-inverted [55]. Because enzymes are stereoselective, L-oligonucleotides are not recognized by the ubiquitous plasma nucleases and escape from subsequent degradation [55]. However, Spiegelmers® cannot be directly selected by the conventional SELEX process. Based on the fundamental principles of stereochemistry, the mirror-image selection methodology was first reported in 1996 [53, 54]. If an in vitro selected D-aptamer binds its natural target, the structural mirror image of this aptamer binds with the same characteristics to the mirror image of the natural target [56]. Spiegelmers have shown promising results in animal models and have good safety profiles in healthy volunteers [55].

Olaptesed pegol (***NOX-A12***) is a pegylated RNA Spiegelmer anti-CXC chemokine ligand (CXCL12), which plays a critical role in the homing and retention of chronic lymphocytic leukemia cells in tissues such as the bone marrow. ***NOX-A12*** binds and neutralizes CXCL12 on the surface of bone marrow stromal cells. As a result, ***NOX-A12*** interferes with the tumor microenvironment and cell mobilization [57] and also increases the tumor-infiltrating T cells [58]. ***NOX-A12*** advanced to Phase IIa studies in patients with relapsed/refractory chronic lymphocytic leukemia and was well tolerated in monotherapy, with no additional toxicity when combined with bendamustine and rituximab [38]. Furthermore, it shows a high response rate of 86% as well as a three-year overall survival rate of >80%, which compares favorably with results achieved with bendamustine and rituximab alone [38]. Additional findings are driving further clinical trials for multiple myeloma [59].

4.4 AS1411

Often, it is considered that unmodified oligonucleotides could never be clinically useful due to their low serum half-lives in the order of minutes [60]. Fortunately, aptamers might have intrinsic stability. It is well established that secondary or tertiary structures can be relatively protected from nuclease attack [60]. G-rich quadruplex-forming aptamers display unusual biophysical and biological properties, and some have recently been investigated as biotherapeutic agents for various human diseases [60, 61].

AS1411 (formerly AGRO100, now also known as ACT-GRO-777) is a 26-mer unmodified G-rich DNA aptamer that contains a G-quadruplex structure [61–63]. It was identified with cancer antiproliferative activity and determined to be an aptamer anti-nucleolin, which is mainly found in the nucleolus but is also present at high levels on the surface of cancer cells [8, 61]. ***AS1411*** showed an excellent safety profile and ability to induce durable responses in some patients with intractable cancers [8]. However, the suboptimal pharmacology and low potency may limit its future development in unselected cancer patients [39]. Its clinical development indicated that ***AS1411*** is safe and can induce durable remissions in a few patients. However, phase I and phase II studies as ***AS1411*** monotherapy in patients with renal cell carcinoma had failed to tolerate standard treatment [39]. ***AS1411*** was also enrolled in a clinical trial in combination with cytarabine in patients with refractory or relapsed acute myeloid leukemia [8]. There are no clinical studies of ***AS1411*** being carried out at present. A better understanding of the mechanisms by which ***AS1411*** is targeted and its cytotoxic effect may stimulate the development of ***AS1411*** as a biotherapeutic [8]. ***AS1411*** has the potential to be conjugated to drugs and nanoparticles as a drug delivery therapy [63–65].

5 Aptamer-Based Immunomodulation

Over the last decade, the study of the regulation of T cell responses has led to Nobel prize-winning to Dr. James P. Allison and Dr. Tasuku Honjo [66, 67]. Immune system is regulated by a group of molecules able to modulate and ensure their correct function [68]. The immune checkpoints are these key regulatory molecules; stimulatory checkpoint pathways promote immune responses, while inhibitory checkpoint pathways inhibit immune responses. From pathogens to tumor cells, they can interact with the immune checkpoints to evade the immune system. In consequence, immune checkpoint therapy has been developed [69].

The suppressive functions of immune checkpoints usually depend on ligand-induced signaling [70]. Particularly, one of the major escape mechanisms of cancer cells is the suppression of T cell activation by PD-1 and/or CTLA-4 [71]. Inhibition of these molecules by immune checkpoint inhibitors can successfully activate the immune system to fight cancer.

To enhance the anti-cancer immunity, an important number of biotherapeutics have been described [69, 70]. In fact, aptamers have emerged as promising molecules with the potential to generate a huge impact in the immunotheranostic field [72, 73].

6 Aptamer in Diagnostics

Aptamer-based diagnostics include several methodologies. Among them, aptamers have been shown as promising tools in molecular imaging. While conventional images provide anatomical and physiological information, molecular imaging is used extensively in medicine to detect, localize, and characterize pathologies as

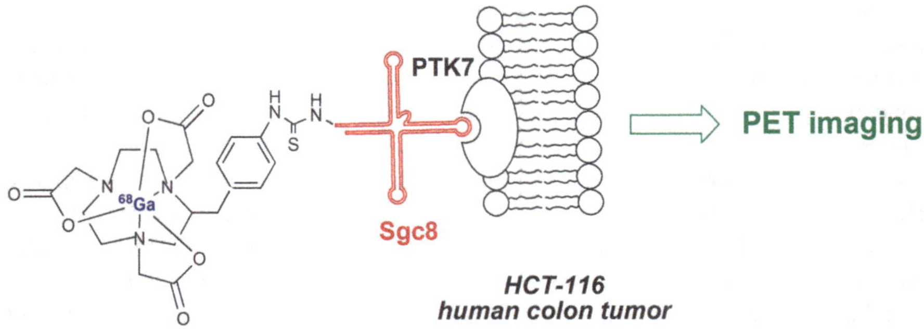

Fig. 2 Scheme of chemical structure of *[^{68}Ga]NOTA-Sgc8* [85]

well as detect molecular alterations, even in real time [74, 75]. Besides, molecular imaging is often needed to further the understanding of disease progression, as well as monitor treatment efficacy.

Thus, using a variety of strategies, aptamers have been evaluated as radiopharmaceuticals to be used in gamma images, Positron Emission Tomography (PET), and Single-photon Emission Computed Tomography (SPECT) [75–77].

On the other hand, Computed Tomography (CT), Magnetic Resonance Imaging (MRI), and Ultrasound (US) are the most well-established conventional modalities in the field of noninvasive medical imaging. Nevertheless, to allow for better delineation of anatomical structures and differentiation between normal and pathological tissues, contrast agents might be used. In fact, imaging research is currently dominated by aptamer-modified contrast media, which overlap molecular imaging with conventional approaches [78]. Aptamers have been used as a targetable contrast agent for MRI, resulting in a high magnetic sensitivity [79–81]. Also, drug-loaded aptamer-combined and gold nanoparticles showed desirable characteristics for CT imaging enhancement in a colon adenocarcimona model [82]. Goncin and colleagues recently developed a targeted microbubble with an anti-P-selectin aptamer to increase the ultrasound molecular imaging signal [83].

The ***Sgc8*** aptamer is currently in clinical trials for the staging of bladder cancer. ***Sgc8*** is an unmodified 88-nt DNA aptamer that binds protein tyrosine kinase-7 [11, 84]. In this sense, ***Sgc8*** was radiolabeled with Gallium-68 for PET/MR evaluation. The trial with ***[^{68}Ga]NOTA-Sgc8*** is currently recruiting volunteers [85] (Fig. 2). This entire aptamer as well as the truncated form, ***Sgc8***-c, is being explored in several imaging modalities [86] and also for theranostic applications [76].

7 Drug Delivery Using Aptamers

In addition to being used as biotherapeutics as such, directly as agonists or antagonists, alternatively, aptamers can be used as targeted drug delivery systems to transport drugs selectively to target sites, reducing off-target effects, decreasing

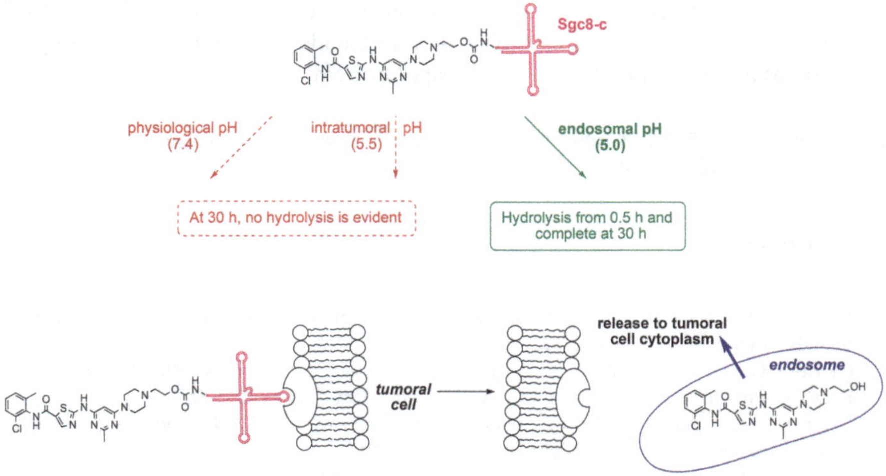

Fig. 3 Scheme of pH-dependent hydrolytic strategy covalently incorporating the anti-tumor agent dasatinib into ***Sgc8-c*** aptamer [20, 94]

toxicities, and thereby enhancing a drug's therapeutic efficacy [87]. Diverse strategies are considered regarding chemical conjugation. Biologically active entities are covalently or noncovalently attached to one or more sites of the aptamer [88]. Aptamer can deliver different biological active entities such as small-molecule drugs, small RNAs (siRNAs), miroRNAs, and toxins, among others [89].

A well-characterized aptamer-drug conjugate was reported by Bagalkot and colleagues [90]. In this study, the A10 2′-fluoropyrimidine RNA aptamer [91] that binds to the prostate-specific membrane antigen with high affinity and specificity was conjugated with doxorubicin (dox) [90]. The same drug was conjugated with other aptamers and using diverse strategies [92]. Later, an aptamer origami was reported to co-deliver dox and antisense oligonucleotides (ASOs) in cancer cells [93].

Moreover, to ensure the release of the drug under physiological conditions, Sicco and colleagues studied the pH-dependent hydrolytic strategy covalently incorporating the anti-tumor agent dasatinib into an aptamer [20, 94] (Fig. 3).

Definitely, the development of aptamers able to delivery drugs has huge potential.

8 Conclusion

Aptamers have emerged as new functional oligonucleotides with high potential for diverse applications. Notably, aptamers are used directly in therapeutic applications as modulatory agents, theranostics, or drug delivery systems. Indeed, aptamers indisputably represent an exciting opportunity in the biotherapeutic field.

Despite research advances, more efforts are needed to maximize their clinical impact. However, their roles as bioactive compounds highlight the versatility of aptamers as a powerful tool in the biotherapeutic field.

Compliance with Ethical Standards This chapter is a review and no humans or animals were involved in the preparation of the chapter.

References

1. Ellington AD, Szostak JW (1990) *In vitro* selection of RNA molecules that bind specific ligands. Nature 346(6287):818–822
2. da Rosa G, Grille L, Calzada V et al (2021) Sequence-dependent structural properties of B-DNA: what have we learned in 40 years? Biophys Rev 13(6):995–1005
3. Riccitelli NJ, Luptak A (2010) Computational discovery of folded RNA domains in genomes and *in vitro* selected libraries. Methods 52(2):133–140
4. Cai S, Yan J, Xiong H et al (2018) Investigations on the interface of nucleic acid aptamers and binding targets. Analyst 143(22):5317–5338
5. Kalra P, Dhiman A, Cho WC et al (2018) Simple methods and rational design for enhancing aptamer sensitivity and specificity. Front Mol Biosci 5:41
6. Zhou W, Saran R, Liu J (2017) Metal sensing by DNA. Chem Rev 117(12):8272–8325
7. Niu C, Ding Y, Zhang C et al (2022) Comparing two cortisol aptamers for label-free fluorescent and colorimetric biosensors. Sens Diagn 1:541–549
8. Bates PJ, Reyes-Reyes EM, Malik MT et al (2017) G-quadruplex oligonucleotide AS1411 as a cancer-targeting agent: uses and mechanisms. Biochim Biophys Acta 1861:1414–1428
9. Tsai YT, Liang CH, Yu JH et al (2019) A DNA aptamer targeting galectin-1 as a novel immunotherapeutic strategy for lung cancer. Mol Ther Nucleic Acids 18:991–998
10. Kim TH, Lee SW (2021) Aptamers for anti-viral therapeutics and diagnostics. Int J Mol Sci 22(8):4168
11. Shangguan D, Li Y, Tang Z et al (2006) Aptamers evolved from live cells as effective molecular probes for cancer study. Proc Natl Acad Sci U S A 103(32):11838
12. Canoura J, Liu Y, Alkhamis O et al (2023) Aptamer-based fentanyl detection in biological fluids. Anal Chem 95(49):18258–18267
13. Sarafraz M, Nakhjavani M, Shigdar S et al (2022) Modelling of mass transport and distribution of aptamer in blood-brain barrier for tumour therapy and cancer treatment. Eur J Pharm Biopharm 173:121–131
14. Doherty C, Wilbanks B, Khatua S et al (2023) Aptamers in neuro-oncology: an emerging therapeutic modality. Neuro-Oncology 26(1):38–54
15. Mufhandu HT, Gray ES, Madiga MC et al (2012) UCLA1, a synthetic derivative of a gp120 RNA aptamer, inhibits entry of human immunodeficiency virus type 1 subtype C. J Virol 86(9): 4989–4999
16. Brown A, Brill J, Amini R et al (2024) Development of better aptamers: structured library approaches, selection methods, and chemical modifications. Angew Chem 63(16):e202318665
17. Mayer G (2009) The chemical biology of aptamers. Angew Chem 48(15):2672–2689
18. Boomer RM, Lewis SD, Healy JM et al (2005) Conjugation to polyethylene glycol polymer promotes aptamer biodistribution to healthy and inflamed tissues. Oligonucleotides 15(3): 183–195
19. Soldevilla MM, Meraviglia-Crivelli de Caso D, Menon AP et al (2018) Aptamer-iRNAs as therapeutics for cancer treatment. Pharmaceuticals (Basel) 11(4):108
20. Sicco E, Cerecetto H, Calzada V et al (2023) Targeted-lymphoma drug delivery system based on the Sgc8-c aptamer. Cancers 15(3):922

21. Tuerk C, Gold L (1990) Systematic evolution of ligands by exponential enrichment: RNA ligands to bacteriophage T4 DNA polymerase. Science 249(4968):505–510
22. Robertson D, Joyce G (1990) Selection *in vitro* of an RNA enzyme that specifically cleaves single-stranded DNA. Nature 344:467–468
23. Silverman SK (2016) Catalytic DNA: scope, applications, and biochemistry of deoxyribozymes. Trends Biochem Sci 41(7):595–609
24. Lee CH, Han SR, Lee SW (2016) Therapeutic applications of aptamer-based riboswitches. Nucleic Acid Ther 26(1):44–51
25. Anandhakumar C, Kizaki S, Bando T et al (2015) Advancing small-molecule-based chemical biology with next-generation sequencing technologies. Chembiochem 16(1):20–38
26. Komarova N, Barkova D, Kuznetsov A (2020) Implementation of high-throughput sequencing (HTS) in aptamer selection technology. Int J Mol Sci 21(22):8774
27. Khabbazian M, Jabbari H (2022) AI-powered aptamer generation. Nat Comput Sci 2:356–357
28. Sefah K, Shangguan D, Xiong X et al (2010) Development of DNA aptamers using cell-SELEX. Nat Protoc 5:1169–1185
29. Kohlberger M, Gadermaier G (2022) SELEX: critical factors and optimization strategies for successful aptamer selection. Biotechnol Appl Biochem 69(5):1771–1792
30. Wang H, Zhang Y, Yang H et al (2018) *In vivo* SELEX of an inhibitory NSCLC-specific RNA aptamer from PEGylated RNA library. Mol Ther Nucl Acids 10:187–198
31. Berezovski MV, Musheev MU, Drabovich AP et al (2006) Non-SELEX: selection of aptamers without intermediate amplification of candidate oligonucleotides. Nat Protoc 1(3):1359–1369
32. Zumrut HE, Ara MN, Maio GE et al (2016) Ligand-guided selection of aptamers against T-cell receptor-cluster of differentiation 3 (TCR-CD3) expressed on Jurkat.E6 cells. Anal Biochem 512:1–7
33. Shigdar S, Schrand B, Giangrande PH et al (2021) Aptamers: cutting edge of cancer therapies. Mol Ther 29(8):2396–2411
34. Kim LA, D'Amore PA (2012) A brief history of anti-VEGF for the treatment of ocular angiogenesis. Am J Pathol 181(2):376–379
35. Khanani AM, Patel SS, Staurenghi G et al (2023) Efficacy and safety of avacincaptad pegol in patients with geographic atrophy (GATHER2): 12-month results from a randomised, double-masked, phase 3 trial. Lancet 402(10411):1449–1458
36. Ismail SI, Alshaer W (2018) Therapeutic aptamers in discovery, preclinical and clinical stages. Adv Drug Deliv Rev 134:51–64
37. Hernández-Jiménez M, Abad-Santos F, Cotgreave J et al (2023) APRIL: a double-blind, placebo-controlled, randomized, Phase Ib/IIa clinical study of ApTOLL for the treatment of acute ischemic stroke. Front Neurol 14:1127585
38. Steurer M, Montillo M, Scarfò L (2019) Olaptesed pegol (NOX-A12) with bendamustine and rituximab: a phase IIa study in patients with relapsed/refractory chronic lymphocytic leukemia. Haematologica 104(10):2053–2060
39. Rosenberg JE, Bambury RM, Van Allen EM et al (2014) A phase II trial of AS1411 (a novel nucleolin-targeted DNA aptamer) in metastatic renal cell carcinoma. Investig New Drugs 32(1): 178–187
40. Menne J, Eulberg D, Beyer D et al (2017) C-C motif-ligand 2 inhibition with emapticap pegol (NOX-E36) in type 2 diabetic patients with albuminuria. Nephrol Dial Transplant 32(2): 307–315
41. Boyce M, Warrington S, Cortezi B et al (2016) Safety, pharmacokinetics and pharmacodynamics of the anti-hepcidin Spiegelmer lexaptepid pegol in healthy subjects. Br J Pharmacol 173(10):1580–1588
42. Ay C, Kovacevic KD, Kraemmer D et al (2023) The von Willebrand factor-binding aptamer rondaptivon pegol as a treatment for severe and nonsevere hemophilia A. Blood 141(10): 1147–1158
43. Kang C (2023) Avacincaptad pegol: First approval. Drugs 83:1447–1453

44. Fleckenstein M, Mitchell P, Freund KB et al (2018) The progression of geographic atrophy secondary to age-related macular degeneration. Ophthalmology 125:369–390
45. Wong WL, Su X, Li X et al (2014) Global prevalence of age-related macular degeneration and disease burden projection for 2020 and 2040: a systematic review and meta-analysis. Lancet Glob Health 2:e10–e16
46. Kim BJ, Mastellos DC, Li Y et al (2021) Targeting complement components C3 and C5 for the retina: key concepts and lingering questions. Prog Retin Eye Res 83:100936
47. Park YG, Park YS, Kim IB (2021) Complement system and potential therapeutics in age-related macular degeneration. Int J Mol Sci 22:6851
48. Jaffe GJ, Westby K, Csaky KG et al (2021) C5 inhibitor avacincaptad pegol for geographic atrophy due to age-related macular degeneration: a randomized pivotal phase 2/3 trial. Ophthalmology 128:576–586
49. Fernández G, Moraga A, Cuartero M et al (2018) TLR4-binding DNA aptamers show a protective effect against acute stroke in animal models. Mol Ther 26:2047–2059
50. Ramirez-Carracedo R, Tesoro L, Hernandez I et al (2020) Targeting TLR4 with ApTOLL improves heart function in response to coronary ischemia reperfusion in pigs undergoing acute myocardial infarction. Biomol Ther 10(8):1167
51. Hernández-Jiménez M, Martín-Vílchez S, Ochoa D et al (2022) First-in-human phase I clinical trial of a TLR4-binding DNA aptamer, ApTOLL: safety and pharmacokinetics in healthy volunteers. Mol Ther Nucl Acids 28:124–135
52. Hernández-Jiménez M, Abad-Santos F, Cotgreave I et al (2023) Safety and efficacy of ApTOLL in patients with ischemic stroke undergoing endovascular treatment: a phase 1/2 randomized clinical trial. JAMA Neurol 80(8):779–788
53. Klussmann S, Nolte A, Bald R et al (1996) Mirror-image RNA that binds D-adenosine. Nat Biotechnol 4(9):1112–1115
54. Nolte A, Klussmann S, Bald R et al (1996) Mirror-design of L-oligonucleotide ligands binding to L-arginine. Nat Biotechnol 14(9):1116–1119
55. Vater A, Klussmann S (2015) Turning mirror-image oligonucleotides into drugs: the evolution of Spiegelmer(®) therapeutics. Drug Discov Today 20(1):147–155
56. Eulberg D, Klussmann S (2003) Spiegelmers: Biostable aptamers. Chembiochem 4(10): 979–983
57. Hoellenriegel J, Zboralski D, Maasch C et al (2014) The Spiegelmer NOX-A12, a novel CXCL12 inhibitor, interferes with chronic lymphocytic leukemia cell motility and causes chemosensitization. Blood 123(7):1032–1039
58. Zboralski D, Hoehlig K, Eulberg D et al (2017) Increasing tumor-infiltrating T cells through inhibition of CXCL12 with NOX-A12 synergizes with PD-1 blockade. Cancer Immunol Res 5(11):950–956
59. Ludwig H, Weisel K, Petrucci MT et al (2017) Olaptesed pegol, an anti-CXCL12/SDF-1 Spiegelmer, alone and with bortezomib-dexamethasone in relapsed/refractory multiple myeloma: a phase IIa study. Leukemia 31(4):997–1000
60. Bates PJ, Laber DA, Miller DM et al (2009) Discovery and development of the G-rich oligonucleotide AS1411 as a novel treatment for cancer. Exp Mol Pathol 86(3):151–164
61. Bates PJ, Kahlon JB, Thomas SD et al (1999) Antiproliferative activity of G-rich oligonucleotides correlates with protein binding. J Biol Chem 274(37):26369–26377
62. Chung WJ, Heddi B, Schmitt E et al (2015) Structure of a left-handed DNA G-quadruplex. Proc Natl Acad Sci USA 112(9):2729–2733
63. Van den Avont A, Sharma-Walia N (2023) Anti-nucleolin aptamer AS1411: an advancing therapeutic. Front Mol Biosci 10:1217769
64. Kawamoto Y, Wu Y, Park S et al (2024) Multivalent dendritic DNA aptamer molecules for the enhancement of therapeutic effects. Chem Commun 60(49):6256–6259. https://doi.org/10.1039/d4cc00578c
65. Puzzo F, Zhang C, Powell Gray B (2023) Aptamer-programmable adeno-associated viral vectors as a novel platform for cell-specific gene transfer. Mol Ther Nucl Acids 31:383–397

66. Freeman GJ, Long AJ, Iwai Y (2000) Engagement of the PD-1 immunoinhibitory receptor by a novel B7 family member leads to negative regulation of lymphocyte activation. J Exp Med 192(7):1027–1034
67. Sharma P, Allison JP (2015) The future of immune checkpoint therapy. Science 348(6230): 56–61
68. Sharpe AH (2017) Introduction to checkpoint inhibitors and cancer immunotherapy. Immunol Rev 276(1):5–8
69. Wei SC, Duffy CR, Allison JP (2018) Fundamental mechanisms of immune checkpoint blockade therapy. Cancer Discov 8(9):1069–1086
70. He X, Xu C (2020) Immune checkpoint signaling and cancer immunotherapy. Cell Res 30:660–669
71. Huang PW, Chang JW (2019) Immune checkpoint inhibitors win the 2018 Nobel Prize. Biomed J 42(5):299–306
72. Garcia MF, Moreno M, Cerecetto H et al (2023) Aptamer-based immunotheranostic strategies. Cancer Biother Radiopharm 38(4):246–255
73. Thomas BJ, Porciani D, Burke DH (2022) Cancer immunomodulation using bispecific aptamers. Mol Ther Nucl Acids 27:894–915
74. Condeelis J, Weissleder R (2010) *In vivo* imaging in cancer. Cold Spring Harb Perspect Biol 2: a003848
75. Calzada V, Moreno M, Newton J et al (2017) Development of new PTK7-targeting aptamer-fluorescent and -radiolabelled probes for evaluation as molecular imaging agents: lymphoma and melanoma *in vivo* proof of concept. Bioorg Med Chem 25(3):1163–1171
76. Sicco E, Baez J, Ibarra M et al (2020) Sgc8-c aptamer as a potential theranostic agent for hemato-oncological malignancies. Cancer Biother Radiopharm 35(4):262–270
77. Song W, Song Y, Li Q et al (2022) Advances in aptamer-based nuclear imaging. Eur J Nucl Med Mol Imaging 49(8):2544–2559
78. Koudrina A, De Rosa MC (2020) Advances in medical imaging: aptamer- and peptide-targeted MRI and CT contrast agents. ACS Omega 5(36):22691–22701
79. Heo D, Ku M, Kim JH et al (2018) Aptamer-modified magnetic nanosensitizer for *in vivo* MR imaging of HER2-expressing cancer. Nanoscale Res Lett 13(1):288
80. Kim ST, Kim HG, Kim YM (2023) An aptamer-based magnetic resonance imaging contrast agent for detecting oligomeric amyloid-β in the brain of an Alzheimer's disease mouse model. NMR Biomed 36(3):e4862
81. Zu G, Cao Y, Dong J et al (2019) Development of an aptamer-conjugated polyrotaxane-based biodegradable magnetic resonance contrast agent for tumor-targeted imaging. ACS Appl Bio Mater 2(1):406–416
82. Alibolandi M, Hoseini F, Mohammadi M (2018) Curcumin-entrapped MUC-1 aptamer targeted dendrimer-gold hybrid nanostructure as a theranostic system for colon adenocarcinoma. Int J Pharm 549(1–2):67–75
83. Goncin U, Curiel L, Geyer CR et al (2023) Aptamer-functionalized microbubbles targeted to P-selectin for ultrasound molecular imaging of murine bowel inflammation. Mol Imaging Biol 25(2):283–293
84. Shangguan D, Tang Z, Mallikaratchy P et al (2007) Optimization and modifications of aptamers selected from live cancer cell lines. Chembiochem 8(6):603–606
85. Ding D, Zhao H, Wei D et al (2023) The first-in-human whole-body dynamic pharmacokinetics study of aptamer. Research (Washington, DC) 6:0126
86. Arévalo AP, Castelli R, Ibarra M et al (2022) *In vivo* evaluation of Sgc8-c aptamer as a molecular imaging probe for colon cancer in a mouse xenograft model. Int J Mol Sci 23(5):2466
87. Zhao Z, Ukidve A, Kim J et al (2020) Targeting strategies for tissue-specific drug delivery. Cell 181(1):151–167
88. Kruspe S, Mittelberger F, Szameit K et al (2014) Aptamers as drug delivery vehicles. ChemMedChem 9(9):1998–2011

89. Gao F, Yin J, Chen Y et al (2022) Recent advances in aptamer-based targeted drug delivery systems for cancer therapy. Front Bioeng Biotechnol 10:972933
90. Bagalkot V, Farokhzad OC, Langer R et al (2006) An aptamer-doxorubicin physical conjugate as a novel targeted drug-delivery platform. Angew Chem 45(48):8149–8152
91. Lupold SE, Hicke BJ, Lin Y et al (2002) Identification and characterization of nuclease-stabilized RNA molecules that bind human prostate cancer cells via the prostate-specific membrane antigen. Cancer Res 62(14):4029–4033
92. Henri JL, Nakhjavani M, McCoombe S et al (2023) Cytotoxic effects of aptamer-doxorubicin conjugates in an ovarian cancer cell line. Biochimie 204:108–117
93. Pan Q, Nie C, Hu Y et al (2020) Aptamer-functionalized DNA origami for targeted codelivery of antisense oligonucleotides and doxorubicin to enhance therapy in drug-resistant cancer cells. ACS Appl Mater Interfaces 12(1):400–409
94. Sicco E, Almeida L, Moreno M et al (2021) Chemical conjugations of Sgc8-c with the lymphoma drug dasatinib to generate selective biotherapeutics. Aptamers 5:15–21

Top Med Chem
https://doi.org/10.1007/7355_2024_173

Published online: 24 December 2024

Gene Therapy

Pedram Moeini and Gloria González-Aseguinolaza

Contents

Abstract Gene therapy is a relatively new medical modality that involves using genetic material or modifying the genetic material of a cell as a therapeutic tool. This chapter will first describe the characteristics of the therapeutic genetic material and the main vectors used for their delivery to the cell. Following this, it will describe the different types of gene therapy modalities according to the mode of administration and mechanism of action, with special attention to new therapeutic strategies based on gene editing. The chapter will conclude by describing the gene therapy products that have reached the market and the limitations encountered in their global distribution.

P. Moeini and G. González-Aseguinolaza (✉)
DNA & RNA Division, Gene Therapy for Rare Diseases Program, Centro de Investigación Medicina Aplicada (CIMA), Universidad de Navarra (UNAV), Pamplona, Navarra, Spain
e-mail: pmoeinigavg@alumni.unav.es; ggasegui@unav.es

Keywords Gene addition, Gene editing, Gene silencing, Gene therapy, Genetic material, Marketed products, Nonviral and viral vectors

1 Introduction

Currently, therapeutic options for the vast majority of diseases of genetic origin are still very limited or absent. Existing treatments usually need to be administered very frequently and throughout the patient's life, or require very invasive procedures, such as organ transplantation [1]. The latter have important additional limitations, such as the reduced number of compatible or available organs, as well as the need to receive immunosuppressive treatments for life. The development of novel, more effective, and safer treatments for these types of diseases is clearly a medical necessity. Among the new therapeutic options, gene therapy (GT), including gene editing, holds significant promise.

GT is a field of medicine that involves using genetic material as a biotherapeutic agent to treat or prevent diseases. GT aims to correct or replace defective genes in order to treat genetic disorders or to provide new functions to cells (Fig. 1) [2]. This

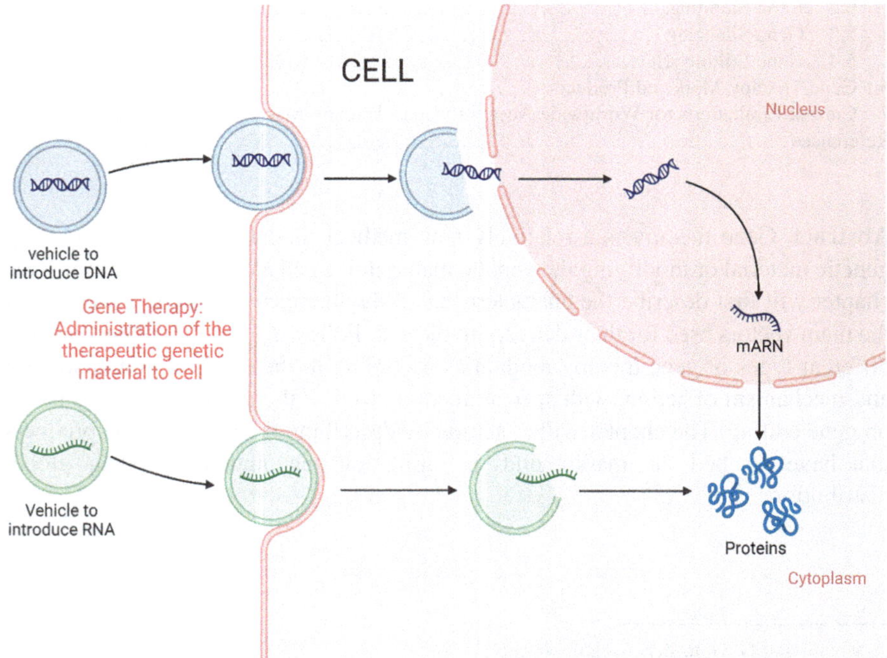

Fig. 1 Representation of the elements involved in the development of a gene therapy treatment: the therapeutic genetic material (DNA or RNA), the cell, and the vehicle that allows the genetic material to enter the cell

chapter will use the definition of GT established by the American Society for Gene and Cell Therapy in 2019 [3]: "GT is the introduction, deletion, or change in the content of a person's genetic code with the aim of treating or curing a disease. The transferred genetic material modifies the levels at which the cell produces a single protein or group of proteins. GT can be used to reduce expression levels of a disease-causing protein, increase the production of disease-fighting proteins, or to produce new or modified proteins." This genetic material includes both DNA and RNA (nucleic acid) molecules, which once inside the cell can perform different functions, always with a therapeutic purpose. GT can be applied to both hereditary diseases and acquired diseases such as cancer, infectious diseases, or neurodegenerative diseases [4]. In fact, most GT clinical trials are indicated for cancer treatment. However, the application of GT to hereditary or genetic diseases is the most intuitive, especially for the treatment of diseases caused by the involvement of a single gene or monogenic diseases. In addition, GT, like organ transplantation, can cure the disease. Among the monogenic diseases that can be treated with GT are many of the so-called rare or minority diseases, which affect a very small number of patients worldwide and, in most cases, lack adequate treatment and a cure [5].

The first GT clinical trial was conducted in the 1990s for the treatment of Adenosine Deaminase (ADA) deficiency, a condition characterized by severe immunodeficiency due to the absence and malfunction of T, B, and NK cells. In this groundbreaking trial, T lymphocytes were extracted from a patient, genetically modified to contain the correct ADA gene, and then reintroduced into the patient. Although the benefits were temporary, this pioneering trial marked a major milestone in the development of GT, demonstrating the feasibility of delivering genetically modified cells to patients and paving the way for further advancements in the field [6].

Following this initial trial, a multitude of clinical trials have been conducted for the treatment of various hereditary diseases, yielding diverse outcomes. These trials have been instrumental in advancing the field of GT by providing valuable insights into its efficacy, safety, and feasibility. The varied outcomes of these trials have contributed to the ongoing refinement and development of GT techniques, paving the way for future innovations in the field. In some of the first GT clinical trials, conducted after very promising results in animal models, a number of adverse effects occurred that reduced expectations about the application of GT. These include the death of a patient with ornithine transcarbamylase (OTC) deficiency [7] and the development of leukemia in patients with severe immunodeficiencies who were treated with bone marrow stem cells transformed with retroviruses carrying the therapeutic gene [8–10]. In the first case, the patient experienced an exaggerated immune reaction to the gene therapy product, which involved a first-generation adenovirus carrying the *OTC* gene. In the second case, the integration of the therapeutic retrovirus genome led to the activation of an oncogene (insertional mutagenesis) [8–10]. Despite these challenges, it is noteworthy that a large number of patients with severe immunodeficiencies who underwent treatment with retroviruses continue to derive benefits from this approach [11, 12]. This underscores the complexity and variability of responses to GT interventions and highlights the need for further research and advancements in optimizing the safety and efficacy of GT for diverse patient populations [11, 12]. Research in GT thereafter focused on developing safer and more effective strategies. Thus, in the last decade, GT has had remarkable successes such as the use of CAR T cells for cancer treatment [13–15], and the arrival of the first GT products to the market for the treatment of hereditary diseases that will be described at the end of this chapter [16–21]. It is expected that the number of GT products reaching to market will grow exponentially in the coming years.

This chapter will cover the essential concepts for understanding how GT works, including the characteristics of the therapeutic genetic material, the vehicles for its administration, and the various modalities. It will focus mainly on GT applied to the treatment of genetic disorders and it will put special emphasis attention in the most recent GT tools based on gene editing. The chapter will conclude by describing the GT products that have reached the market and the current limitations for worldwide access.

2 Therapeutic Genetic Material

The treatment of a genetic deficiency requires, on the one hand, a highly efficient transfer of the genetic material to the cell and, on the other hand, a sustained expression of the therapeutic gene, ideally, throughout the patient's life. To this end, it is very important to carefully select all the genetic elements necessary for the appropriate expression of therapeutic product [22, 23]. First of all, it is critical for the appropriate selection of the DNA sequence that will regulate expression, the promoter region (Fig. 2a). This sequence can have very different features and will be selected based on the characteristics of the indication to be treated. The promoter can be either ubiquitously expressed (in any type of cell in the body), or specific to a certain tissue or cell type [22]. A ubiquitous expression promoter is chosen when the disease affects various organs/tissues, while a cell-specific promoter is preferred when the disease impacts a particular cell type, and expression in other cells may be harmful. Promoters can also be constitutively expressed, providing stable protein expression, or inducible, responding to specific stimuli, either endogenous or exogenous. Expression can be either induced or repressed when needed (Fig. 2b, c). Inducible expression is utilized when precise control over protein levels is necessary, either to avoid toxicity or for optimal functionality [24–27].

In addition to the promoter region, other elements can be included to modulate the expression of the therapeutic gene such as expression enhancer sequences, intronic sequences that provide stability to the messenger, or sequences that allow the inhibition of expression in specific cell types such as target sequences for microRNAs (repressor) [27] (Fig. 2a). The inclusion of microRNA target sequences prevents the expression of the therapeutic protein in antigen-presenting cells, thereby reducing the risk of developing immune responses to the therapeutic product [27].

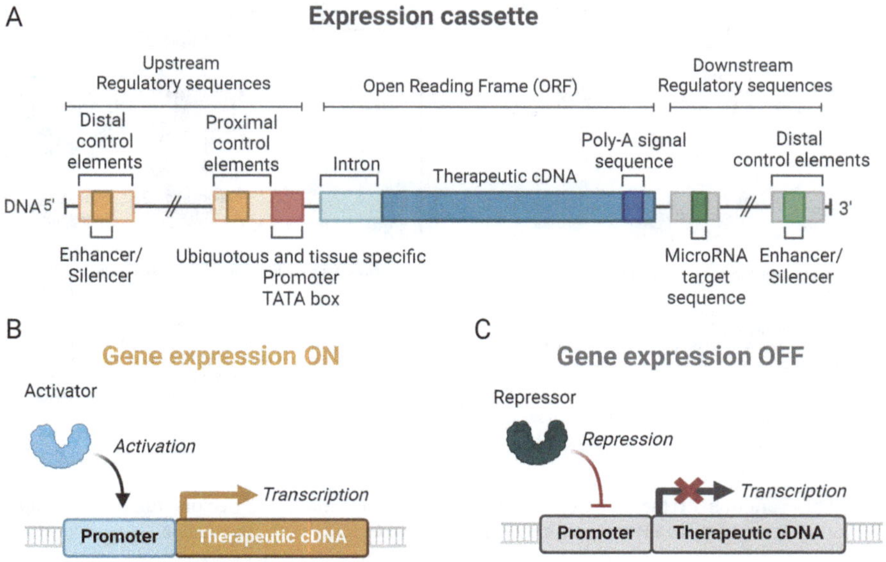

Fig. 2 Genetic elements that must be taken into account when correctly designing a gene therapy vector. (**a**) Upstream regulatory elements comprising distal and proximal transcriptional control elements containing the promoter sequence with its TATA box to indicate the initiation of transcription, and the sequence that modulates its activity as stimulatory sequences (enhancer) or repressors (silencer). The open reading frame contains the coding sequence (therapeutic cDNA) in which introns can be introduced to provide stability to the messenger RNA and the poly-A signal sequence. Finally, downstream regulatory sequences can be added to modulate the expression. (**b**, **c**) Inducible expression systems. (**b**) The therapeutic protein is induced by the administration of a specific molecule that will act over the promoter activity (could be a protein or a drug) and levels of expression can be modulated. (**c**) Protein expression is active and can be repressed by the administration of a specific molecule

As it will be seen throughout the chapter, with the evolution of GT, the use of new genetic materials has been incorporated, such as small RNA molecules to silence the expression of certain proteins or the use of messenger RNA molecules.

3 Gene Transfer Vectors

The key step for the proper functioning of GT is to introduce the genetic material inside the target cell, and in most cases to reach its nucleus where the necessary machinery for the biotherapeutic, nucleic acid, to be active is located. To enter the cell and its nucleus, it is necessary to cross a membrane that surrounds the cell and another that surrounds the nucleus (Fig. 1). Both membranes are composed of lipids and proteins and have a highly hydrophobic character. Nucleic acids are negatively charged molecules that make it difficult for them to pass through the membranes. Although it is true that the administration of "naked" DNA or RNA can cross these

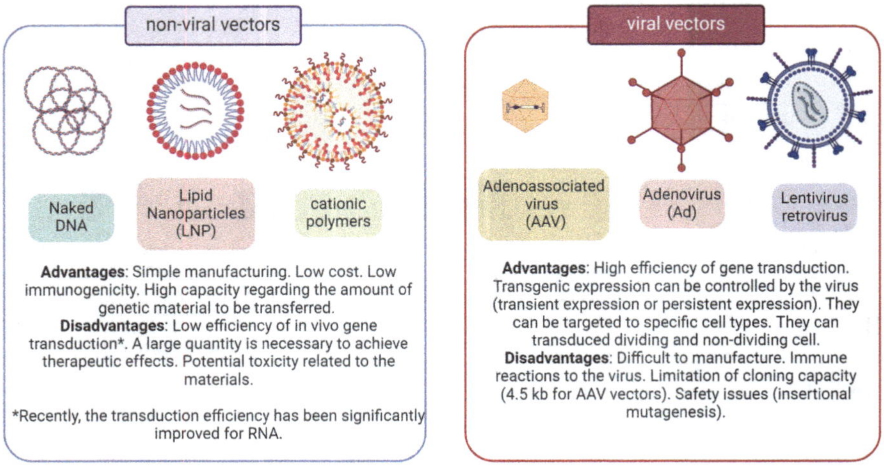

Fig. 3 Description of the most commonly used nonviral and viral vectors for gene therapy, indicating some of their advantages and disadvantages. Created with BioRender

barriers and exert its effect, its efficacy is very low. Therefore, it is necessary to use vehicles that both protect the genetic material from degradation and serve as transporters to deliver the genetic material into the cell [2, 28].

There are two main families of vehicles or vectors depending on their origin, viral and nonviral vectors (Fig. 3). Viral vectors take advantage of the natural ability of viruses to introduce their genetic material into the cell and produce their own proteins hijacking the cellular machinery. Nonviral vectors attempt to mimic the characteristics of viral vectors in the absence of the potential pathogenic effects of the latter.

3.1 Nonviral Gene Transfer Vectors

Nonviral vectors are mostly composed of polymeric or lipid particles that package the genetic material inside and protect it as it passes into the cell [28–30]. The use of nonviral vectors has several advantages over the use of viral vectors, such as their ease of production, unlimited size of the genetic material to be transferred, and theoretically a better safety profile. The limitations of nonviral vectors include the stability of these particles, their cellular uptake, and, so far, a limited ability to achieve durable transgenic expression [25].

3.1.1 Administration of Naked Genetic Material

The simplest form of nonviral vector is the administration of naked DNA or RNA, but as mentioned, its efficiency in entering the cell is very low. This low efficacy can be corrected by using different methods. One of the simplest methods used so far is hydrodynamic injection, which is capable of transferring genetic material with great efficiency to the liver or muscle [31, 32]. This technique involves the injection of a high volume of a solution containing DNA or RNA at a relatively high speed. The rapid injection of a large volume of fluid induces an increase in pressure in the organ and facilitates the entry of genetic material into the cell. This technique is most often used in mice for proof-of-concept studies and has also been tested in large animal models such as pigs, sheep, or monkeys [31]. Another technique for introducing naked genetic material into the cell is the use of ultrasound either alone or in combination with microbubbles [33, 34]. However, one of the most widely used techniques is electroporation which increases the permeability of the cellular plasma membrane by applying an oscillating electric field to the surface of the cells, allowing the entry of genetic material [35].

3.1.2 Lipidic or Polymeric Nanoparticles

Liposomes[1] are very similar in composition to cell membranes. They are made up of lipids with amphiphilic properties (having a hydrophilic part and a hydrophobic part) that when dispersed in an aqueous medium spontaneously form spherical structures. Their size can range from 20 nm to several tens of μm. The interior of liposomes is hydrophilic and can encapsulate many different compounds, including nucleic acids, both DNA and RNA [36]. The use of liposomes in GT has experienced an important resurgence with the use of these vehicles for the administration of interfering RNAs and messenger RNAs. The use of RNA makes it possible to obtain a therapeutic effect without the need to introduce the genetic material into the nucleus, since the messenger RNA (mRNA) gives rise to the therapeutic protein in the cell cytoplasm. In recent years and thanks to improvements in the chemistry of the synthesis of more stable and large quantities, several studies have been carried out using mRNA contained in liposomes for the treatment of hereditary diseases with very promising results; and the first clinical trials [37].

Cationic polymers form nanoparticles with nucleic acids through electrostatic interactions that allow nucleic acids to be transferred into the cell. One of the first cationic polymers explored was polylysine (PLL). However, if the genetic material entered, it remained retained in the PLL-nanoparticle, not being able to reach the cytoplasm or nucleus of the cell. In the early 1990s, several cationic polymers, lipopolyamines, and polyamidoamines such as polyethylenimine (PEI), were found to be capable of mediating high transfection efficiencies [36]. PEI is a synthetic cationic polymer with high amine density and high buffering capacity. The entry of PEI particles into the endosome causes a large increase in ionic concentration within the endosome resulting in osmotic swelling of the endosome and its rupture, termed the proton sponge effect. The use of PEI nanoparticles makes it possible to protect

[1] See Chapter "Encapsulation and Structural Modifications of Biotherapeutics".

DNA from lysosomal nucleases and the arrival of genetic material to the nucleus of the cell [37].

Both liposomes and cationic polymers can be modified on their surface by binding ligands that allow their targeting to specific cell types. These modifications have facilitated the delivery of genetic material to organs like the liver or heart [38].

3.2 *Viral Gene Transfer Vectors*

To generate a viral GT vector, typically, the genes necessary for replication or involved in their pathogenicity are removed from the viral genome and replaced with the desired genetic material. However, in some cases, maintaining the virus's replicative capacity is beneficial, such as with oncolytic viruses that are designed to target and kill tumor cells [2, 25].

Viral vectors commonly used in GT for inherited diseases include those based on adeno-associated viruses, adenoviruses, retroviruses, and lentiviruses (Fig. 3). Each type of vector offers unique characteristics and advantages suited to different therapeutic goals and target diseases.

3.2.1 Adeno-Associated Virus (AAV)-Based Vectors

AAVs belong to the Parvoviridae family and the genus Dependovirus [39]. Their genome is a 4.7 kb single-stranded DNA containing two genes: the rep gene, which encodes proteins involved in replication, and the cap gene, encoding the proteins for virus capsid formation. At the genome ends are fork-shaped inverted terminal repeats (ITRs), containing replication initiation and termination sequences, and the packaging signal (Fig. 4a).

AAVs are naturally deficient in replication and rely on coinfection with a helper virus like adenovirus to complete their life cycle. This dependency on a helper virus is reflected in their name "adenoassociated" and their classification within the dependovirus genus [39]. In the absence of coinfection with the helper virus, the AAV genome is integrated into chromosome 19 of the cell, at the so-called AAV locus, where it remains latent. Another important characteristic of AAVs is that no pathology has been associated with their infection [40].

For the production of GT vectors based on AAVs, the rep and cap genes are removed from the virus genome, and are replaced by the therapeutic genetic material, and only the ITR sequences are maintained (Fig. 4a) [39, 40]. The absence of rep in the recombinant genome causes the virus to lose its ability to integrate into host DNA, and once inside the nucleus, the genome of the virus remains mostly episomal, i.e., not integrated, reducing the risk of insertional mutagenesis. In addition to these characteristics, most AAVs hardly induce inflammatory responses, which allow the infection to go unnoticed and the cell not to be eliminated. This helps maintain the expression of the therapeutic material for a long time [40–42]. For

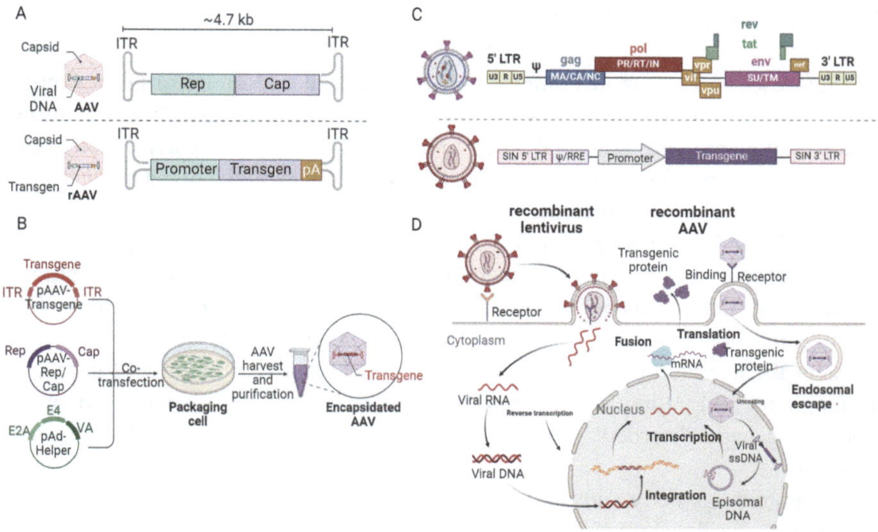

Fig. 4 (**a**) Graphical representation of the structure of the genome of wild type (WT) and recombinant AAV genomes. The WT genome contains two genes: the rep gene, which codes for proteins involved in AAV replication, and the cap gene, which produces the capsid proteins. At both ends of the genome are the inverted terminal repeats (ITRs), which contain the packaging signal and the initiation and termination signals for replication. In the recombinant vector, these two genes are removed and replaced by the therapeutic expression cassette flanked by the ITRs. (**b**) Graphical representation of the recombinant AAV production process, in which the rep and cap genes and the elements necessary from the helper virus are provided in trans to the recombinant genomes in the packaging cells. (**c**) Graphical representation of the structure of the genome of WT and recombinant lentiviral genomes, where only the long terminal repeats (LTRs) are required for the production of the recombinant lentivirus. A 400-nucleotide deletion in the LTR has been introduced to eliminate its promoter activity (SIN LTR) and increase safety. (**d**) Graphical representation of the viral cycle of recombinant lentivirus and AAVs. Created with BioRender

all these reasons, the AAVs used as GT vectors show a priori a very high safety profile.

For the production of recombinant AAVs, it is necessary to provide the *rep* and *cap* genes and the adenovirus genes to the packing cell. This makes it possible to use different *cap* genes in their production, giving rise to capsids with different properties in terms of their seroreactivity and their ability to infect different cell types which represents an additional very attractive feature (Fig. 4b) [42].

The primary limitations of AAV vectors for broad application are their limited cloning capacity (< 4.5–4.7 kb) and the presence of preexisting humoral immunity to AAV. A significant portion of the population has been naturally exposed to AAV, leading to the development of antibodies that can neutralize AAV infections. This preexisting immunity poses a challenge to the effectiveness of AAV-based GT in a considerable number of individuals [43, 44].

Currently, AAVs are the most widely used GT vectors in the in vivo treatment of hereditary disorders.

3.2.2 Adenoviral (Ad) Vectors

Adenoviruses (Ad) are more complex than AAVs, with a 36 kb linear DNA genome containing several structural and nonstructural genes. Similar to AAVs, the adenoviral genome remains episomal after infecting the cell. The early recombinant Ad-based vectors lacked genes involved in viral replication but retained a large portion of the genome, allowing the introduction of exogenous genetic material between 5 and 7 kb in size [45]. This manipulation prevented viral replication but maintained the expression of structural proteins, which triggered a significant inflammatory and immune response as previously mentioned [7]. This robust immune reaction resulted in the rapid elimination of infected cells, leading to a transient therapeutic effect. While these vectors have proven effective for vaccine development and antitumor treatments, they are less suitable for the treatment of inherited genetic disorders, which typically require a more sustained therapeutic effect [46].

The immunogenicity and limited cargo capacity of early-generation Ad vectors have led to the development of newer, more advanced adenoviral vector designs aimed at improving their safety and efficacy profile for gene therapy applications. Recently, Ad vectors have been engineered with all viral genes removed, retaining only the essential sequences for packaging. These innovative adenoviruses, known as High-Capacity Adenoviruses (HC-Ad), demonstrate highly efficient modification of specific cell types, such as hepatocytes. They enable long-term stable expression of recombinant proteins and offer a substantial cloning capacity of up to 36 kb [47].

However, a current limitation in the widespread use of HC-Ad vectors is the challenge of large-scale production at clinical grade. Addressing this production obstacle is crucial for advancing the clinical application of these advanced adenoviral vectors in GT [47].

3.2.3 Retroviral (RV) and Lentiviral (LV) Vectors

Due to the lack of active integration into the host cell genome, both AAV and HC-Ad vectors become diluted after cell division, leading to a loss of therapeutic effect. This limitation currently restricts their application in pediatric patients. In contrast, RVs and LVs integrate into the genome of the target cell, enabling the transfer of genetic information from cell to cell during division.

RVs and LVs are enveloped viruses with a single-stranded RNA genome. The viral RNA encodes three key proteins: gal, pol, and env. In vector production, these proteins are replaced with therapeutic genetic material, typically with a capacity of 7–8 kb (Fig. 4c) [48, 49]. They possess their own reverse transcriptase, which facilitates the conversion of their RNA into a double-stranded DNA known as provirus. This provirus then integrates into the host cell's genome.

Initially, clinical trials utilized retroviruses, specifically the moloney leukemia virus derived from mice, before transitioning to lentiviruses derived from the human immunodeficiency virus. Through genetic engineering efforts with these viruses, safer vectors have been developed that avoid integration into the host genome, preventing alterations in cellular protein expression previously reported with RV [11, 12]. Additionally, these viruses, which naturally target specific cell types, can be modified to alter the envelope proteins, creating vectors with diverse tropisms [50].

These vectors are generally used for ex vivo gene therapy strategies, for this reason these vectors were initially used for the correction of hematological disorders. However, although still in preclinical phases, groups are working on the systemic administration of direct injection of LV to achieve the permanent transformation of organs such as the brain or liver [50, 51]. For example, systemic administration of an LV to adult mice and dogs resulted in stable expression of factor IX in the liver [52]. However, mild acute toxicity and low efficacy were observed in dogs as a result of LV uptake by macrophages. Strategies to prevent uptake by these cells are currently being investigated in order to improve hepatocyte transduction and reduce systemic toxicity. For instance, the incorporation of the human phagocytosis inhibitor CD47 into lentiviral particles resulted in a significant increase in transgenic expression and a reduction in toxicity in monkeys [53]. Furthermore, pseudotyping with the baculovirus GP-64 protein has been shown to allow liver endothelial cell targeting resulting in hemophilia A correction in mice [54].

4 Gene Therapy Modalities by Mode of Administration

GT can be divided into two main modalities based on the form of administration: ex vivo and in vivo (Fig. 5).

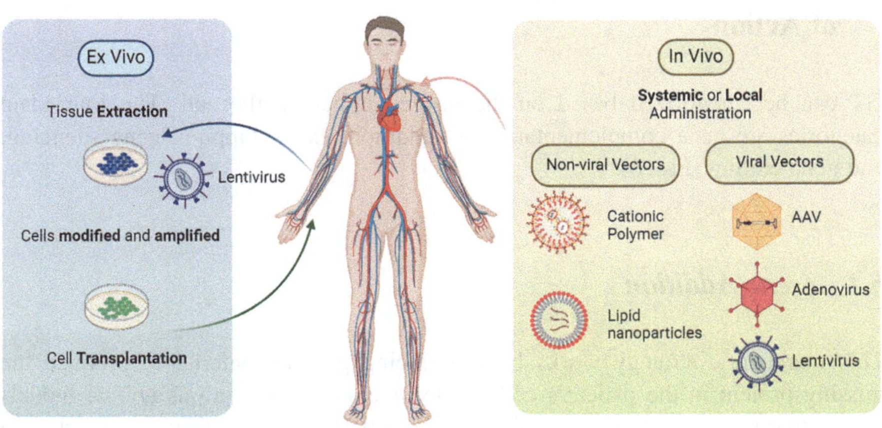

Fig. 5 Graphical representation of the GT modalities according to the mode of administration and the gene transfer vehicles more commonly used. Created with BioRender

4.1 Ex Vivo GT

In ex vivo GT, cells are extracted from the patient, cultured and amplified, and then modified using a vector of choice. For this therapy to have a long-lasting effect, the modified cells must be able to divide once administered to the patient without losing the therapeutic genetic material. Typically, hematopoietic stem cells (CD34+) are used, as their genome can be permanently altered through the use of gene transfer vehicles that integrate the therapeutic cargo into the cell's genome. Ex vivo GT is primarily used to treat metabolic diseases of hematopoietic origin, such as genetic deficiencies causing severe immunodeficiencies or various types of anemia [2, 8–12, 48–50]. The most commonly used vectors for ex vivo GT have been integrative viruses like retrovirus or lentiviruses (Fig. 4d). However, with the development of highly efficient and increasingly secure gene editing tools, many groups are now exploring these approaches.

4.2 In Vivo GT

In in vivo GT, the therapeutic agent is directly administered to the patient using various routes of administration, which can be systemic or local. A range of vector types are used, but the ones demonstrating the most promising results are those based on AAVs [2, 5, 42, 43]. As previously indicated, AAVs deliver the recombinant genome inside the cell that remained episomal (not integrated) (Fig. 4d).

5 Gene Therapy Modalities According to Mechanism of Action

GT can be categorized based on the desired effect on the cell. The four main categories are gene complementation or, addition, exon skipping, gene silencing, and gene editing (Fig. 6).

5.1 Gene Addition

This therapeutic strategy works by introducing genetic material to produce the missing protein in the patient's cells. Extensive studies in various animal models have demonstrated the efficacy of this approach in targeting different organs and many different indications [5, 42, 43, 48, 49]. Gene addition can be achieved ex vivo by modifying the patient's cells using lentiviruses and then readministering them to the patients, or in vivo by administering viral or nonviral vectors.

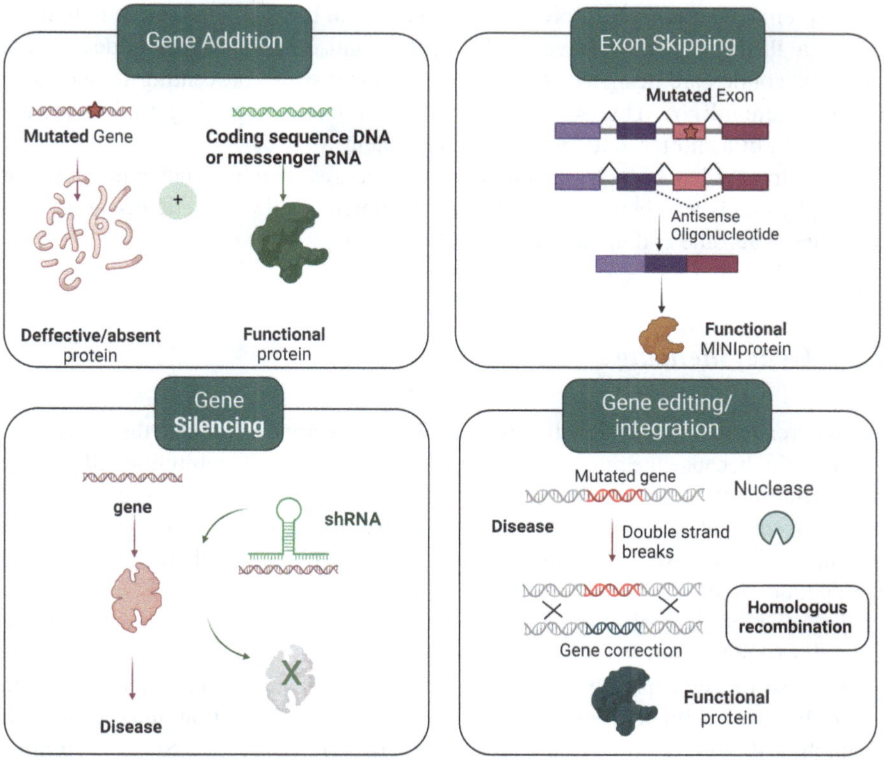

Fig. 6 Schematic representation of TG modalities according to mechanism of action. Created with BioRender

For this strategy to work effectively, meticulous selection of all regulatory elements controlling the expression of the therapeutic gene is crucial, with particular focus on the promoter region (Fig. 6). This is because the promoter plays a pivotal role in determining both the expression levels of the protein and the specific cell type in which it is expressed, as previously indicated (Fig. 2) [22].

Traditionally, gene addition strategies involved the use of nonviral vectors carrying DNA or viral vectors that are capable of expressing the therapeutic protein. However, recent advancements in stabilizing RNA molecules have paved the way for the initiation of clinical trials administering messenger RNA, marking a significant evolution in gene therapy approaches [55, 56].

5.2 *Exon Skipping*

In some hereditary diseases, the problem lies in mutations in one of the exons (coding regions) of the gene, resulting in the introduction of premature stop codons.

These premature stop codons prevent the synthesis of the functional protein. In these cases, a therapeutic alternative is the use of antisense oligonucleotides. These oligonucleotides are designed to bind to the mutant exon, preventing its inclusion in the mature mRNA. This process, known as "exon skipping" (Fig. 6), leads to the production of a shorter, but still functional, protein.

This strategy has been successfully used in the treatment of Duchenne muscular dystrophy and is also being studied for the treatment of other genetic diseases, such as Usher syndrome and epidermolysis bullosa [57].

5.3 Gene Silencing

For the treatment of some genetic diseases, it is necessary to eliminate the expression of a protein, because the mutated protein is toxic or because its inhibition reduces the production of the substrate that must be metabolized by the missing enzyme and whose accumulation is noxious. This is the case, for example, in some autosomal dominant diseases or in diseases where substrate reduction therapy achieves a therapeutic effect.

To achieve this goal, different strategies can be employed: Antisense Oligonucleotides (ASOs)[2] and RNA Interference (RNAi) [58–61].

The use of antisense oligonucleotides (ASOs) that, by complementarity, form heteroduplexes with a target mRNA, preventing its translation into protein or promoting its destruction by ribonuclease H. The initial use of ASOs was limited by the short half-life of the oligonucleotide, but advances in introducing modified nucleotides that protect them from degradation and significantly extend their plasma half-life have allowed several of these molecules to reach the patient and become therapeutic products [58, 59].

RNA interference (RNAi) is a highly evolutionarily conserved mechanism whereby double-stranded RNA (dsRNA) molecules silence the posttranscriptional expression of specific genes [60, 61]. This phenomenon was first discovered in plants and in the *Caenorhabditis elegans* nematode. In 2001, the existence of similar processes in mammalian cells was demonstrated, leading to the emergence of new tools to study the function of different genes and identify their potential use in therapeutic applications. Two main types of RNAi molecules are used: microRNAs (miRNAs) and small interfering RNAs (siRNAs). miRNAs are small, noncoding dsRNAs that are naturally transcribed by the genome. miRNAs are synthesized from a precursor with a hairpin structure (pre-miRNA) and are processed in the nucleus and cytoplasm to form a silencing complex that binds to and blocks the translation of mRNA [59]. On the other hand, siRNAs are dsRNA molecules, 21 to 23 base pairs (bp) in length, specifically designed to silence target gene expression. They can be introduced exogenously into the cell or organism in their short form (21–23 bp) or as

[2] See Chapter "Antisense Oligonucleotides (ASOs)".

longer dsRNA molecules, which are then processed by the cellular machinery to form gene-silencing complexes. The difference between miRNAs and siRNAs is that miRNAs have partial complementarity with the target sequence, while siRNAs require almost perfect complementarity [59].

ASOs, microRNAs, or siRNAs can be administered directly using nonviral vectors or expressed inside the cell using viral vectors. Therapies based on the use of siRNAs have reached the market for the treatment of several indications with excellent results [60, 61].

5.4 Gene Editing

The field of GT has recently been revolutionized with the entry of tools that enable the editing of the cell's genome. This strategy is based on introducing the correct gene into the cell's genome, correcting the mutation that causes the disease, or silencing the expression of a specific gene [62]. This method uses "tailor-made" nucleases capable of editing DNA in very specific sequences, and there are four different types of nucleases to edit the cell's DNA: Zn-finger nucleases, transcription activator-like effector nuclease (TALEN) nucleases, and bacterial complex-based nucleases consisting of Clustered Regularly Interspaced Short Palindromic Repeats (CRISPR) and an associated nuclease, Cas9 (Fig. 7). These complexes can bind to a specific region in the DNA, and once bound, the nuclease element in each system creates a double-strand break (DSB) in the DNA at the target location. This DSB triggers the cell's natural DNA repair mechanisms, which can lead to either nonhomologous end joining (NHEJ) or homology-directed repair (HDR). NHEJ often results in small insertions or deletions (indels) at the site of the break, while HDR can be used to introduce specific genetic changes or edits (Fig. 7). By harnessing this mechanism, these systems enable precise and targeted modifications to the genome [62–64].

5.4.1 Zinc-Finger Nucleases (ZFNs)

The first successful platform used for genome editing in human somatic cells was ZFNs, which are artificial proteins consisting of zinc finger DNA-binding domains fused to nuclease domains derived from FokI (endonuclease originally derived from *Flavobacterium okeanokoites*). Each ZFN domain is designed to recognize a specific trinucleotide sequence [65]. Once adjacent DNA sites are bound by the two ZFN domains, the FokI nuclease domains dimerize and cleave the DNA at a precise location within the target sequence (Fig. 7). This dimerization requirement improves the accuracy while decreasing off-target effects [65, 66]. Despite the specificity and efficiency, their complex structure makes the design process time-consuming (~3 months) and the delivery challenging. Furthermore, due to their size, this approach requires a complex delivery system like the use of three distinct AAV

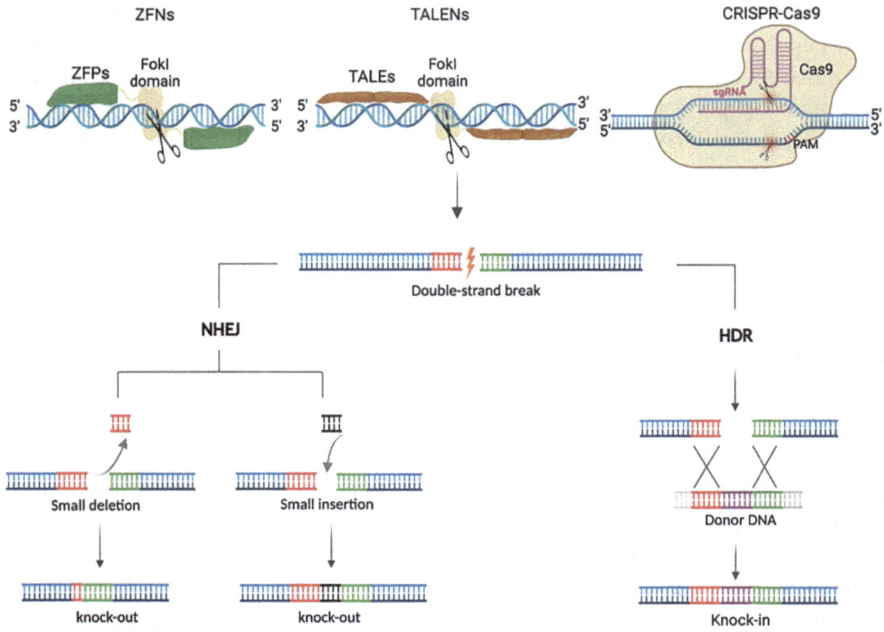

Fig. 7 Platforms for targeting specific DNA sequences: ZFNs, zinc-finger nucleases; TALENs, transcription activator-like effector nucleases; CRISPR, clustered regularly interspaced short palindromic repeats-associated nuclease systems. These nucleases can introduce double-strand breaks (DSBs) at the target locus, which can be repaired via homologous-directed repair (HDR) if a donor template is provided, or through nonhomologous end joining (NHEJ), leading to indel formation and potential gene silencing. This figure illustrates the various platforms available for targeted DNA sequence modification and the mechanisms by which these nucleases introduce genetic changes, highlighting their potential applications in gene editing and therapeutic interventions. Created with BioRender

vectors, two vectors carrying the ZFNs and a third one carrying the DNA template. This limits the efficacy of the treatment since the three vectors have to transduce the same cell; additionally, it complicates the production and regulatory authorization [64].

5.4.2 Transcription Activator-like Effector Nucleases (TALENs)

TALENs, are DNA-binding proteins first identified in the plant pathogen *Xanthomonas*. Each TALE repeat contains a highly conserved 33–34 amino acid sequence. Within each repeat, two hypervariable residues (also known as RVDs for repeat variable residues) at positions 12 and 13 confer DNA specificity. TALENs, like ZFNs, are fused to the catalytic domain of the FokI endonuclease to produce TALENs [66–68]. They also exhibit high specificity in targeting DNA sequences, both generating 5′ overhangs. They are easier to design than ZFNs, primarily due to their modular structure and the direct correspondence between TALE repeats and

DNA base pairs, simplifying the design process (~2 weeks). However, one potential limitation of TALENs is their larger size, which can pose important challenges for in vivo delivery.

5.4.3 Clustered Regularly Interspaced Short Palindromic Repeats (CRISPR)-Associated Nuclease (CRISPR-Cas)

CRISPR-Cas, originally discovered as a bacterial defense mechanism against viruses, has revolutionized genetic engineering and therapy [69]. Its adaptation for use in human cells is achieved through harnessing the Cas9 enzyme's ability to precisely target and edit DNA. The simplicity of CRISPR-Cas9 design has played a significant role in its rapid emergence as a widely used genome editing tool, overcoming the main limitation of previous nuclease-based platforms and opening up vast possibilities in biomedicine [70, 71]. From correcting disease-causing mutations to modulating gene expression, CRISPR-Cas technologies offer promising therapeutic applications for a wide range of genetic disorders [71].

CRISPR-Cas involves a two-component system: the guide RNA (gRNA) and the nuclease Cas protein. The gRNA is designed to be complementary to a specific target sequence within the genome sequence of interest. When the gRNA binds to its complementary sequence, the Cas protein is recruited to the target site and creates DSBs [66, 70, 71].

Over the years, numerous Cas enzymes have been discovered and characterized. There are over 20 known types of natural Cas proteins, each with its own unique characteristics and properties. In the last years, many research groups have been working on the development of in silico approaches to engineer Cas enzymes with tailored properties, enabling the development of highly efficient and specific genome editing tools for diverse applications [72].

The most frequently used Cas enzymes in CRISPR-based genome editing applications include Cas9 from *Staphylococcus pyogenes* (SpCas9) or from *Staphylococcus aureus* (SaCas9) characterized by their efficiency, versatility, and ease of use; Cas12a (Cpf1) from *Acidaminococcus* sp. BV3L6 and *Lachnospiraceae* bacterium ND2006 are commonly used alternatives to Cas9; Cas13 enzymes, such as Cas13a (C2c2) and Cas13b, are used primarily for RNA-targeting applications, including RNA editing and RNA interference, Cas12b from *Francisella novicida* and *Bacillus hisashii*, have been explored for their potential in genome editing and other applications [72]. These Cas enzymes offer diverse capabilities and can be tailored to specific experimental needs, making them valuable tools in molecular biology and biotechnology research [73, 74].

Furthermore, Cas9 enzymes can be genetically modified to alter their properties. Cas9 is a multifunctional protein with several domains [73, 75]. The recognition (REC) domain is responsible for binding to the guide RNA (gRNA) and forming the Cas9-gRNA complex it interacts with both the crRNA (CRISPR RNA) and the tracrRNA (trans-activating CRISPR RNA) components of the gRNA. The nuclease (NUC) domain which is responsible for cleaving the target DNA strand complementary to the crRNA consists of two subdomains: the RuvC-like domain and the

HNH domain. A bridge helix connects the REC and NUC domains and plays a role in transmitting conformational changes upon RNA binding to the catalytic site, thus activating the nuclease activity. The Protospacer Adjacent Motif (PAM)-Interacting Domain interacts with the PAM sequence located adjacent to the target DNA sequence and is necessary for Cas9 recognition and cleavage of the target DNA. The introduction of inactivating mutations in one of the NUC domains results in a Cas9 protein with nickase (nCas9) activity, meaning it only cuts one of the DNA strands [75, 76]. Inactivation of both NUC domains leads to the generation of "dead" Cas9 (dCas9), which can bind to the target sequence but cannot modify it [77]. The fusion of nCas9 or dCas9 with various enzymes has resulted in the development of highly attractive tools for genome modification, like Base editors and prime editors [77–82] (Fig. 8).

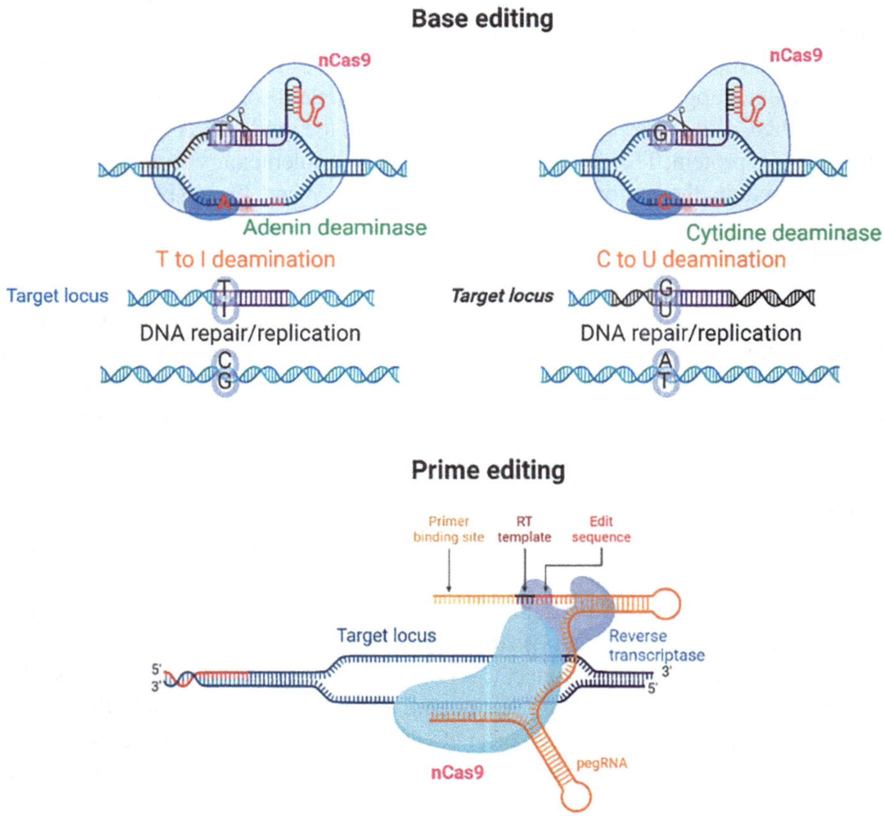

Fig. 8 Adaptability of CRISPR/Cas9: alternative applications. The CRISPR/Cas9 system has shown remarkable adaptability, leading to the development of innovative applications beyond gene editing. These include base editing: this approach utilizes adenine or cytidine deaminases to introduce precise base changes without requiring double-strand breaks (DSBs) or donor templates. Prime editing: this technique combines the benefits of HDR and base editing, allowing for precise insertions, deletions, and all possible base-to-base conversions without DSBs or donor templates. Created with BioRender

In base editing, nCas9 is fused with a single-stranded DNA-specific cytidine deaminase or adenine deaminase, enabling the catalytic modification of the host DNA sequence. This modification converts C•G base pairs to T•A base pairs (resulting in cytosine base editors, CBEs) or A•T base pairs to G•C base pairs (resulting in adenine base editors, ABEs) [80]. The key advantage of base editors is their potential for highly efficient gene correction compared to HDR strategies. However, meticulous characterization of the guide RNA is crucial to prevent off-target base editing within the intended region. In the last few years, many different variants of base editors enzymes have been developed to improve their efficacy and fidelity such as BE4, BEmax, ABEmax, transformed BE (tBE), etc. [83–86]. Due to the large size of base editors, delivery methods such as adenovirus, dual AAV systems, hydrodynamic tail vein injection of naked DNA, and LNPs have been utilized in initial mouse studies [83–86].

Prime editors comprise an engineered reverse transcriptase fused to Cas9 nickase, which induces a nick at the target DNA site. The prime editing guide RNA (pegRNA) contains a 3′ extension that binds to the nicked target DNA strand, forming a primer-template complex for the reverse transcriptase. This enzyme then synthesizes the desired sequence onto the nicked DNA based on a template encoded in the pegRNA. Subsequent cellular DNA repair processes integrate the edited DNA permanently into the genome. Prime editing offers versatility and has shown high efficiency in cellular systems [87, 88].

While prime editing has successfully corrected disease-causing mutations in vitro and ex vivo, its large size presents challenges for in vivo applications. Various delivery systems have been explored in mouse models, including dual AAVs, HC-Ad, and hydrodynamic tail vein injection [87, 88].

6 Gene Therapy-Marketed Products

There are currently a significant number of approved GT products worldwide. These products utilize various GT biotherapeutics and vehicles, including siRNA, miRNA, CRISPR/Cas9, and viral and nonviral vectors like AAV, AdV, or LNPs [13–21, 89, 90].

The first one was ***gendicine*** (China 2003), an Ad vector carrying the p53 tumor-suppressor gene for head and neck squamous cell carcinoma, followed by ***oncorine*** (China, 2005), an oncolytic adenovirus for head and neck cancer. ***Rexin-G*** (Philippines, 2007) is a retrovirus carrying a cytocidal cyclin G1 gene for pancreatic cancer and bone and soft tissue sarcoma. In 2012, the first GT product was approved in Europe, ***glybera***, an AAV vector carrying the LPL gene for lipoprotein lipase deficiency. Was the first GT product approved for a rare disorder but was withdrawn in 2017 due to a lack of demand and high price. The next product approved after ***glybera*** was ***imlygic*** (US, 2015) consisting of a modified herpes simplex virus type 1 for melanoma treatment. Followed by the first GT product for a severe hereditary immunodeficiency, ***strimvelis*** (Europe, 2016), an ex vivo stem cell GT for

ADA-SCID, in which CD34+ cells were transformed with retrovirus expressing ADA. One year later, two CAR T-cell products were approved; ***kymriah*** (US, 2017), a CAR T-cell therapy for B-cell acute lymphoblastic leukemia and diffuse large B-cell lymphoma, and ***yescarta*** (US, 2017), a CAR T-cell therapy for large B-cell lymphoma. In both cases, lentivirus was used to introduce the CAR molecule into the T cells. Followed by ***abecma*** (US, 2021), ***breyanzi*** (US, 2021) and ***tecartus*** (US, 2020) are CAR T-cell therapies for multiple myeloma, large B-cell lymphoma, and mantle cell lymphoma.

It took 5 years for the approval of the second AAV-based product, ***luxturna*** (US, 2017), an AAV serotype 2 vector carrying the RPE65 gene for inherited retinal dystrophy followed by ***zolgensma*** (US, 2019), an AAV9 vector GT for spinal muscular atrophy.

Some other examples are: ***Zynteglo*** (Europe, 2019), an ex vivo stem cell gene therapy for beta-thalassemia; ***upstaza*** (Europe, 2022), an AAV2 vector carrying the AADC gene for aromatic L-amino acid decarboxylase deficiency; ***roctavian*** (Europe, 2022), an AAV5 vector carrying the B-domain deleted human factor VIII gene for severe hemophilia A; ***hemgenix*** (US 2022), an AAV5 vector carrying the human factor VIII gene for severe hemophilia B; ***elevidys*** (US, 2023), an AAV9 carrying the microdystrophin gene for the treatment of Duchenne Muscular dystrophy (DMD).

Several ex vivo stem cell gene therapy products have been also recently approved: ***zynteglo*** (Europe, 2019) for transfusion-dependent beta-thalassemia; ***libmeldy*** (Europe, 2020) for metachromatic leukodystrophy; ***skysona*** (Europe, 2022) for cerebral adrenoleukodystrophy.

In addition, several siRNA therapeutics products have been commercialized: ***leqvio*** (Europe, 2020), small interfering RNA (siRNA) therapy for hypercholesterolemia; ***patisiran*** (US 2018), hereditary transthyretin-mediated amyloidosis; ***givosiran*** (US 2019), for acute hepatic porphyria; ***lumasiran*** (US 2020), for primary hyperoxaluria type 1; ***inclisiran*** (US 2021), for primary hypercholesterolemia; ***vutrisiran*** (US and Europe 2022), for hereditary transthyretin-mediated amyloidosis; ***nedosiran*** (US 2023), for primary hyperoxaluria type 1.

Finally, very recently the first gene editing therapeutic product for a rare disease has been approved, ***casgevy*** (US 2023) indicated for the treatment of sickle cell disease and transfusion-dependent beta-thalassemia, a cellular gene therapy that consists of autologous CD34+ hematopoietic stem cells (HSCs) edited with CRISPR/Cas9 technology on the BCL11A gene, a transcription factor that represses the expression of γ-globin and fetal hemoglobin (HbF) in erythroid cells. Downregulation of bcl11a expression resulted in the upregulation of HbF.

7 Current Limitations for Worldwide Access to Gene Therapy-Marketed Products

As it was indicated at the beginning of this chapter, most hereditary diseases currently lack a satisfactory therapeutic alternative. Many of these diseases affect children who either do not survive the disease or do so with serious sequelae. Recent significant advances in terms of the technological development of vectors and the knowledge of the molecular mechanisms involved in the development of the disease invite optimism and allow venturing that in the coming years, gene therapy could become the go-to treatment for these diseases. In this regard, the efforts made in recent years by various groups have culminated in the approval of several therapies and the start of promising ongoing clinical trials.

The current limitations for worldwide access to GT products that have reached the market include aspects like the price, production, distribution, etc. described below [91–96]. GT treatments are often extremely expensive, with some therapies costing hundreds of thousands to millions of dollars per patient, making them unaffordable for many healthcare systems and patients, especially in low- and middle-income countries. Many health insurance plans do not cover the high costs of GT, and reimbursement policies can vary significantly between countries, limiting patient access even in countries where the therapies are approved and available. Additionally, different countries have varying regulatory requirements for the approval of GT products, which can delay or prevent the availability of these therapies in certain regions due to complex and lengthy approval processes [96].

The production of GT products is complex and requires highly specialized facilities and expertise, limiting the ability to scale up production and distribute these therapies widely. Furthermore, administering GT products often requires specialized medical infrastructure and trained personnel, which may not be available in many regions, including the need for facilities capable of handling GT vectors and managing potential side effects. Finally, there are ethical concerns and social acceptance issues related to GT, particularly regarding germline modifications and long-term effects, which can influence regulatory decisions and public acceptance.

Currently, GTs are approved for a relatively small number of rare genetic disorders, limiting their overall impact and availability to a broader patient population. There is still a need for more long-term data on the safety and efficacy of GTs, which can affect regulatory approvals and healthcare providers' willingness to adopt these treatments widely. Addressing these limitations requires coordinated efforts among stakeholders, including governments, regulatory agencies, pharmaceutical companies, healthcare providers, and patient advocacy groups, to ensure broader and more equitable access to GT worldwide.

Compliance with Ethical Standards This chapter is a review and no humans or animals were involved in the preparation of the chapter.

References

1. National Institutes of Health. National Center for Advancing Translational Sciences. https://ncats.nih.gov/research/research-activities/TRND. Accessed July 30th 2024
2. Friedmann T (1992) A brief history of gene therapy. Nat Genet 2:93–98
3. American Society of Gene and Cell Therapy (2019) Gene and cell therapy FAQs. https://www.asgct.org/education/more-resources/gene-and-cell-therapy-faqs. Accessed July 30th 2024

4. Gene therapy clinical trials worldwide. J Gene Med. https://a873679.fmphost.com/fmi/webd/GTCT. Accessed July 30th 2024
5. Baylot V, Le TK, Taïeb D et al (2024) Between hope and reality: treatment of genetic diseases through nucleic acid-based drugs. Commun Biol 7(1):489
6. Culver KW, Anderson WF, Blaese RM (1991) Lymphocyte gene therapy. Hum Gene Ther 2: 107–109
7. Wilson JM (2009) Lessons learned from the gene therapy trial for ornithine transcarbamylase deficiency. Mol Genet Metab 96(4):151–157
8. Hacein-Bey-Abina S, von Kalle C, Schmidt M et al (2003) LMO2-associated clonal T cell proliferation in two patients after gene therapy for SCID-X1. Science 302(5644):415–419
9. Hacein-Bey-Abina S, Garrigue A, Wang GP et al (2008) Insertional oncogenesis in 4 patients after retrovirus-mediated gene therapy of SCID-X1. J Clin Invest 118:3132–3142
10. Hacein-Bey-Abina S, von Kalle C, Schmidt M et al (2003) A serious adverse event after successful gene therapy for X-linked severe combined immunodeficiency. N Engl J Med 348: 255–256
11. Gaspar HB, Cooray S, Gilmour KC et al (2011) Long-term persistence of a polyclonal T cell repertoire after gene therapy for X-linked severe combined immunodeficiency. Sci Transl Med 3(97):97ra79
12. Hacein-Bey-Abina S, Le Deist F, Carlier F et al (2002) Sustained correction of X-linked severe combined immunodeficiency by ex vivo gene therapy. N Engl J Med 346:1185–1193
13. Rose S (2017) First-ever CAR T-cell therapy approved in U.S. Cancer Discov 7(10):OF1
14. Calmes-Miller J (2018) FDA approves second CAR T-cell therapy. Cancer Discov 8:5–6
15. Majzner RG, Mackall CL (2019) Clinical lessons learned from the first leg of the CAR T cell journey. Nat Med 25(9):1341–1355
16. Ma CC, Wang ZL, Xu T et al (2020) The approved gene therapy drugs worldwide: from 1998 to 2019. Biotechnol Adv 40:107502
17. Shahryari A, Saghaeian Jazi M, Mohammadi S et al (2019) Development and clinical translation of approved gene therapy products for genetic disorders. Front Genet 10:868
18. Ylä-Herttuala S (2016) ADA-SCID gene therapy endorsed by European Medicines Agency for marketing authorization. Mol Ther 24:1013–1014
19. Iglesias-Lopez C, Agustí A, Vallano A et al (2021) Current landscape of clinical development and approval of advanced therapies. Mol Ther Methods Clin Dev 23:606–618
20. Shchaslyvyi AY, Antonenko SV, Tesliuk MG et al (2023) Current state of human gene therapy: approved products and vectors. Pharmaceuticals 16(10):1416
21. U.S. Food and Drug Administration (FDA) Approved Cellular and Gene Therapy Products. Available at: https://www.fda.gov/vaccines-blood-biologics/cellular-gene-therapy-products/approved-cellular-and-gene-therapy-products. Accessed July 31th 2024
22. Benskey MJ, Sandoval IM, Miller K et al (2019) Basic concepts in viral vector-mediated gene therapy. Methods Mol Biol 1937:3–26
23. Herrera-Carrillo E, Liu YP, Berkhout B (2017) Improving miRNA delivery by optimizing miRNA expression cassettes in diverse virus vectors. Hum Gene Ther Methods 28(4):177–190
24. Lipinski DM (2019) A comparison of inducible gene expression platforms: implications for recombinant adeno-associated virus (rAAV) vector-mediated ocular gene therapy. Adv Exp Med Biol 1185:79–83
25. Ingusci S, Verlengia G, Soukupova M et al (2019) Gene therapy tools for brain diseases. Front Pharmacol 10:724
26. Vilaboa N, Voellmy R (2006) Regulatable gene expression systems for gene therapy. Curr Gene Ther 6(4):421–438
27. Geisler A, Fechner H (2016) MicroRNA-regulated viral vectors for gene therapy. World J Exp Med 6(2):37–54
28. Arjmand B, Larijani B, Sheikh Hosseini M et al (2020) The horizon of gene therapy in modern medicine: advances and challenges. Adv Exp Med Biol 1247:33–64

29. Hayat SMG, Farahani N, Safdarian E et al (2019) Gene delivery using lipoplexes and polyplexes: principles, limitations and solutions. Crit Rev Eukaryot Gene Expr 29(1):29–36
30. Jiang C, Chen J, Li Z et al (2019) Recent advances in the development of polyethylenimine-based gene vectors for safe and efficient gene delivery. Expert Opin Drug Deliv 16(4):363–376
31. Sendra L, Herrero MJ, Aliño SF (2018) Translational advances of hydrofection by hydrodynamic injection. Genes 9(3):136
32. Le Guen YT, Le Gall T, Midoux P et al (2020) Gene transfer to skeletal muscle using hydrodynamic limb vein injection: current applications, hurdles and possible optimizations. J Gene Med 22(2):e3150
33. Decker RE, Lamantia ZE, Emrick TS et al (2020) Sonodelivery in skeletal muscle: current approaches and future potential. Bioengineering 7(3):107
34. Omata D, Unga J, Suzuki R et al (2020) Lipid-based microbubbles and ultrasound for therapeutic application. Adv Drug Deliv Rev 154-155:236–244
35. Sokołowska E, Błachnio-Zabielska AU (2019) A critical review of electroporation as a plasmid delivery system in mouse skeletal muscle. Int J Mol Sci 20(11):2776
36. Wahane A, Waghmode A, Kapphahn A et al (2020) Role of lipid-based and polymer-based non-viral vectors in nucleic acid delivery for next-generation gene therapy. Molecules 25(12): 2866
37. Shi B, Zheng M, Tao W et al (2017) Challenges in DNA delivery and recent advances in multifunctional polymeric DNA delivery systems. Biomacromolecules 18(8):2231–2246
38. Jogdeo CM, Siddhanta K, Das A et al (2024) Beyond lipids: exploring advances in polymeric gene delivery in the lipid nanoparticles era. Adv Mater:e2404608
39. Berns KI (1974) Molecular biology of the adeno-associated viruses. Curr Top Microbiol Immunol 65:1–20
40. Wang D, Tai PWL, Gao G (2019) Adeno-associated virus vector as a platform for gene therapy delivery. Nat Rev Drug Discov 18(5):358–378
41. El Andari J, Grimm D (2021) Production, processing, and characterization of synthetic AAV gene therapy vectors. Biotechnol J 16(1):e2000025
42. Li C, Samulski RJ (2020) Engineering adeno-associated virus vectors for gene therapy. Nat Rev Genet 21(4):255–272
43. Zabaleta N, Unzu C, Weber ND et al (2023) Gene therapy for liver diseases – progress and challenges. Nat Rev Gastroenterol Hepatol 20(5):288–305
44. Pupo A, Fernández A, Low SH et al (2022) AAV vectors: the Rubik's cube of human gene therapy. Mol Ther 30(12):3515–3541
45. Gao J, Mese K, Bunz O et al (2019) State-of-the-art human adenovirus vectorology for therapeutic approaches. FEBS Lett 593(24):3609
46. Chen W, Hu Y, Ju D (2020) Gene therapy for neurodegenerative disorders: advances, insights and prospects. Acta Pharm Sin B 10(8):1347–1359
47. Cantore A, Ranzani M, Bartholomae CC et al (2015) Liver-directed lentiviral gene therapy in a dog model of hemophilia B. Sci Transl Med 7(277):277ra28
48. Milani M, Annoni A, Moalli F et al (2019) Phagocytosis-shielded lentiviral vectors improve liver gene therapy in nonhuman primates. Sci Transl Med 11(493):eaav7325
49. Milani M, Canepari C, Assanelli S et al (2024) GP64-pseudotyped lentiviral vectors target liver endothelial cells and correct hemophilia a mice. EMBO Mol Med 16(6):1427–1450
50. Taghdiri M, Mussolino C (2024) Viral and non-viral systems to deliver gene therapeutics to clinical targets. Int J Mol Sci 25(13):7333
51. Song X, Liu J, Chen T et al (2024) Gene therapy and gene editing strategies in inherited blood disorders. J Genet Genomics S1673-8527(24):00180–00182
52. Rodrigues M, Yokota T (2018) An overview of recent advances and clinical applications of exon skipping and splice modulation for muscular dystrophy and various genetic diseases. Methods Mol Biol 1828:31–55
53. Gheibi-Hayat SM, Jamialahmadi K (2021) Antisense oligonucleotide (AS-ODN) technology: principle, mechanism and challenges. Biotechnol Appl Biochem 68(5):1086–1094

54. Katzmann JL, Packard CJ, Chapman MJ et al (2020) Targeting RNA with antisense oligonucleotides and small interfering RNA: JACC state-of-the-art review. J Am Coll Cardiol 76(5): 563–579
55. Setten RL, Rossi JJ, Han SP (2019) The current state and future directions of RNAi-based therapeutics. Nat Rev Drug Discov 18(6):421–446
56. Yin W, Rogge M (2019) Targeting RNA: a transformative therapeutic strategy. Clin Transl Sci 12(2):98–112
57. Ernst MPT, Broeders M, Herrero-Hernandez P et al (2020) Ready for repair? Gene editing enters the clinic for the treatment of human disease. Mol Ther Methods Clin Dev 18:532–557
58. Uddin F, Rudin CM, Sen T (2020) CRISPR gene therapy: applications, limitations, and implications for the future. Front Oncol 10:1387
59. Wang D, Zhang F, Gao G (2020) CRISPR-based therapeutic genome editing: strategies and in vivo delivery by AAV vectors. Cell 181(1):136–150
60. Lombardo A, Genovese P, Beausejour CM et al (2007) Gene editing in human stem cells using zinc finger nucleases and integrase-defective lentiviral vector delivery. Nat Biotechnol 25: 1298–1306
61. Zhang HX, Zhang Y, Yin H (2019) Genome editing with mRNA encoding ZFN, TALEN, and Cas9. Mol Ther 27(4):735–746
62. Li T, Huang S, Jiang WZ et al (2011) TAL nucleases (TALNs): hybrid proteins composed of TAL effectors and FokI DNA-cleavage domain. Nucleic Acids Res 39:359–372
63. Miller JC, Tan S, Qiao G et al (2011) A TALE nuclease architecture for efficient genome editing. Nat Biotechnol 29:143–148
64. Jinek M, Chylinski K, Fonfara I et al (2012) A programmable dual-RNA-guided DNA endonuclease in adaptive bacterial immunity. Science 337:816–821
65. Doudna JA, Charpentier E (2014) Genome editing. The new frontier of genome engineering with CRISPR-Cas9. Science 346:1258096
66. Kotagama OW, Jayasinghe CD, Abeysinghe T (2019) Era of genomic medicine: a narrative review on CRISPR technology as a potential therapeutic tool for human diseases. Biomed Res Int 2019:1369682
67. Koonin EV, Makarova KS, Zhang F (2017) Diversity, classification and evolution of CRISPR-Cas systems. Curr Opin Microbiol 37:67–78
68. Komor AC, Badran AH, Liu DR (2017) CRISPR-based technologies for the manipulation of eukaryotic genomes. Cell 168:20–36
69. Sahu S, Poplawska M, Lim SH et al (2023) CRISPR-based precision medicine for hematologic disorders: advancements, challenges, and prospects. Life Sci 333:122165
70. Mali P, Aach J, Stranges PB et al (2013) CAS9 transcriptional activators for target specificity screening and paired nickases for cooperative genome engineering. Nat Biotechnol 31:833–838
71. Ran FA, Hsu PD, Lin CY et al (2013) Double nicking by RNA-guided CRISPR Cas9 for enhanced genome editing specificity. Cell 154:1380–1389
72. Qi LS, Larson MH, Gilbert LA et al (2013) Repurposing CRISPR as an RNA-guided platform for sequence-specific control of gene expression. Cell 152:1173–1183
73. Villiger L, Joung J, Koblan L et al (2024) CRISPR technologies for genome, epigenome and transcriptome editing. Nat Rev Mol Cell Biol 25(6):464–487
74. Hess GT, Tycko J, Yao D et al (2017) Methods and applications of CRISPR-mediated base editing in eukaryotic genomes. Mol Cell 68:26–43
75. Levy JM, Yeh WH, Pendse N et al (2020) Cytosine and adenine base editing of the brain, liver, retina, heart and skeletal muscle of mice via adeno-associated viruses. Nat Biomed Eng 4:97–110
76. Anzalone AV, Koblan LW, Liu DR (2020) Genome editing with CRISPR-Cas nucleases, base editors, transposases and prime editors. Nat Biotechnol 38(7):824–844
77. Jiang T, Zhang XO, Weng Z et al (2021) Deletion and replacement of long genomic sequences using prime editing. Nat Biotechnol 40:227–234

78. Choi J, Chen W, Suiter CC et al (2022) Precise genomic deletions using paired prime editing. Nat Biotechnol 40:218–226
79. Villiger L, Grisch-Chan HM, Lindsay H et al (2018) Treatment of a metabolic liver disease by in vivo genome base editing in adult mice. Nat Med 24:1519–1525
80. Anderson CD, Arthur JA, Zhang Y et al (2023) Non-viral in vivo cytidine base editing in hepatocytes using focused ultrasound targeted microbubbles. Mol Ther Nucleic Acids 33:733–737
81. Packer MS, Chowdhary V, Lung G et al (2022) Evaluation of cytosine base editing and adenine base editing as a potential treatment for alpha-1 antitrypsin deficiency. Mol Ther 30:1396–1406
82. Chadwick AC, Wang X, Musunuru K (2017) In vivo base editing of PCSK9 (Proprotein Convertase Subtilisin/Kexin Type 9) as a therapeutic alternative to genome editing. Arterioscler Thromb Vasc Biol 37:1741–1747
83. Musunuru K, Chadwick AC, Mizoguchi T et al (2021) In vivo CRISPR base editing of PCSK9 durably lowers cholesterol in primates. Nature 593:429–434
84. Rothgangl T, Dennis MK, Lin PJC et al (2021) In vivo adenine base editing of PCSK9 in macaques reduces LDL cholesterol levels. Nat Biotechnol 39:949–957
85. Davis JR, Wang X, Witte IP et al (2022) Efficient in vivo base editing via single adeno-associated viruses with size-optimized genomes encoding compact adenine base editors. Nat Biomed Eng 6:1272–1283
86. Kasiewicz LN, Biswas S, Beach A et al (2023) GalNAc-lipid nanoparticles enable non-LDLR-dependent hepatic delivery of a CRISPR base editing therapy. Nat Commun 14:2776
87. Zhi S, Chen Y, Wu G et al (2022) Dual-AAV delivering split prime editor system for in vivo genome editing. Mol Ther 30:283–294
88. Wang Q, Liu J, Janssen JM et al (2021) Broadening the reach and investigating the potential of prime editors through fully viral gene-deleted adenoviral vector delivery. Nucleic Acids Res 49: 11986–12001
89. Gene Therapy Net. Gene Therapy Products on the Market. https://www.genetherapynet.com/gene-therapy-products-on-the-market.html. Accessed July 31th 2024
90. Kalland ME, Pose-Boirazian T, Palomo GM et al (2024) Advancing rare disease treatment: EMA's decade-long insights into engineered adoptive cell therapy for rare cancers and orphan designation. Gene Ther 31(7–8):366–377
91. Darrow JJ (2019) Luxturna: FDA documents reveal the value of a costly gene therapy. Drug Discov Today 24:949–954
92. Harlow EM, Adair JE (2024) Make gene therapies more available by manufacturing them in lower-income nations. Nature 631(8021):502–504
93. van der Loo JC, Wright JF (2016) Progress and challenges in viral vector manufacturing. Hum Mol Genet 25:R42–R52
94. Moradi V, Khodabandehloo E, Alidadi M et al (2024) Progress and pitfalls of gene editing technology in CAR-T cell therapy: a state-of-the-art review. Front Oncol 14:1388475
95. Lee NK, Chang JW (2024) Manufacturing cell and gene therapies: challenges in clinical translation. Ann Lab Med 44(4):314–323
96. Wang J, Griffiths E, Tounekti O et al (2023) Canadian regulatory framework and regulatory requirements for cell and gene therapy products. Adv Exp Med Biol 1430:91–116

Top Med Chem
https://doi.org/10.1007/7355_2024_174

Published online: 31 December 2024

Nucleic Acids Production

Natalia Sol Adler and Ana Sol Peinetti

Contents

Abstract Nucleic acid drugs are emerging as a powerful therapeutic class, with a growing number of approved biotherapeutics demonstrating their potential. However, their transition from treating rare diseases to more common afflictions presents new challenges in production scalability and environmental impact. This chapter explores current oligonucleotide production techniques, highlighting their strengths and limitations, and ongoing efforts to overcome the latter. Additionally, it discusses the structural modifications required for oligonucleotide biotherapeutics and how to achieve them using available synthetic methods.

Keywords Chemical oligo synthesis, Enzymatic oligo synthesis, Liquid-phase oligo synthesis (LPOS), Solid-phase oligo synthesis (SPOS), Structural chemical oligo modification

---

N. S. Adler and A. S. Peinetti (✉)
INQUIMAE (CONICET) – Departamento de Química Inorgánica, Analítica y Química Física, Facultad de Ciencias Exactas y Naturales, Universidad de Buenos Aires, Ciudad Universitaria, Buenos Aires, Argentina
e-mail: apeinetti@qi.fcen.uba.ar

1 Introduction

The ability to sequentially synthesize oligonucleotides, one nucleotide at a time, enables precise control over the composition and size of nucleic acid. Millions of synthetic oligonucleotides (ONs) are prepared every year for use in laboratories around the world. For most applications, very small quantities of nucleic acids are required. For biochemical and biological experiments, oligonucleotide synthesis is performed mainly on the 25–100 nmol scale, and for biophysical studies a larger amount (10 μmol or more) can be prepared. However, syntheses of multi-kilogram quantities of oligonucleotides are needed for some industrial applications such as biotherapeutic agents.

As discussed in the previous sections of this book,[1] biotherapeutic oligonucleotides represent a powerful drug approach with the potential to treat a wide range of diseases. Currently, more than 20 oligonucleotide biotherapeutics have reached the market [1–5], and hundreds more are in clinical trials and preclinical development. This growth indicates an impending need to produce oligonucleotides at a scale of metric tons per year and at lower costs [6, 7]. Moreover, biotherapeutic oligonucleotides have transitioned from specialized treatments for rare diseases to therapies for more common disorders, placing unprecedented demands on the synthetic capacity to manufacture these complex molecules at scale, in greater purity, quantity, and at reduced cost [8].

In addition to the need for producing greater quantities of sequences and scaling up to kilogram levels, biotherapeutic oligonucleotides involve a wide range of synthetic demands in terms of size, sequence, and chemical composition. For example, antisense oligonucleotides are short and fully modified oligos. Instead, mRNA vaccines need the production of extended oligonucleotides, spanning several thousand nucleotides. Moreover, each of these ONs necessitates multiple chemical modifications to effectively demonstrate their biological activity.

Thus, to unlock the full clinical potential of oligonucleotide therapies, these manufacturing challenges must be addressed. This chapter will describe well-established nucleic acid synthetic methods based on phosphoramidite chemistry, as well as new synthetic approaches that involve new chemistry, such as phosphorus (V) (P(V))-based synthesis, and enzymatic synthesis of oligonucleotides. It will also summarize new strategies that can overcome the limitations in terms of synthesis scale, length of the nucleic acid synthesized, and stereoselectivity. Finally, it will focus on how to incorporate the chemical modifications that are needed for therapeutic applications during the synthesis of oligonucleotides, discussing advantages and limitations of the different synthetic approaches.

[1] See Chapters "Antisense Oligonucleotides (ASOs), Small Interferring RNAs (siRNAs), Aptamers, and Gene Therapy".

2 Nucleic Acid Production

2.1 *Chemical Oligonucleotide Synthesis*

Oligo synthesis has a long history, beginning in academic labs in the 1950s, followed by automation and commercialization in the 1980s and progressing into high-throughput array-based methods in the 1990s [9]. The first synthetic oligos were synthesized using phosphodiester [10], H-phosphonate [11], and phosphotriester [12] chemistries. The phosphoramidite method was pioneered by Marvin Caruthers in the early 1980s. The development of solid-phase technology and automation further boosted this method, which is now established as the method of choice [13]. However, emerging techniques based on P(V) monomers are steadily gaining ground [14].

2.1.1 Solid-Phase Chemical Synthesis

Solid-phase chemical oligo synthesis consists of a cycle that adds bases one at a time to a growing oligo chain attached to a solid support held between filters, in columns that enable all reagents and solvents to pass through freely. The support election and development are essential for the synthesis. Before going into details of the oligo synthesis cycle, we will focus on the support that makes the synthesis possible.

Many types of solid support have been used, but controlled pore glass (CPG) was widely adopted because of its ability to solve swelling issues encountered with resin-based supports. Therefore, oligonucleotide synthesis employing CPG and its modifications were extensively developed [15].Glass supports with 50 nm pores are mechanically robust and are used routinely in the synthesis of short oligonucleotides. However, synthesis yields fall off dramatically for sequences longer than 40 bases, since the pores become blocked by the growing oligonucleotide and the diffusion of the reagents through the column is reduced. Although larger-pore CPG is more fragile, 100 nm pore supports have proven to be satisfactory for the synthesis of oligonucleotides up to 100 bases long [13].

However, the growing demand for synthetic oligonucleotides, particularly related to biotherapeutic applications, and the need to scale up the synthetic process spotlighted two limitations of CPG support. First, the unfavorable loading capacity of CPG restricts the amount of product that can be synthetized [15]. Second, in a scaled-up synthesis, a large excess of reactants (typically between 10- and 50-fold) is needed, because of the poor solvent diffusion of the CPG-attached substrates by less-polar solvents typically used in chemical synthesis. Thus, to achieve single-step efficiency above 99%, multiple excess reactants must be applied, exponentially increasing the chemical cost for large-scale synthesis [15].

For all of the above, multiple efforts are being made to design new materials for solid-phase oligo synthesis (SPOS). Nowadays, the most frequent reports on SPOS supports, alternative to unmodified CPG support, relate to copolymers of styrene and

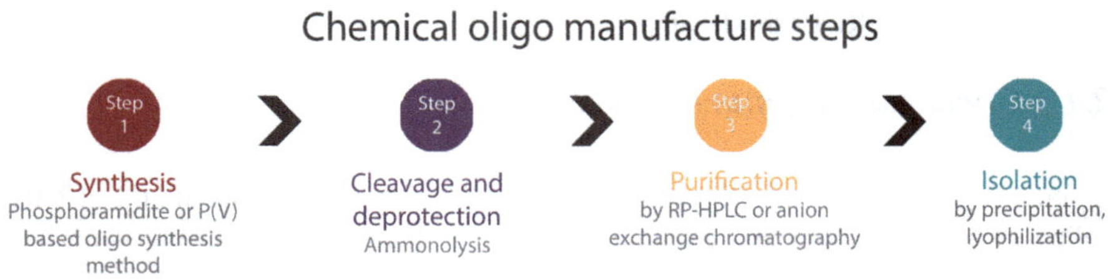

Fig. 1 The four steps involved in all chemical oligonucleotide manufacture

divinylbenzene as crosslinking agents, with different amino functions [16]. Even though swelling of polymer resins is a major concern, recent work has addressed this issue [17]. Different modifications of these supports have been proposed allowing for higher yields and purity [18, 19] as well as costs and synthesis time reductions [20]. Derivatization and functionalization of CPG with low-molecular mass compounds to improve the synthesis process have also been reported [21–23]. Finally, efforts to develop new materials called hybrid supports, consisting of CPG particles coated with a very thin crosslinked polymer film based on polystyrene, have been made [24]. These supports have high loading capacity, no swelling issues, sufficient yields even for longer chains, easy isolation, purification, and automation. Though few hybrid supports are currently available, there is significant potential for development in this field.

Building on the previously described supports, the chemical oligonucleotide synthesis takes place. The chemical manufacture of oligonucleotides typically consists of four key steps (Fig. 1): (1) oligonucleotide synthesis on the solid support; (2) cleavage from the solid support and removal of protecting groups; (3) purification by preparative chromatography; and (4) isolation by precipitation to desalt and lyophilization.

1. Synthesis

Phosphoramidite-Based Oligo Synthesis

Phosphoramidite-based oligo synthesis occurs in a column charged with a solid support, with the first nucleotide or a general linker molecule attached. The latter are referred to as universal supports and allow for the incorporation of modified/labeled nucleotides in the 3′-end of the oligos, since columns with modified bases attached may not be readily available. The synthesis typically proceeds in the 3′- to 5′--direction (opposite to the 5′- to 3′-direction of DNA biosynthesis in DNA replication) in a four-step cycle where one nucleotide per cycle is added and every step is followed by a solvent wash. The synthesis is computer-controlled and fully automated, and the reagents flow through the column sequentially to perform the reactions on the solid support.

The complete phosphoramidite DNA synthesis steps, for the case in which the first nucleotide is preloaded in the column, are outlined in Fig. 2a. The support-bound nucleoside has a 5′-(4,4′-dimethoxytrityl) (5’-DMT) protecting group to

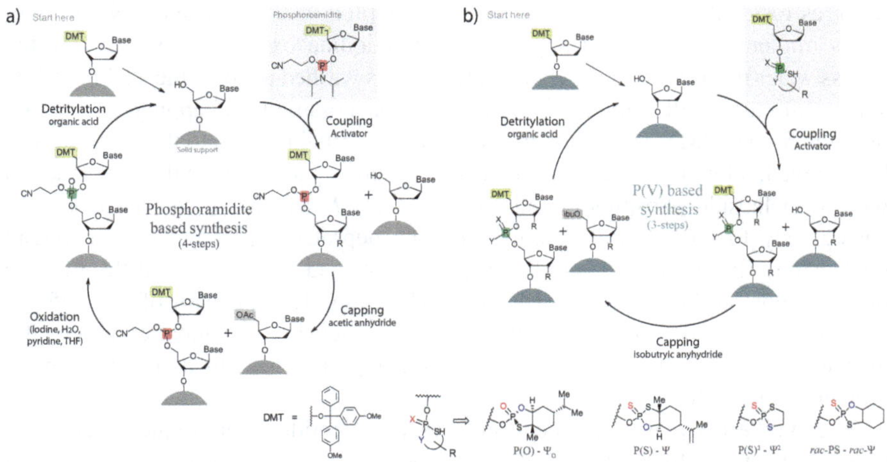

Fig. 2 Scheme of different chemical oligonucleotide synthesis. (**a**) Four-step phosphoramidite-based oligo synthesis. (**b**) Three-step P(V)-based oligo synthesis. The oxidation state of P is highlighted in green (P(III)) and red (P(V))

prevent polymerization during resin functionalization. Thus, the first action in the synthesis process is to deprotect the nucleoside through a *detritylation reaction*, where the DMT group on the 5′-hydroxyl group is removed with a solution of an organic acid, typically trichloroacetic acid in dichloromethane.

Following detritylation, the next nucleoside phosphoramidite in the sequence is incorporated in the *Coupling step*. The DMT-protected phosphoramidite monomer is added in a large excess to favor reaction yield, and mixed with an acidic activator employing acetonitrile as reaction solvent. The diisopropylamino group of the nucleoside phosphoramidite is protonated by the activator and is thereby converted to a good leaving group. It is rapidly displaced by attack of the 5′-hydroxyl group of the growing oligonucleotide chain and a new phosphorus-oxygen bond is formed, creating a phosphite triester. During the *Coupling step*, some 5′ hydroxyl groups may remain unreacted leading, if left unchecked, to the formation of oligonucleotides with deletion mutations. Thus, in order to prevent further extension of coupling failures and minimize *N*-1 impurities, a *Capping step* is incorporated after the coupling reaction to block the remaining 5′ hydroxyl groups. In this step, an acetic anhydride-*N*-methylimidazole (NMI) solution in tetrahydrofuran, with the addition of a small quantity of pyridine, is added to the synthesis column. The electrophilic mixture rapidly acetylates the 5′ hydroxyl groups, and the pyridine ensures that the pH remains basic to prevent detritylation of the other nucleoside phosphoramidites.

The phosphite-triester (P(III)) formed in the *Coupling step* is unstable in acidic conditions and must be converted to a stable P(V) species prior to the next *Detritylation step*. Therefore, an *Oxidation step* is performed where a phosphodiester group is formed by iodine oxidation in the presence of water and pyridine. The resultant DNA backbone is protected with a 2-cyanoethyl group to prevent undesirable reactions during subsequent synthesis cycles. Moreover, in order to dry the resin, as residual water can inhibit the next coupling reaction, some DNA synthesizers implement a second Capping step after iodine oxidation. In this step, the excess water reacts with the acylating agent and is washed away in the THF/pyridine solvent mixture. Following the coupling of phosphoramidites, capping, and oxidation, the 5′-end DMT protecting group of the DNA chain bound to the support must be removed, allowing the exposed hydroxyl group to participate in the next reaction with a nucleoside phosphoramidite. Thus, a new *Detritylation step* is performed analogously to the process described for the deprotection of the support-bound nucleoside. The reaction is rapid and quantitative and can be tracked through the orange color produced by the cleaved DMT group. The intensity of the 495 nm absorbance is used to determine the coupling efficiency and most commercially available DNA synthesizers measure and record trityl yield to monitor synthesis efficiency in real time.

The synthesis cycle is repeated until every nucleotide in the sequence is incorporated and the fully protected oligonucleotide is assembled on the support. All reactions in the cycle are rapid: a single four-step cycle takes approximately 45 min and a full synthesis of a 20-mer[2] is complete in less than 24 h [7].

P(V)-Based Oligo Synthesis

While nature builds DNA relying solely on the P(V) oxidation state, unlike the P (III) state found in phosphoramidite chemistry, early attempts to mimic this P (V) approach in the lab were abandoned in the 1970s due to the superior coupling efficiency of P(III)-based methods [14, 25, 26]. However, from a fundamental standpoint, a P(V) approach offers several advantages, as it eliminates the need for an oxidation step, simplifies protecting group strategies, and potentially allows for the use of more stable building blocks. This suggests that revisiting P(V) methods with modern advancements could be a promising avenue for DNA synthesis, and the past few years have seen a growing push toward developing a new P(V) platform for oligonucleotide synthesis [21, 27].

Inspired by the pioneering work of Stec and colleagues on P(V)-based oxathiaphospholane (OTP) heterocycles [28–30], and through a systematic evaluation of key components such as the OTP backbone, phosphorus source, and leaving group, the Ψ (PSI, Phosphorus Sulfur Incorporation) platform was developed by Knouse and colleagues [27]. With Ψ reagents, a synthesis cycle similar to the previously described phosphoramidite cycle can be proposed, relying on direct PO bond formation during the *Coupling Step*, as well as the requisite of *Deprotection* and *Capping Steps* as the sequence grows, but avoiding the *Oxidation Step* required in P(III) chemistry (Fig. 2b). Moreover, almost the same capping and deblocking reagents listed for the phosphoramidite cycle can be employed here but, notably, the only activator needed during the *Coupling Step* is the inexpensive amine base

[2]Denotes the length of the oligonucleotide (from Greek meros: "part"). The way to annotate is by first indicating the number of nucleotides followed by a hyphen and finally the "mer" particle. For example, an oligonucleotide of 35 nucleotides would usually be called a "35-mer".

1,8-diazabicyclo[5.4.0]undec-7-ene (DBU). Remarkably, the Ψ platform was derived from inexpensive limonene, which is readily available on a commodity scale in either enantiomer.

Since the initial study did not explore the incorporation of various oligonucleotide chemical modifications or its compatibility with modern automated synthesizer protocols, Huang and colleagues [21] later expanded the platform by introducing three new reagent systems (rac-Ψ, Ψ_O, and Ψ^2, Fig. 2b). This extension offered a unified P(V) methodology, moving away from traditional P(III)-based oligonucleotide synthesis, and enabled precise, controlled synthesis of specific chimeric oligonucleotides (Fig. 2b). They also provided a comprehensive protocol for using these reagents in commercial synthesizers with a single coupling process for all Ψ reagents, achieving full reaction conversion in under two minutes, comparable to the industry-standard P(III) protocol. While decades of continuous refinement have cemented P(III)-based oligonucleotide synthesis as the high-yielding standard, this recently developed P(V) chemistry offers a compelling alternative to be explored. Notably, P(V) methods excel at generating homogeneous, stereopure oligonucleotides—a feat currently challenging for P(III) approaches. This advantage has not been lost on the pharmaceutical industry, with P(V) platforms experiencing rapid adoption for diverse applications [31–33].

2. Cleavage and Deprotection

Once the synthesis is complete, the full-length product must be cleaved from the solid support. The most commonly used linker to attach the 3′-end of the oligonucleoside to the support is the succinyl link [13] which is stable in all conditions employed during the synthesis process but can be easily cleaved by treatment with concentrated ammonium hydroxide at room temperature. The backbone and base protecting groups are then generally removed by heating the resulting aqueous ammonium hydroxide solution. Conditions regarding the removal of different heterocyclic basis protecting groups can be found elsewhere [13]. Once cleaved from the solid support, the crude oligonucleotide can be analyzed for concentration, identity, and purity; crude purities of greater than 80% are not uncommon [7].

In the case of popular universal support with the UnyLinker [34] molecule attached, concentrated ammonium hydroxide can also be employed to cleave the oligo from the column, but extended cleavage times and higher temperatures are required. Universal linkers containing a urea bridge to the support allowing for significantly milder reaction conditions [35]. This is particularly advantageous when working with monomers that have sensitive functional groups.

3. Purification

Oligonucleotide purification can be achieved through various preparative chromatography methods, with reversed-phase high-performance liquid chromatography (RP-HPLC) and anion exchange (AEX) chromatography being the most prevalent. The hydrophobic 5’-DMT protecting group of the oligonucleotide often remains intact during purification, as its properties facilitate chromatographic separation.

The stationary phase employed in RP-HPLC typically consists of a polystyrene-divinylbenzene copolymer or C18-modified silica. The oligonucleotide is retained through interaction with the stationary phase via its DMT group. Impurities lacking the DMT group are washed out using a medium-strength eluent, followed by elution of the product at higher solvent strength. The DMT group can then be cleaved in solution using the previously described acidic treatment.

AEX chromatography employs a quaternary ammonium-functionalized stationary phase that attracts the negatively charged oligonucleotides at low salt concentrations. Here, the interaction of DMT groups with the somewhat hydrophobic resin can be exploited: a wash step using an intermediate salt concentration can remove some DMT-off impurities followed by on-column detritylation achieved with an acidic wash and product elution using an increasing salt gradient. Alternatively, the entire purification can be performed in DMT-off mode, relying solely on the salt gradient to separate impurities.

4. Isolation

Following purification, the obtained solution undergoes desalting to eliminate eluent buffer components and is concentrated before further processing or final isolation. Two main desalting methods are often employed: precipitation and tangential flow filtration (TFF).

Oligonucleotides can be precipitated from aqueous solution by ethanol or 2-propanol addition, either at a reduced temperature or using salt promoters [36]. This process separates the desired oligonucleotide from inorganic salts, which remain in the discarded supernatant. The remaining oligonucleotide pellet is redissolved in water to the required concentration.

TFF utilizes a membrane to desalt the oligonucleotide solution [37]. Smaller molecules like water, salts, and organic compounds pass through the membrane (permeate), while larger oligonucleotides are retained (retentate) and concentrated. Water is introduced into the retentate to flush out small molecules and facilitate desalting. This process, commonly referred to as ultrafiltration/diafiltration (UF/DF), significantly reduces inorganic salt concentration while increasing oligonucleotide concentration.

Following desalting, the next steps depend on the final biotherapeutic oligonucleotide structure. Double-stranded oligonucleotides, like small interfering RNAs (siRNAs), require an annealing step to join the complementary strands. This typically involves mixing and heating solutions of both strands, followed by cooling to promote hybridization. Finally, the active pharmaceutical ingredient (API) is isolated by lyophilization (freeze-drying) to obtain a low-density, hygroscopic, amorphous solid, usually as the sodium salt. This solid form is typically stored frozen and shipped to the biotherapeutic product facility for formulation into the final drug product.

The outlined manufacturing process is a significant accomplishment, capable of synthesizing, purifying, and isolating oligonucleotides with exceptional yield (~50%) and purity (~90%). This process is versatile, capable of producing a wide range of nucleic acid derivatives, including modified, single and double-stranded oligonucleotides, aptamers, and conjugates. This is particularly impressive given the size and complexity of the molecules and the numerous chemical transformations involved [7]. Moreover, the process can be scaled up from gram to kilogram quantities to produce oligonucleotides for clinical and commercial use. However, large-scale manufacturing (>100-kg batches) is not yet accessible through this process.

Besides the success of this method for producing oligos for a wide variety of applications, there are some limitations when thinking on therapeutic application, most of them related with the need of scaling-up. First, monomer excess is a must. An excess of heavily protected monomers is required for efficient coupling, which reduces atom efficiency and generates by-products that must be removed through washing steps. Coupling efficiency is crucial, since small decreases (0.1–0.5%) have a significant impact on the overall synthesis yield because of the cumulative effect of a series of poor couplings. On top of that, final products require chromatographic purification, which necessitates the use of prohibitively large volumes of acetonitrile (1,000 kg per kilogram of oligonucleotide). Therefore, this achievement comes at an environmental cost in terms of waste, chemical hazards, and energy efficiency and this burden will only grow as volumes of therapeutic oligonucleotides increase [7]. Thus, several approaches are being pursued in order to make oligonucleotide production more sustainable. Early efforts focused on reducing or eliminating the use or generation of toxic materials, solvents, and reagents, and the development of methodologies that allowed the reuse of valuable materials such as amidites, solid-support, and protecting groups, consequently improving the atom economy and cost-efficiency of oligonucleotide synthesis [38]. In addition, purification stage is responsible for approximately half of the material used in the manufacturing process; therefore, improvements to this stage would make a significant contribution to creating a greener process [7], so there are ongoing efforts to develop HPLC-free purification methods for oligonucleotides based on fluorous chemistry [39], biotin–avidin interaction [40–42], orthogonal chemical reactions [43–45], oximation reaction [46, 47], and polymerization reactions [48, 49].

2.1.2 Liquid-Phase Synthesis

To date, conventional solid-phase synthesis has been a successful method for producing oligonucleotide-based drugs, as the initial development of this promising therapeutic approach was limited to the treatment of rare diseases. However, the recent U.S. Food & Drug Administration (FDA) approval of the biotherapeutic ***inclisiran*** as an adjunct to statin therapy for patients with clinical atherosclerotic cardiovascular disease [50], a condition affecting over 500 million people worldwide, marks a shift in the field from low-volume drugs to therapies for widespread diseases. Therefore, large-scale manufacturing of oligonucleotides has become a high priority for process chemists, and attention has been drawn to the liquid-phase oligonucleotide synthesis (LPOS) as a possible alternative strategy to meet the scale-up challenge.

In the LPOS technique, the synthesis occurs in the same manner as described for SPOS, adding one nucleotide per cycle, but the growing oligonucleotide is attached to a soluble anchor macromolecule, and the reactions occur under homogeneous conditions ensuring even distribution of reactants as well as higher accessibility of the reagents for reaction. Thus, the main advantage of LPOS is that its performance is not affected by the scale of the synthesis, since limitations related to steric hindrance are eliminated, and can be carried out in conventional batch reactors used in the production of current APIs. Moreover, the homogeneous nature of the protocol allows for the reaction to be performed at higher concentrations, enabling monitoring and optimization of each step, which helps reduce the use of excess reagents and solvents. Furthermore, by increasing crude purity, purification processes can be simplified or eliminated, contributing to a more sustainable methodology. Even though di- and tri-nucleotides have been successfully obtained in solution without a support [51], longer oligonucleotides require a soluble support for efficient purification. The properties of the macromolecular supports allow separation of the growing ON from excess reagents through techniques like precipitation, extraction, or chromatography.

While numerous soluble-supports protocols have been reported in the last decades and were thoroughly reviewed by Molina & Sanghvi [52], two stand out for their potential for large-scale implementation. On the one hand, a strategy designed by the Livingston group based on organic solvent nanofiltration (OSN) technology [53], and, on the other hand, a recently developed alkyl-chain-soluble support (ACSS) with anchor molecules such as benzyl or diphenylmethyl alcohol or amine, but with longer than C18 alkyl chains (trade name AJIPHASE®) [54, 55].

The first strategy is a membrane-based method that efficiently separates the ON-linked polyethylene glycol (PEG) (acting as the soluble polymeric support) from small molecule reagents and impurities after each step. This novel approach utilizes diafiltration to isolate the PEG-bound oligonucleotide based primarily on size and shape-related features of the molecules in the reaction mixture. In order to apply this technology to a wide range of compounds, various membranes having a specific molecular weight cut-off can be employed.

On the other hand, the AJIPHASE support macromolecular design enables easy purification of the anchored oligonucleotide by simple precipitation in acetonitrile or methanol. Remarkably, this protocol requires only one precipitation step for each nucleotide extension cycle, which results in significantly higher overall yield and reduction of solvent consumption compared to other LPOS protocols. This approach is scalable up to 200 kg and has already successfully produced multiple 10-kg batches of a phosphorodiamidate morpholino oligomer (PMO) under good manufacturing practices (GMP) for clinical trials, with considerably less excess monomer waste compared to solid-supported PMO synthesis. The FDA has also approved the manufacture of an undisclosed commercial oligonucleotide using AJIPHASE technology [56]. Furthermore, they propose that oligonucleotides can be constructed using phosphotriester or H-phosphonate chemistry, which might be useful in lowering the cost of starting material [15].

Another interesting strategy is the solution-phase ONs synthesis where di- and trinucleotide building blocks (blockmer) are used instead of conventional amidites, named convergent LPOS. This technique allows for faster synthesis times and improves stepwise and overall yields. In addition, the use of a blockmer approach is expected to furnish full-length ON that is devoid of n−1 truncated products which translates into easier purification. Convergent LPOS dinucleotide and trinucleotide synthesis was successfully carried out using phosphoramidite [57] and H-phosphonate chemistry [58].

In conclusion, LPOS holds significant promise for large-scale (>100 kg) oligonucleotide production due to recent advancements. However, scaling LPOS for commercial production hinges on overcoming current limitations. This includes developing a broader range of soluble scaffolds with improved solubility and cost-effectiveness: most likely, customizable supports tailored to each specific oligonucleotide sequence will be required. Another key challenge for LPOS is its current excessive use of reagents and solvents. Strategies employed in SPOS, such as recovering and reusing amidites and protecting groups, hold promise for adaptation to LPOS, potentially leading to a more sustainable process. Thus, building on the knowledge gained from SPOS, LPOS is rapidly progressing thanks to ongoing research efforts in both industry and academia.

2.2 *Enzymatic Oligonucleotide Synthesis*

Today, almost every strand of DNA manufactured outside of a cell is produced via chemical synthesis strategies described in the previous section. These methods have been highly successful, but new applications in biotherapeutics have revealed their limitations. Besides the limitations in terms of scalability and sustainability that were discussed, chemical synthesis has another limitation: the length of sequences is limited to 120–150 nucleotides. It is known that the efficiency of direct synthesis steadily drops off beyond ~200-mers, with about 50% product yield for 120 mers. Thus, today longer nucleic acids, such as gene length, must be gradually constructed in stages by assembling smaller strands [59, 60]. Furthermore, the chemical oligo synthesis involves several purification and isolation steps as described, a workflow that is labor-intensive. Thus, a synthesis approach that provides higher purity oligonucleotide is desirable.

An alternative to this chemical synthesis is enzymatically driven nucleic acid synthesis, which harnesses the catalyzing power of enzymes to assemble the four nucleotide bases into nucleic acid strands. In this section, we describe the effort in enzymatic oligonucleotides synthesis to generate de novo nucleic acid sequences, as well as methods that enable the production of large quantities of nucleic acid. The enzymatically driven synthesis shows a great potential to tackle the challenge of obtaining longer nucleic acid.

2.2.1 Template-Independent Enzymatic Oligonucleotide Synthesis (TiEOS)

A short strand of nucleic acid synthesized on a solid support can be extended by DNA polymerases using deoxynucleoside 5′-triphosphates (dNTPs). However, DNA polymerases generally require a template DNA strand for base pairing, which limits their ability to generate de novo DNA sequences. Therefore, an alternative enzyme is required to efficiently elongate polynucleotide chains in the absence of a template strand. In 1959, F. J. Bollum described the first ssDNA polymerase, terminal deoxynucleotidyl transferase (TdT), capable of template-independent (primer-free) synthesis [61]. TdT can catalyze the binding between the 3′ OH of ssDNA and the 5′ triphosphate groups of the dissociative nucleotide, independent of a complementary template [62, 63], opening up the possibility of tailoring this reaction to reach a de novo template-independent oligonucleotide synthesis (TiEOS) [64].

TiEOS is schematized in Fig. 3 and involves a two steps cycle that occurs in a solid substrate for retaining the polynucleotide during synthesis. The solid substrate contains an initiator strand (>3 nucleotides) for TdT enzyme binding and polymerization. This initiator strand is synthesized by chemical synthesis. Then, TdT is able to elongate oligonucleotides in the 5′ to 3′ direction, accepting any of the four canonical nucleotides, i.e., without specificity. TdT can generate long polymers (>1 kb), using dNTPs without any modification [65]. Thus, for stepwise nucleotide

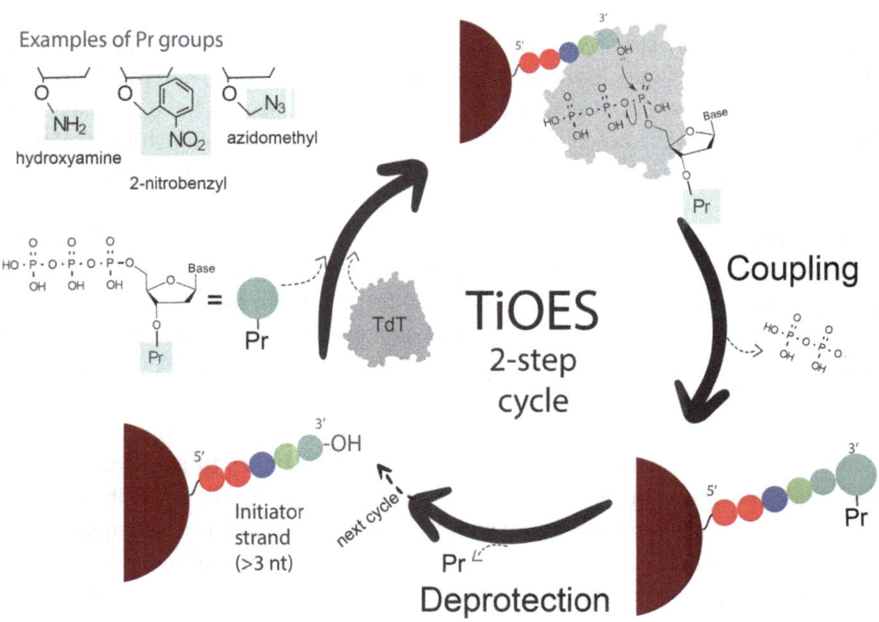

Fig. 3 Scheme of template-independent enzymatic oligonucleotide synthesis. Pr represents the protecting group. Examples of the most commonly used Pr are included

addition and a controlled oligonucleotide synthesis, dNTPs are protected from further extension until it is selectively converted to the reactive hydroxyl group ($dNTP_{Pr} \rightarrow dNTP$, where Pr is a protecting group); as such, $dNTP_{Pr}$ is also termed the reversible terminator (RT). After multiple cycles of extension and deblocking, the full-length single strand sequence is released from the solid support and isolated for subsequent use. To this end, initiator strand incorporates a highly specific deoxyuridine cleavage site at its 3′ end which is enzymatically labile. Upon completion of the synthesis, this site is cleaved by uracil DNA glycosylase, releasing the assembled sequence [8].

The success of enzymatic DNA synthesis depends on two distinct elements: an effective polymerization biocatalyst (enzyme) and the right reversible terminator ($dNTPs_{Pr}$). In terms of reversible terminators, the blocking groups are typically affixed at the 3′-hydroxyl moiety to prevent further nucleophilic attack on the α-phosphorous of incoming nucleoside triphosphates. Different suitable protecting groups have been developed (Fig. 3), such as hydroxyamine (3′-ONH_2) [66, 67], azidomethyl [68], or 2-nitrobenzyl terminators [69, 70]. 2-Nitrobenzyl group may be of particular interest for the TiEOS application because it is a photocleavable group [71] and this could be very beneficial in terms of high-throughput synthesis automation. Another promising candidate for TiEOS is the 3′-ONH_2 due to its advantageous characteristics such as mild reagent deblocking conditions employing aqueous sodium nitrite, and its small size from a steric chemistry standpoint. Other Pr are attempted for the protection of 3′-OH [69, 71–74], however, TdT must be able to accommodate the protected dNTPs in its active site, which limits the choice of Pr. Indeed, the native form of TdT appears to allow some incorporation of these 3′-blocked monomers but has a relatively small active cavity [75], making it more difficult to incorporate sugar-modified nucleotides than natural nucleotides; resulting in the slow incorporation of protected dNTPs. Thus, the challenge has been to balance the choice of a 3′-reversible terminator that is both an efficient substrate for TdT and is rapidly removable under mild conditions.

With this constraint, different approaches have been developed. On the one hand, extensive re-engineering of TdT is applied [76–78]. For instance, 1,000-fold higher catalytic activity with 3′-ONH_2-dNTPs has been achieved by re-engineering TdT, compared to its commonly used form [78]. However, the engineered polymerase currently requires about 5 min to add a modified monomer, whereas native TdT, when using unblocked nucleotides, can add a monomer in just 10–20 s [63, 79]. This disparity indicates that there is still significant room for optimization. Consequently, many companies are actively working on improving TdT efficiency.

On the other hand, Palluk and colleagues conceived a different approach wherein each polymerase molecule is site-specifically labeled with a tethered nucleoside triphosphate. When the tethered dNTP is incorporated into a primer, it remains covalently attached to the 3′ end, blocking further elongation by other polymerase-dNTP conjugates [80]. Based on this technology, ANSA Biotechnologies, a pioneer company in enzymatic DNA synthesis, has de novo synthesized a DNA sequence with 1,005 nucleotides successfully in 2023, which is currently the longest single synthesis of DNA oligonucleotides and an important milestone in the field of enzymatically driven DNA synthesis [81]. This strategy requires more enzyme, but it remains cost-effective because it eliminates the need for high concentrations of expensive modified nucleotides. Moreover, the utilization of TdT–dNTP conjugates allows for a reduction in reaction time to 10–20 s, offering a much faster synthesis.

The second key element for the success of enzymatic DNA synthesis is TdT. While TdT is well-suited for enzymatic synthesis, several important limitations arise from its features. We have already discussed the challenge of using this enzyme to incorporate $dNTPs_{Pr}$. Additionally, this enzyme shows a preference for incorporating certain nucleotides over others, which can affect the sequence fidelity and lead to deletions during synthesis [64, 82]. Moreover, TdT works only on single-stranded DNA. TdT contains loop 1, a Lariat-like conformation, which clamps down on the initiator via hydrogen bonding and precludes dsDNA from the active site [63, 83]. Consequently, the efficiency of the synthesis is reduced if the strand under construction begins to form secondary structures.

Furthermore, with a similar two-step cycle synthesis, de novo RNA strands have been reported. Church and colleagues reported the enzymatic synthesis of natural RNA oligonucleotide sequences in an aqueous based-format using 3'-*O*-allyl ether reversible terminators and an enzyme capable of their incorporation [84]. They evaluated the performance of 24 unique NTP building blocks, and prepared mutant variants of CID1 Poly(U) Polymerase (PUP). This mutant has an increased ability to incorporate NTP bases compared to wild-type enzymes.

Although TiEOS is currently in the early stages of its technology lifecycle, several startups are proving the viability of enzymatic synthesis as a rapid, accurate, and efficient alternative to conventional chemical oligonucleotide synthesis [85]. For instance, DNA Script recently introduced the first commercial benchtop DNA synthesizer based on enzymatic synthesis [86]. Novel hardware and high-performance enzymes with long shelf-life are all part of the future development of enzymatic oligonucleotide synthesis. The potential of recycling reactants [87] for multiple cycles could have a dramatic impact on the costs of such synthesis, with a drop of the cost in three orders of magnitude, reaching even an order of magnitude less than phosphoramidite-based synthesis [64]. For instance, TdT can be recycled by covalently attaching the enzymes to a solid support [87]. Currently, this opportunity does not exist for chemical synthesis.

2.2.2 Template-Dependent Polymerase-Based Oligonucleotide Synthesis

Using templates during the enzymatic oligonucleotide synthesis could substantially improve the speed and extension yields of this oligonucleotide synthesis [88]. To reach de novo enzymatic oligonucleotide synthesis, Hoff and colleagues have reported an approach involving the use of DNA polymerases (or reverse transcriptases) to synthesize single-stranded oligonucleotides controlled by transient hybridization to neighboring strands (Fig. 4a). They designed a "universal template" that contains all possible three base combinations and includes spacer poly-T sequences

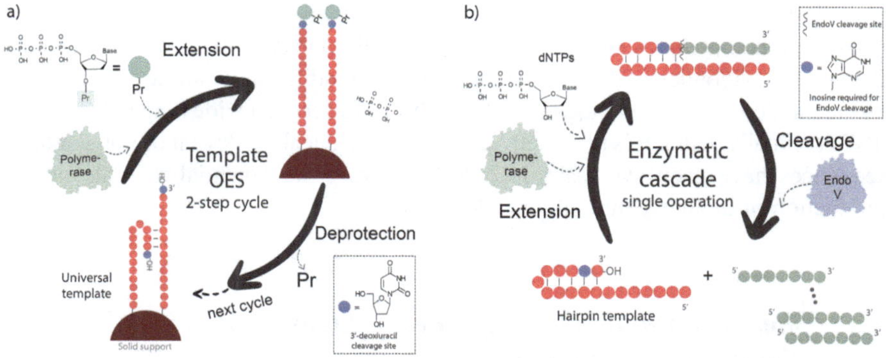

Fig. 4 Two different template-dependent polymerase-based approaches for oligonucleotide synthesis. (**a**) De novo synthesis of single-stranded oligonucleotides controlled by transient hybridization to neighboring strands, in a step-by-step synthetic approach. (**b**) Enzymatic cascade approach for the synthesis of single-stranded oligonucleotides, by extension and cleavage of a catalytic self-priming hairpin template in a single operation

at the 3′ and 5′ ends as well as a 3′ deoxyuracil [88]. Transient hybridization of the template to itself through hairpin formation or to a neighboring template molecule is enough to enable polymerase-mediated addition of a 3′-protected nucleotide to the template's 3′-end. The synthesis cycle is analogous to TiEOS, containing a first extension step using a polymerase in the presence of the $dNTP_{Pr}$ building blocks which contain a 3′-O-azidomethyl reversible terminator. After nucleotide addition, excess dNTP is removed and a second step of deprotection is performed using tris (2-carboxyethyl)phosphine (TCEP). The final sequence is cleaved from the template at a deoxyuracil site positioned at the 3′-end of the starting template, using Uracil-DNA glycoylase in combination with an apurinic/apyrimidic endonuclease. Using this strategy, a 20 nucleotide ssDNA oligonucleotide was synthesized [88].

Recently, in a substantially different approach, iterative rounds of chain extension and deprotection are replaced by a biocatalytic platform to synthesize oligonucleotides from unprotected nucleoside triphosphate (dNTP) building blocks in a single operation (Fig. 4b) [89]. This approach functions in aqueous conditions and does not require solid supports. This strategy, named enzymatic cascade, uses a DNA polymerase and dNTP building blocks to extend a catalytic self-priming hairpin template [89]. An endonuclease V (EndoV) then selectively cleaves a single strand of the resulting duplex DNA downstream of an inosine base built into the hairpin sequence, releasing the product and regenerating the template. The larger size of the polymerase relative to the DNA makes it impossible for the endonuclease to bind to the template until the amplification is complete and the polymerase is released, thus avoiding premature cleavage of the sequences. This allows for both processes to occur in a single operation. Although this method is not suitable for de novo oligo synthesis, it is specifically tailored for the large-scale production of short, modified oligonucleotides typically used as therapeutics. Moreover, it is interesting to notice that the template, generated by chemical synthesis, is added to the reaction in a catalytic amount representing less than 1% of the DNA generated. Furthermore, therapeutic oligonucleotides containing different modifications are accessible with this technology, as polymerases have been shown to accept a wide range of modified dNTPs. Finally, using this new method, a range of clinically relevant oligonucleotide sequences have been synthesized [2, 89], showing the potential to scale up the production of biotherapeutic oligonucleotides.

2.3 *Chemoenzymatic Oligo Synthesis: Ligase-Based Oligonucleotide Synthesis*

The capability to synthesize long nucleic acid sequences could pave the way for next-generation therapies with more complex combinations of chemical modifications . So far, synthesis of these longer sequences is restricted mainly to chemoenzymatic strategies. In this strategy, chemically synthesized short oligonucleotide fragments (typically 3–8 nucleotides long) are assembled into longer sequences using DNA or RNA ligases. Shorter oligonucleotides are easier to produce with high purity using chemical methods, as byproducts like truncated sequences from incomplete coupling reactions tend to accumulate with each nucleotide addition. Shorter fragments are also easier to purify using HPLC [90].

Ligases activate a 5′-monophosphate group of one oligonucleotide fragment using adenosine triphosphate (ATP) or nicotinamide adenine dinucleotide (NAD). Then, the 3′-hydroxyl of a second fragment attacks the adenylated intermediate, ligating the fragments with concomitant release of adenosine monophosphate (AMP). Different ligation-based approaches have been reported, where the accuracy and number of oligonucleotide molecules that can be combined in a single step have been improved.

On the one hand, an approach reported by GSK involves annealing multiple fragments to a complementary DNA template to control the order of assembly and then using a double-strand ligase to form the desired sequence (Fig. 5a) [91, 92]. Then, upon heating, the product detaches from the support-bound template and can be isolated by filtration, achieving sufficient purity to eliminate the need for chromatography [2]. Moreover, the use of a template ensures a high-purity product, as fragments with deletions or insertions do not anneal to the template and are therefore not incorporated into the final product.

Alternative, template-independent ligation strategies have been reported. Ajinomoto and Almac reported a method to synthesize double-strand siRNA by complementary base pairing between overlapping fragments (Fig. 5b). Using this approach a ligation of up to six fragments, including multiple oligonucleotide modifications, in one-pot reaction has been demonstrated [93]. Furthermore, a template-independent ligation strategy toward single-stranded sequences has been developed and involves the use and engineering of RNA ligases (Fig. 5c) [94]. In

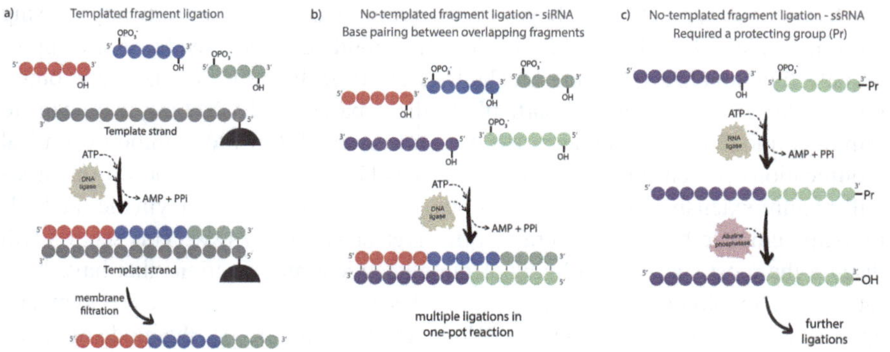

Fig. 5 Representation of different ligase-based oligonucleotide synthesis

this case, the donor 5′-monophosphorylated fragment is synthesized with an additional 3′-phosphate protecting group to prevent uncontrolled polymerization. After ligation reaction, the 3′-phosphate can subsequently be removed from the ligated sequence using an alkaline phosphatase enabling further ligations [92, 94, 95].

Thus chemoenzymatic ligase-based oligo synthesis approaches permit the production of not only short, clinically relevant oligonucleotides, but also much larger fragment (>150 nucleotides) and heavily modified sequences (>50 chemically modified nucleotides) which are difficult to obtain through conventional chemical or enzymatic methods.

3 Structure Modifications in Biotherapeutic Oligonucleotides

Oligonucleotides offer a powerful approach for drug discovery, as their design can be guided solely by the target gene sequence, enabling rapid identification of highly specific lead compounds [96]. This contrasts with traditional small-molecule drugs, which often require extensive and iterative screening followed by significant medicinal chemistry optimization [97]. However, unmodified DNA and RNA exhibit minimal therapeutic activity because they are rapidly degraded, exhibit poor cellular uptake and/or are filtered out of blood in biological environments. Thus, enormous efforts have been made to address the delivery challenge and the most commonly used strategies include chemical modification to improve “drug-likeness”, covalent conjugation to cell-targeting or cell-penetrating moieties, and nanoparticle formulation [98]. Chemical modifications of the nucleic acid backbone, the ribose sugar moiety and the nucleobase itself have been extensively employed in order to improve oligonucleotide pharmacokinetics, pharmacodynamics, and biodistribution [99]. Moreover, most ONs approved to date lacks an additional delivery vehicle and so are dependent on structure modification alone to facilitate tissue delivery.

Diverse modifications have been introduced site-specifically using phosphoramidite-based and Ψ-based oligonucleotide chemical synthesis by appropriately functionalized building blocks [14, 21, 100]. In contrast, enzymatic oligonucleotide synthesis methods, particularly those based on TdTs, face challenges to introduce modified oligos due to the poor tolerance of TdTs to common chemical modifications present in therapeutic sequences [101]. Addressing these limitations will require extensive enzyme engineering. On the other hand, enzymatic methods utilizing template-based polymerases offer greater versatility [89]. This section will discuss the most successful oligonucleotide chemical modifications that have been introduced in biotherapeutic oligonucleotides (Fig. 6) focusing on the synthetic approaches available to obtain them. While it will briefly discuss the mode of action of each modification, a more detailed description of the rationale behind as well as the biological activity of these modifications can be found in previous chapters of this book and elsewhere [1, 102, 103]. Finally, it will briefly describe two bioconjugation methods that commercially available therapeutic ONs incorporated and can be introduced during the ONs synthesis process.

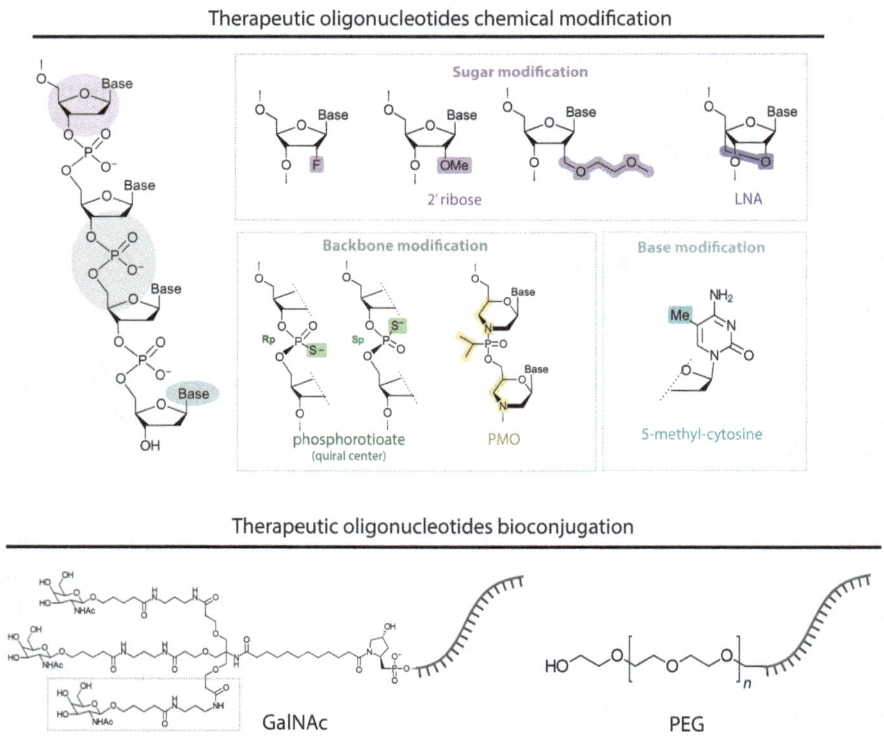

Fig. 6 Chemical modifications and conjugations that are used in approved biotherapeutic oligonucleotides. Me: methyl, LNA: locked nucleic acid, PMO: phosphorodiamidate morpholino oligomer, GalNAc: moiety derived from *N*-acetylgalactosamine, PEG: moiety derived from polyethylene glycol

3.1 Backbone Modification

Phosphorothioate (PS) modification, where a non-bridging oxygen atom in the phosphate backbone is replaced with sulfur, has become a cornerstone of therapeutic oligonucleotides. Initially introduced for their nuclease resistance, compared to standard phosphate linkages, PS modifications were later found to enhance cellular uptake [104, 105] and in vivo bioavailability [106]. This dual benefit has led to their widespread adoption, making PS modifications a standard feature in most biotherapeutic oligonucleotides today (at least 10 of the 20 FDA-approved therapeutic ONs contain this modification). PS backbone modifications are effective in creating ONs for antisense knockdowns, and also in downregulating gene expression by hybridizing to a target mRNA [107]. Even though PS backbone modifications often reduce the binding affinity of an ON for its target [108], this limitation can be overcome by incorporating additional types of modifications to the molecule [1].

PS modification can be easily incorporated in standard phosphoramidite chemistry by employing a sulfurizing agent such as 3-((dimethylamino-methylidene) amino)-3*H*-1,2,4-dithiazole-3-thione (DDTT) in the *Oxidation Step* described in the SPOS section. However, the introduction of the sulfur atom in the PS linkage results in the generation of a chiral center at each modified phosphorus atom (Fig. 6), and since this technique accounts for no stereoselectivity the resulting product is a mixture of up to 2^n individual therapeutic molecules, where n equals the number of modified linkages.

Stereochemical variation at a single center in small molecules is well-known to profoundly impact their pharmacological properties [107, 109, 110]. This call for the speculation that the racemic mixtures of therapeutic ONs currently employed might contain a large number of inactive stereoisomers that dilute the overall potency of the drug, alongside a small number of highly effective molecules. Identification of these hyperfunctional stereoisomers could be a game-changer in biotherapeutic oligonucleotide development, allowing for lower doses with more efficacious compounds. Interestingly, some researchers argue that a mix of stereoisomers is required to achieve a balance between good activity and stability [111]. For instance, recently Østergaard and colleagues reported a systematic analysis of the effect of controlling phosphorothioate chirality in the DNA gap for enhancing ASO potency and therapeutic profile, confirming that stereorandom PS linkages in the DNA gap region offer the optimal balance of activity and metabolic stability [112].

For all of the above, multiple efforts have been made to develop stereoselective synthetic approach [113], but have encountered different obstacles and thus, no practical and scalable methods for the synthesis of stereopure PS oligonucleotides, other than all-*S*p or all-*R*p centers, having pharmaceutical-grade purity were developed with phosphoramidite ONs synthesis until recently. In 2017, Iwamoto and colleagues [110] reported a method for stereocontrolled synthesis of PS ASOs that uses nucleoside 3′-oxazaphospholidine derivatives as monomers, termed SOSICS (stereocontrolled oligonucleotide synthesis with iterative capping and sulfurization). This strategy employs a stereochemically pure nucleoside oxaphospholidine monomer for the coupling step, resulting in ONs with exceptional stereochemical and chemical purity due to both high conversion and clean inversion of stereochemical configuration.

In addition, the recent demonstration of new reagents and the introduction of new chemistry with Ψ-based chemical synthesis have enabled the simple, at-will and controlled synthesis of specific chimeric oligonucleotides. Figure 2 shows the reagents available to obtain the different stereoisomers (Ψ, Ψ^2, Ψ_o, *rac*-Ψ).

In terms of enzymatic synthesis, PS-modified oligonucleotides with TiOES have not been reported yet. However, the incorporation of modified nucleoside thiotriphosphate by TdT has been demonstrated by the Hollenstein group [101]. This is expected because most DNA and RNA polymerases readily incorporate phosphorothioate-modified triphosphates [82]. Moreover, Church group found that α-PS modified NTPs can be enzymatically incorporated by their engineering PUP enzyme; however, deblocking with the allyl ether Pd/TPPTS chemistry resulted in the formation of reduction side products. Despite attempts to prevent PS degradation during deblocking, the efforts were unsuccessful, preventing the formation of the desired N + 2* product [84]. Thus, further investigations are required to incorporate PS modification in the TiOES cycle.

Moreover, polymerase generally incorporates only the *S*p-PS-dNTPs stereoisomers with inversion of stereochemistry [114, 115], thus, it is possible that enzymatic synthesis might allow some degree of stereocontrol. Actually, the enzymatic cascade oligo synthesis method described previously [89] has demonstrated the successful production of PS-modified oligonucleotides. In this case, no activity toward *R*p-PS–modified dNTPs was observed with any of the polymerases tested in their work, resulting in all-*R*p single stereoisomer oligonucleotide as final product. Nonetheless, the authors envision that the development of polymerases that incorporate the activity toward *R*p-PS-dNTP will enable production of libraries of stereochemically defined PS-modified oligonucleotides.

In contrast to negatively charged PS oligonucleotides, PMOs contain a modified backbone in which the phosphodiester linkage and the ribose sugar ring are replaced with morpholine rings and noncharged phosphodiamidate linkages (Fig. 6). PMOs show high specificity, good water solubility and are resistant to nuclease degradation [116]. However, they suffer from a decreased binding affinity for serum proteins, which leads to rapid blood clearance and limited tissue distribution [103], so different strategies have been implemented in order to enhance PMOs bioavailability [117]. PMO backbone linkages also contain chiral centers, but the effects of defined PMO stereochemistry have not been as explored as in the PS backbone case [98]. Nonetheless, recent studies have shown dramatic difference in safety profile among stereoisomers in which only one phosphorus stereochemistry differs, highlighting the need for development and characterization of stereopure compounds [118].

So far, this PMO modification has been incorporated to the FDA-approved biotherapies ***eteplirsen***, ***golodirsen***, ***viltolarsen***, and ***casimersen*** [103]. PMOs are currently synthesized with phosphoramidate P(V) chemistry on a solid support using chlorophosphoramidate building blocks, proceeding in a 5′ to 3′ direction. This approach differs from most other oligonucleotide chemistries, where synthesis typically occurs in the 3′ to 5′ direction, which severely limits the ability to synthesize chimeras having PMOs and other analogs. Furthermore, the active chlorophosphoramidate synthon used is less reactive and therefore leads to significantly lower yields than the phosphoroamidite (P(III)) oligonucleotide synthesis strategy [119]. Given these challenges, Cartuhers and Paul recently developed a synthetic route for PMOs employing a P(III) phosphoramidite approach, and have successfully prepared several morpholino phosphorodiamidite building blocks to synthesize many novel PMO compounds [119].

Besides the PS and PMO backbone modifications, there are many other kinds of backbone modifications available, such as boranophosphates, phosphorodithioates, phosphorothiolates phosphoramidates, and amides [120, 121]. However, these modifications have not been incorporated yet in any of FDA-approved therapeutics ONs.

3.2 *Modifications of Nucleobase*

Oligonucleotide base modifications are often used to create stronger Watson-Crick base pairing in order to induce higher affinity of the ON for its target, thus enhancing its therapeutic activity [110, 122, 123]. However, it is worth noting that the increased binding affinity could also lead to off-target binding increasing the risk of adverse effects. Additionally, base modifications can act to strengthen the 3D structure of aptamers [124].

The most common modification used on therapeutic ONs is the addition of a methyl group in the 5-position of pyrimidines [120]. This modification is well known to stabilize DNA/RNA duplex [125] and increase the oligonucleotide melting temperature by ~0.5°C per substitution, which is believed to be caused from the ability of the methyl group to stack between the planar nucleobases in the major groove [99]. Methylated cytosine bases (m^5C) have been incorporated in at least three of the FDA-approved oligonucleotide biotherapeutics: ***nusinersen***, ***tofersen,*** and ***eplontersen***, and are also present in other ASOs under development [98]. This modification can be easily introduced during phosphoramidite-based synthesis by employing the corresponding commercially available methylated cytosine phosphoramidite, and during P(V)-based synthesis with the corresponding methylated cytosine Ψ_o reagent.

In addition, 5-substituted pyrimidine dNTPs/NTPs are mostly good to excellent substrates for DNA/RNA polymerases since those substituents point out to the major groove of the nucleic acid [126, 127]. Thus, it is expected that the enzymatic synthesis approach does not have major limitations to include m^5C-modified nucleotides. For instance, it has been reported that TdT is capable of extending ssDNAs with unnatural nucleoside triphosphates dNaMTP and dTPT3TP, which have been most extensively employed to produce unnatural base pair-containing DNAs to expand the genetic alphabet [90]. However, no reports of TiOES m^5C-modified oligo synthesis are available. On the other hand, the synthesis of ***pegaptanib*** biotherapeutic containing several m^5C modifications was demonstrated with the enzymatic cascade method.

Following the success of 5-methyl substitution, several analogs with 5-substitutions such as propynyl [128] and thiazolyl [129] were evaluated and in every case these modifications resulted in further improved duplex stability. However, Terrazas and Kool [130] recently showed that despite the strong stabilizing effects of the 5-propynyl modification, this bulky substitution caused negative effects in RNAi activity, probably due to disruption of interactions in the major groove of the RISC complex [130]. Up to date, none of these other 5-substitution modifications have been incorporated to any of FDA approved oligonucleotides.

3.3 Modifications of Sugar

Extensive research has been conducted on chemical modifications of the furanose sugar moiety within nucleic acids in order to enhance binding affinity and reduce susceptibility toward nuclease cleavage [131]. Notably, even small structural changes to the sugar structure can have a profound influence on the biophysical and biological properties of oligonucleotides, which can be illustrated by the sharp contrast between DNA and RNA, differing only by the presence of a hydroxyl group at the 2′ position of the furanose ring. This seemingly minor modification ultimately influences their distinct roles within biological systems. This remarkable effect is explained by the presence of an electron-withdrawing group at the 2′ position and its influence in the conformational structure of the ribose ring [132]. This conformational equilibrium changes the distance between the 3′- and 5′-phosphodiester moieties, which has a deep impact on the overall structure and stability of nucleic acid duplexes [99]. In general, RNA duplexes are more stable than DNA duplexes, with mixed strands (RNA/DNA) falling in between. Leveraging this principle, chemists have designed numerous sugar-modified analogs to mimic the RNA-like conformation as a general strategy to enhance the RNA-binding affinity of ASOs [133].

One of the common sugar modifications on therapeutic oligonucleotides are 2′-*O*-alkyl modifications, which have also been shown to increase oligonucleotide nuclease resistance leading to improved stability in plasma, increased tissue half-lives and, consequently, prolonged drug effects [134]. 2′-*O*-methyl (2′-OMe) is a naturally occurring modification and was one of the earliest sugar modifications employed in antisense medicinal chemistry. 2′-OMe remains today one of the most applied modification in several strategies, including the FDA approved ***pegaptanib***, ***patisiran***, ***givosiran***, ***lumasiran***, ***avacincaptad pegol***, ***nedosiran***, ***vutrisiran,*** and ***inclisiran*** biotherapeutics [4, 5, 103, 135]. The appealing properties of 2′-OMe oligonucleotides inspired researchers to explore other alkyl substituents [136], and 2′-*O*-methoxyethyl (2′MOE) emerged as particularly effective, offering both increased nuclease resistance and improved binding affinity [133]. This modification can be found in the FDA-approved drugs ***mipomersen***, ***nusinersen***, ***inotersen***, ***tofersen,*** and ***eplontersen*** [5, 103].

Another extensively investigated 2′ modification is the incorporation of a fluorine atom to replace the 2′-OH group. This 2'-F modification appears to have a privileged role for enhancing the drug-like properties of siRNA [137] and has also been incorporated to the FDA-approved oligonucleotides ***pegaptanib***, ***givosiran***, ***lumasiran***, ***inclisiran***, ***vutrisiran***, ***nedosiran,*** and ***avacincaptad pegol*** [4, 5, 103, 135]. Even though 2'-F-modified oligonucleotides have shown enhancement of duplex stability without compromising base-pair specificity, the phosphodiester form of these compounds was not resistant to nucleases. However, the 2'-F-modified phosphorothioate oligonucleotides were highly nuclease-resistant and retained exceptional binding affinity to the RNA targets [138].

All the modifications mentioned above can be readily incorporated during phosphoramidite-based oligo chemical synthesis employing the commercially available 2′ modified phosphoramidites. Related to the incorporation of these modifications in the 2′ position of the furanose ring through enzymatic synthesis, it is known that nucleotides displaying sugar modifications are often poor substrates for polymerases and require the help of engineered polymerases to be successfully incorporated into DNA. Recently, Church group reported the first example of 10 synthetic cycles of TiOES achieving an RNA sequence that incorporates several NTPs with 2'-F or 2'-OMe modification. Although the overall yield of synthesis was low, it was a remarkable step.

The previously discussed modifications aimed to shift the conformational equilibrium of the ribose ring toward the RNA-like form. Another well-studied sugar modification is the locked nucleic acid (LNA). LNA is a nucleic acid analog that is conformationally locked in the favorable C3-endo sugar conformation by an O2' to C4' methylene linkage (Fig. 6). LNA exhibits exceptionally high binding affinity and sequence specificity, as well as high stability and low toxicity in biologic systems [139, 140]. LNA-enhanced binding affinities enable short-sequence ASOs to function as potent therapeutics, as long as an appropriately specific sequence can be found. ***Miravirsen***, a now discontinued 15-mer ASO that reached phase II clinical trials, exemplifies this concept, with 8 LNA and 7 DNA units, demonstrating the promise of shorter, LNA-containing ASOs for therapeutic applications [141].

LNA contains natural phosphodiester linkages and, therefore, can be obtained by automated synthesis using conventional phosphoramidite chemistry with the corresponding commercially available modified monomers. It is worth mentioning that LNA monomers are compatible with other modified monomers and allow for PS linkages along with other backbone modifications. Moreover, LNA modifications are also compatible with the Ψ platform for P(V)-based ON synthesis.

Initial studies of the possibility of including LNA-nucleotides by TiOES have been reported. Hollestain group synthesized LNA nucleoside triphosphates with various 3′-*O*-blocking groups, such as phosphate, esters, and ethers [142]. They discovered that ether protecting groups can be incorporated into DNA by template-independent synthesis using TdT and PUP polymerase. However, the efficiency of incorporating N + 1 products rarely exceeded 60%–70%, making it impractical for de novo LNA synthesis. Interestingly, although both polymerases accepted several3'-*O*-blocked LNA nucleotides as substrates, they stalled after incorporating a single LNA nucleotide, whether protected or unprotected [101, 142, 143].

3.4 Bioconjugation

Conjugating specific molecules to oligonucleotides is a promising therapeutic approach that enhances the delivery of nucleic acid-based biotherapeutics and, thus, its presence is increasing in the pharmaceutical development pipeline. Bioconjugates are distinct, homogeneous, single-component molecular entities with precise stoichiometry, which allows for simple, high-scale synthesis and well-defined pharmacokinetic properties. Moreover, these covalently conjugated molecules may play one or more roles in recognition, targeting, and cellular internalization.

PEG attachment is a method to increase solubility of oligonucleotides and prevent the rapid elimination, thus increasing tissue distribution [144]. However, the incorporation of long PEG chains can negatively influence the pharmacodynamic effect by reducing hybridization efficiency. Therefore, care must be taken when incorporating this modification into antisense agents [145]. Up to date, only the two FDA-approved therapeutic aptamers, ***pegaptanib*** [103] and ***avacincaptad pegol*** [5], possess PEG conjugation.

The GalNAc (*N*-acetylgalactosamine) modification, introduced by Nair and colleagues [146], leads to hydrophilic glycoconjugates, increasing the cellular internalization in the liver due to binding to the ASialoGlycoProtein Receptor (ASGPR), a receptor predominantly expressed on hepatocytes. The FDA-approved ON biotherapeutics ***inclisiran***, ***givosiran***, ***lumasiran***, ***vutrisiran***, ***nedosiran***, and ***eplontersen*** are all GalNAc conjugates [4, 5, 103, 135].

Both modifications can be easily incorporated at any position of the ONs through classic phosphoramidite synthesis, as modified "ready-to-use" building blocks and solid supports are commercially available. Other conjugates, such as lipids [147], peptides [148], aptamers [149], and antibodies [150] have been studied, but none have yet been incorporated into approved therapies and have been revised elsewhere [151].

4 Summary and Future Directions

Oligonucleotide biotherapeutics are now recognized as the third major drug discovery platform, in addition to small-molecule drugs and antibodies. This impact increased the demand for large volumes of complex modified oligonucleotide products, thus pushing advances in oligonucleotide synthesis that involve improvement in already established technologies and the development of new chemistries (Ψ platform), new enzymatic processes (biocatalytics), and new manufacturing technologies (liquid phase).

Phosphoramidite chemistry has been the gold standard in ONs manufacturing, but its severe limitations regarding sustainability, facing the increasing demand of therapeutic ONs, are a major concern. Moreover, the difficulty of generating stereochemically defined oligonucleotides is a critical issue. In this regard, the development of the Ψ platform derived from inexpensive limonene and with strict stereocenter control represents a significant breakthrough.

SPOS is, to date, the protocol of choice due to its speed, efficiency, batch-to-batch consistency, and scalability from gram to kilogram quantities. However, SPOS struggles with cost-effective production in hundreds of kilograms at fast turnaround times. LPOS then emerges as a promising candidate for the upcoming ONs demand, as its homogeneous nature enables higher reaction concentrations, minimal reagent waste, and the desired scalability.

Enzymatic oligonucleotide synthesis has also appeared as a promising synthetic method. On the one hand, TiOES approach has shown high success for the de novo synthesis of long, unmodified DNA and RNA sequences. However, many challenges appear when modifications, particularly in the nucleobase and sugar need to be incorporated, due to no-template polymerases (TdT, PUP) limitations. Hence, advances in polymerases engineering are necessary to enhance this approach, which can be achieved through either rational design methods or evolutionary experiments [115, 152]. On the other hand, the enzymatic cascade method developed by Moody and colleagues shows the potential to scale up the production of many short and highly modified oligonucleotides. While this method is not capable of de novo oligonucleotide synthesis, the chemically obtained template is only required in catalytic quantities ($< 1\%$).

To unlock the full therapeutic potential of oligonucleotides, a unified approach that includes diverse synthesis methods, combining chemical and biocatalytic strategies, and bringing together beneficial aspects of each technique is called for.

Compliance with Ethical Standards This chapter is a review and no humans or animals were involved in the preparation of the chapter.

References

1. Crooke ST, Baker BF, Crooke RM et al (2021) Antisense technology: an overview and prospectus. Nat Rev Drug Discov 20:427–453
2. Obexer R, Nassir M, Moody ER et al (2024) Modern approaches to therapeutic oligonucleotide manufacturing. Science 384:eadl4015
3. A review on commercial oligonucleotide drug products – ScienceDirect. https://www.sciencedirect.com/science/article/abs/pii/S0022354924001527. Accessed 2 July 2024
4. Al Musaimi O, Al Shaer D, Albericio F et al (2023) 2022 FDA TIDES (peptides and oligonucleotides) harvest. Pharmaceuticals 16:336
5. Al Shaer D, Al Musaimi O, Albericio F et al (2024) 2023 FDA TIDES (peptides and oligonucleotides) harvest. Pharmaceuticals 17:243

6. Kim JF, Gaffney PRJ, Valtcheva IB et al (2016) Organic solvent nanofiltration (OSN): a new technology platform for liquid-phase oligonucleotide synthesis (LPOS). Org Process Res Dev 20:1439–1452
7. Andrews BI, Antia FD, Brueggemeier SB et al (2021) Sustainability challenges and opportunities in oligonucleotide manufacturing. J Org Chem 86:49–61
8. Hoose A, Vellacott R, Storch M et al (2023) DNA synthesis technologies to close the gene writing gap. Nat Rev Chem 7:144–161
9. Kosuri S, Church GM (2014) Large-scale de novo DNA synthesis: technologies and applications. Nat Methods 11:499–507
10. Michelson AM, Todd AR (1955) Nucleotides part XXXII. Synthesis of a dithymidine dinucleotide containing a 3′: 5′-internucleotidic linkage. J Chem Soc 0:2632–2638
11. Hall RH, Todd A, Webb RF (1957) 644. Nucleotides. Part XLI. Mixed anhydrides as intermediates in the synthesis of dinucleoside phosphates. J Chem Soc 0:3291–3296
12. Khorana HG, Razzell WE, Gilham PT et al (1957) Synthesis of dideoxyribonucleotides. J Am Chem Soc 79:1002–1003
13. ATDBio – Solid-phase oligonucleotide synthesis. https://atdbio.com/nucleic-acids-book/Solid-phase-oligonucleotide-synthesis. Accessed 1 July 2024
14. Knouse KW, Flood DT, Vantourout JC et al (2021) Nature chose phosphates and chemists should too: how emerging P(V) methods can augment existing strategies. ACS Cent Sci 7: 1473–1485
15. Brzezinska J, Trzciński S, Strzelec J et al (2023) From CPG to hybrid support: review on the approaches in nucleic acids synthesis in various media. Bioorg Chem 140:106806
16. Yang J, Stolee JA, Jiang H et al (2018) Solid-phase synthesis of phosphorothioate oligonucleotides using sulfurization byproducts for in situ capping. J Org Chem 83:11577–11585
17. Kim SW, Cho YI, Jung KE (2021) Avoiding high-pressure problem for modified RNA -attached polystyrene support by pre-swelling using toluene in the oligonucleotide synthesis. Bull Korean Chem Soc 42:1296–1303
18. Bhadra J, Kundu J, Ghosh KC et al (2015) Synthesis of phosphorodiamidate morpholino oligonucleotides by H-phosphonate method. Tetrahedron Lett 56:4565–4568
19. Bhadra J, Kundu J, Ghosh KC et al (2018) Corrigendum to "Synthesis of phosphorodiamidate morpholino oligonucleotides by H-phosphonate method" [Tetrahedron Lett. 56 (2015) 4565–4568]. Tetrahedron Lett 59:1974–1975
20. Li C, Callahan AJ, Simon MD et al (2021) Fully automated fast-flow synthesis of antisense phosphorodiamidate morpholino oligomers. Nat Commun 12:4396
21. Huang Y, Knouse KW, Qiu S et al (2021) A P(V) platform for oligonucleotide synthesis. Science 373:1265–1270
22. Grajkowski A, Takahashi M, Cawrse BM et al (2020) An expedient process for reducing the formation of process-related impurities during solid-phase synthesis of potential nucleic acid-based drugs. Bioorg Med Chem 28:115779
23. Grajkowski A, Takahashi M, Cawrse BM et al (2021) An improved PEG-linked solid support for minimizing process-related impurities during solid-phase synthesis of DNA and RNA sequences. Curr Protoc 1:e108
24. Rothstein ML, Rothstein DM, Lee DP (2016) Hybrid solid supports useful for oligonucleotide production. US Patent 9,273,087 B2, 1 Mar 2016
25. Reese CB, Saffhill R (1968) Oligonucleotide synthesis via phosphotriester intermediates: the phenyl-protecting group. Chem Commun 13:767–768
26. Letsinger RL, Lunsford WB (1976) Synthesis of thymidine oligonucleotides by phosphite triester intermediates. J Am Chem Soc 98:3655–3661
27. Knouse KW, deGruyter JN, Schmidt MA et al (2018) Unlocking P(V): reagents for chiral phosphorothioate synthesis. Science 361:1234–1238
28. Stec WJ, Grajkowski A, Kobylanska A et al (1995) Diastereomers of nucleoside 3'-o-(2-thio-1,3,2-oxathia(selena)phospholanes): building blocks for stereocontrolled synthesis of oligo (nucleoside phosphorothioate)s. J Am Chem Soc 117:12019–12029

29. Stec WJ, Karwowski B, Boczkowska M et al (1998) Deoxyribonucleoside 3′-o-(2-thio- and 2-oxo-"spiro"-4,4-pentamethylene-1,3,2-oxathiaphospholane)s: monomers for stereocontrolled synthesis of oligo(deoxyribonucleoside phosphorothioate)s and chimeric PS/PO oligonucleotides. J Am Chem Soc 120:7156–7167
30. Stec WJ, Grajkowski A, Koziolkiewicz M et al (1991) Novel route to oligo (deoxyribonucleoside phosphorothioates). Stereocontrolled synthesis of P-chiral oligo (deoxyribonucleoside phosphorothioates). Nucleic Acids Res 19:5883–5888
31. Shi X, Kiesman WF, Antia F et al (2021) Convergent liquid phase syntheses of oligonucleotides. World Patent 2020227618 A3, 12 Nov 2020
32. Albaek N, Funder E, Katkevica D et al (2022) Novel process for preparing nucleotide P (V) monomers. European Patent 3956340A1, 23 Feb 2022
33. Schmidt MA, Zheng B, Cohen BM et al (2020) Process for the preparation of a cyclic dinucleotide. World Patent 2020117739, 11 Jun 2020
34. Ravikumar VT, Kumar RK, Olsen P et al (2008) UnyLinker: an efficient and scaleable synthesis of oligonucleotides utilizing a universal linker molecule: a novel approach to enhance the purity of drugs. Org Process Res Dev 12:399–410
35. Yagodkin A, Weisel J, Azhayev A (2011) Versatile solid supports for oligonucleotide synthesis that incorporate urea bridge. Nucleos Nucleot Nucl 30:475–489
36. Moore MN, Arthur JC, Vansooy K et al (2002) Processes of purifying oligonucleotides. World Patent 2002100873A1, 19 Dec 2002
37. Introduction to tangential flow filtration for laboratory and process development applications. https://cdn.cytivalifesciences.com/api/public/content/S13cdYQ2lkiBbGm8_R533Q-original?v=643951f5 Accessed 03 July 2024
38. Sanghvi YS, Ravikumar VT, Scozzari AN et al (2001) Applications of green chemistry in the manufacture of oligonucleotide drugs. Pure Appl Chem 73:175–180
39. Pearson WH, Berry DA, Stoy P et al (2005) Fluorous affinity purification of oligonucleotides. J Org Chem 70:7114–7122
40. Fang S, Bergstrom DE (2004) Reversible 5′-end biotinylation and affinity purification of synthetic RNA. Tetrahedron Lett 45:7987–7990
41. Franzini RM, Biendl S, Mikutis G et al (2015) "Cap-and-catch" purification for enhancing the quality of libraries of DNA conjugates. ACS Comb Sci 17:393–398
42. Pinto A, Chen SX, Zhang DY (2018) Simultaneous and stoichiometric purification of hundreds of oligonucleotides. Nat Commun 9:2467
43. Igata Y, Saito-Tarashima N, Matsumoto D et al (2017) A 'catch and release' strategy towards HPLC-free purification of synthetic oligonucleotides by a combination of the strain-promoted alkyne-azide cycloaddition and the photocleavage. Bioorg Med Chem 25:5962–5967
44. Agustin E, Okai PNA, Khan I et al (2016) A fast click–slow release strategy towards the HPLC-free synthesis of RNA. Chem Commun 52:1405–1408
45. Grajkowski A, Cieslak J, Beaucage SL (2016) Solid-phase purification of synthetic DNA sequences. J Org Chem 81:6165–6175
46. Grajkowski A, Cawrse BM, Takahashi M et al (2022) An improved process for the release of synthetic DNA sequences from a solid-phase capture support. Tetrahedron Lett 106:154077
47. He M, Wu X, Mao S et al (2021) Bio-orthogonal chemistry enables solid phase synthesis and HPLC and gel-free purification of long RNA oligonucleotides. Chem Commun 57:4263–4266
48. Fang S, Fueangfung S, Lin X et al (2011) Synthetic oligodeoxynucleotide purification by polymerization of failure sequences. Chem Commun 47:1345–1347
49. Pokharel D, Yuan Y, Fueangfung S et al (2014) Synthetic oligodeoxynucleotide purification by capping failure sequences with a methacrylamide phosphoramidite followed by polymerization. RSC Adv 4:8746–8757
50. Albosta MS, Grant JK, Taub P et al (2023) Inclisiran: a new strategy for LDL-C lowering and prevention of atherosclerotic cardiovascular disease. Vasc Health Risk Manag 19:421–431
51. Suchsland R, Appel B, Müller S (2018) Synthesis of trinucleotide building blocks in solution and on solid phase. Curr Protoc Nucleic Acid Chem 75:e60

52. Molina AG, Sanghvi YS (2019) Liquid-phase oligonucleotide synthesis: past, present, and future predictions. Curr Protoc Nucleic Acid Chem 77:e82
53. Gaffney PRJ, Kim JF, Valtcheva IB et al (2015) Liquid-phase synthesis of 2′-methyl-RNA on a homostar support through organic-solvent nanofiltration. Chem 21:9535–9543
54. Takahashi D, Inomata T, Fukui T (2017) AJIPHASE®: a highly efficient synthetic method for one-pot peptide elongation in the solution phase by an fmoc strategy. Angew Chem Int Ed 56: 7803–7807
55. 平井邦博, 片山智 (2012) Method for producing oligonucleotide. World Patent 2012157723A1, 22 Nov 2012
56. FDA approves first drug using AJIPHASE production process. In: Contract Pharma. https://www.contractpharma.com/contents/view_breaking-news/2020-10-07/fda-approves-first-drug-using-ajiphase-production-process/. Accessed 27 May 2024
57. Chen C-H, Chen W-Y, Chen Y-C et al (2010) Convergent solution phase synthesis of chimeric oligonucleotides by a 2+2 and 3+3 phosphoramidite strategy. Aust J Chem 63:227–235
58. Reese CB, Yan H (2002) Solution phase synthesis of ISIS 2922 (Vitravene) by the modified H-phosphonate approach. J Chem Soc Perkin Trans 1:2619–2633
59. Gibson DG, Benders GA, Andrews-Pfannkoch C et al (2008) Complete chemical synthesis, assembly, and cloning of a mycoplasma genitalium genome. Science 319:1215–1220
60. Gibson DG, Glass JI, Lartigue C et al (2010) Creation of a bacterial cell controlled by a chemically synthesized genome. Science 329:52–56
61. Bollum FJ (1959) Thermal conversion of non priming deoxyribonucleic acid to primer. J Biol Chem 234:2733–2734
62. Deibel MR, Coleman MS (1980) Biochemical properties of purified human terminal deoxynucleotidyl transferase. J Biol Chem 255:4206–4212
63. Motea EA, Berdis AJ (2010) Terminal deoxynucleotidyl transferase: the story of a misguided DNA polymerase. Biochim Biophys Acta 1804:1151–1166
64. Jensen MA, Davis RW (2018) Template-independent enzymatic oligonucleotide synthesis (TiEOS): its history, prospects, and challenges. Biochemistry 57:1821–1832
65. Tjong V, Yu H, Hucknall A et al (2011) Amplified on-chip fluorescence detection of DNA hybridization by surface-initiated enzymatic polymerization. Anal Chem 83:5153–5159
66. Chen F, Gaucher EA, Leal NA et al (2010) Reconstructed evolutionary adaptive paths give polymerases accepting reversible terminators for sequencing and SNP detection. Proc Natl Acad Sci USA 107:1948–1953
67. Hutter D, Kim MJ, Karalkar N et al (2010) Labeled nucleoside triphosphates with reversibly terminating aminoalkoxyl groups. Nucleos Nucleot Nucl 29:879–895
68. Guo J, Xu N, Li Z et al (2008) Four-color DNA sequencing with 3′-O-modified nucleotide reversible terminators and chemically cleavable fluorescent dideoxynucleotides. Proc Natl Acad Sci USA 105:9145–9150
69. Metzker ML, Raghavachari R, Richards S et al (1994) Termination of DNA synthesis by novel 3′-modifieddeoxyribonucleoside 5′-triphosphates. Nucl Acids Res 22:4259–4267
70. Stemple DL, Fraser AG, Mankowska S et al (2019) Compositions and methods for template-free geometric enzymatic nucleic acid synthesis. World Patent 2019140353A1, 18 Jul 2019
71. Mathews AS, Yang H, Montemagno C (2016) Photo-cleavable nucleotides for primer free enzyme mediated DNA synthesis. Org Biomol Chem 14:8278–8288
72. Bowers J, Mitchell J, Beer E et al (2009) Virtual terminator nucleotides for next-generation DNA sequencing. Nat Methods 6:593–595
73. Wu W, Stupi BP, Litosh VA et al (2007) Termination of DNA synthesis by N6-alkylated, not 3'-O-alkylated, photocleavable 2′-deoxyadenosine triphosphates. Nucleic Acids Res 35: 6339–6349
74. Wojciechowski F, Ybert T (2021) Method for preparing 3′-O-amino-2′-deoxyribonucleoside-5′-triphosphates. US Patent 20210300961A1, 30 Sep 2021

75. Gouge J, Rosario S, Romain F et al (2013) Structures of intermediates along the catalytic cycle of terminal deoxynucleotidyl transferase: dynamical aspects of the two-metal ion mechanism. J Mol Biol 425:4334–4352
76. Chua JPS, Go MK, Osothprarop T et al (2020) Evolving a thermostable terminal deoxynucleotidyl transferase. ACS Synth Biol 9:1725–1735
77. Barthel S, Palluk S, Hillson NJ et al (2020) Enhancing terminal deoxynucleotidyl transferase activity on substrates with 3′ terminal structures for enzymatic de novo DNA synthesis. Genes 11:102
78. Lu X, Li J, Li C et al (2022) Enzymatic DNA synthesis by engineering terminal deoxynucleotidyl transferase. ACS Catal 12:2988–2997
79. Yu M, Tang X, Li Z et al (2024) High-throughput DNA synthesis for data storage. Chem Soc Rev 53:4463–4489
80. Palluk S, Arlow DH, De Rond T et al (2018) De novo DNA synthesis using polymerase-nucleotide conjugates. Nat Biotechnol 36:645–650
81. Li S, Tan W, Jia X et al (2024) Recent advances in the synthesis of single-stranded DNA in vitro. Biotechnol J 19:2400026
82. Sarac I, Hollenstein M (2019) Terminal deoxynucleotidyl transferase in the synthesis and modification of nucleic acids. Chembiochem 20:860–871
83. Delarue M, Boulé JB, Lescar J et al (2002) Crystal structures of a template-independent DNA polymerase: murine terminal deoxynucleotidyltransferase. The EMBO J 21:427–439
84. Wiegand DJ, Rittichier J, Meyer E et al (2023) A Platform for controlled template-independent enzymatic synthesis of RNA oligonucleotides and therapeutics. bioRxiv 2023.06.29.547106
85. Eisenstein M (2020) Enzymatic DNA synthesis enters new phase. Nat Biotechnol 38:1113–1115
86. SYNTAX System. In: DNA Script. https://www.dnascript.com/products/syntax/. Accessed 1 Jul 2024
87. Efcavitch JW (2020) Reusable initiators for synthesizing nucleic acids. US Patent 10683536B2, 16 Jun 2020
88. Hoff K, Halpain M, Garbagnati G et al (2020) Enzymatic synthesis of designer DNA using cyclic reversible termination and a universal template. ACS Synth Biol 9:283–293
89. Moody ER, Obexer R, Nickl F et al (2023) An enzyme cascade enables production of therapeutic oligonucleotides in a single operation. Science 380:1150–1154
90. Wang G, He C, Zou J et al (2022) Enzymatic synthesis of DNA with an expanded genetic alphabet using terminal deoxynucleotidyl transferase. ACS Synth Biol 11:4142–4155
91. Crameri A, Hill ML, Lovelock SL et al (2018) Novel processes for the production of oligonucleotides. World Patent 2018011067A2, 18 Jan 2018
92. Van Giesen KJD, Thompson MJ, Meng Q et al (2023) Biocatalytic synthesis of antiviral nucleosides, cyclic dinucleotides, and oligonucleotide therapies. JACS Au 3:13–24
93. Paul S, Gray D, Caswell J et al (2023) Convergent biocatalytic mediated synthesis of siRNA. ACS Chem Biol 18:2183–2187
94. Biocataltic Oligonucleotide Synthesis Technology. In: Knowledge Centre. https://www.almacgroup.com/knowledge/library/biocataltic-oligonucleotide-synthesis-technology/. Accessed 30 Jun 2024
95. Kikuchi Y, Sakaguchi K (1978) Enzymatic synthesis of a segment of bacteriophage Qβ coat protein gene. Nucleic Acids Res 5:591–598
96. Moumné L, Marie A-C, Crouvezier N (2022) Oligonucleotide therapeutics: from discovery and development to patentability. Pharmaceutics 14:260–284
97. Hoelder S, Clarke PA, Workman P (2012) Discovery of small molecule cancer drugs: successes, challenges and opportunities. Mol Oncol 6:155–176
98. Roberts TC, Langer R, Wood MJA (2020) Advances in oligonucleotide drug delivery. Nat Rev Drug Discov 19:673–694
99. Wan WB, Seth PP (2016) The medicinal chemistry of therapeutic oligonucleotides. J Med Chem 59:9645–9667

100. Flemmich L, Bereiter R, Micura R (2024) Chemical synthesis of modified RNA. Angew Chem 136:e202403063
101. Flamme M, Hanlon S, Iding H et al (2021) Towards the enzymatic synthesis of phosphorothioate containing LNA oligonucleotides. Bioorg Med Chem Lett 48:128242
102. Kulkarni JA, Witzigmann D, Thomson SB et al (2021) The current landscape of nucleic acid therapeutics. Nat Nanotechnol 16:630–643
103. Bost JP, Barriga H, Holme MN et al (2021) Delivery of oligonucleotide therapeutics: chemical modifications, lipid nanoparticles, and extracellular vesicles. ACS Nano 15:13993–14021
104. Liang X, Sun H, Shen W et al (2015) Identification and characterization of intracellular proteins that bind oligonucleotides with phosphorothioate linkages. Nucleic Acids Res 43: 2927–2945
105. Wang S, Allen N, Vickers TA et al (2018) Cellular uptake mediated by epidermal growth factor receptor facilitates the intracellular activity of phosphorothioate-modified antisense oligonucleotides. Nucleic Acids Res 46:3579–3594
106. Sands H, Gorey-Feret LJ, Cocuzza AJ et al (1994) Biodistribution and metabolism of internally 3H-labeled oligonucleotides. I. Comparison of a phosphodiester and a phosphorothioate. Mol Pharmacol 45:932–943
107. Eckstein F (2014) Phosphorothioates, essential components of therapeutic oligonucleotides. Nucleic Acid Ther 24:374–387
108. Freier SM, Altmann KH (1997) The ups and downs of nucleic acid duplex stability: structure-stability studies on chemically-modified DNA:RNA duplexes. Nucleic Acids Res 25:4429–4443
109. Research C for DE and (2020) Development of new stereoisomeric drugs. https://www.fda.gov/regulatory-information/search-fda-guidance-documents/development-new-stereoisomeric-drugs. Accessed 30 May 2024
110. Iwamoto N, Butler DCD, Svrzikapa N et al (2017) Control of phosphorothioate stereochemistry substantially increases the efficacy of antisense oligonucleotides. Nat Biotechnol 35:845–851
111. Wan WB, Migawa MT, Vasquez G et al (2014) Synthesis, biophysical properties and biological activity of second generation antisense oligonucleotides containing chiral phosphorothioate linkages. Nucleic Acids Res 42:13456–13468
112. Østergaard ME, De Hoyos CL, Wan WB et al (2020) Understanding the effect of controlling phosphorothioate chirality in the DNA gap on the potency and safety of gapmer antisense oligonucleotides. Nucleic Acids Res 48:1691–1700
113. Guga P, Stec WJ (2003) Synthesis of phosphorothioate oligonucleotides with stereodefined phosphorothioate linkages. Curr Protoc Nucleic Acid Chem 14:1–28
114. Griffiths AD, Potter BVL, Eperon IC (1987) Stereospecificity of nucleases towards phosphorothioate-substituted RNA: stereochemistry of transcription by T7 RNA polymerase. Nucleic Acids Res 15:4145–4162
115. Arangundy-Franklin S, Taylor AI, Porebski BT et al (2019) A synthetic genetic polymer with an uncharged backbone chemistry based on alkyl phosphonate nucleic acids. Nat Chem 11: 533–542
116. Summerton J, Weller D (1997) Morpholino antisense oligomers: design, preparation, and properties. Antisense Nucleic Acid Drug Dev 7:187–195
117. Amantana A, Moulton HM, Cate ML et al (2007) Pharmacokinetics, biodistribution, stability and toxicity of a cell-penetrating peptide-morpholino oligomer conjugate. Bioconjug Chem 18:1325–1331
118. Kanatsu K, Takahashi Y, Sakaguchi T et al (2024) Discovery and characterization of stereodefined PMO-gapmers targeting tau. bioRxiv. https://doi.org/10.1101/2024.05.09.591947
119. Paul S, Caruthers MH (2023) Synthesis of backbone-modified morpholino oligonucleotides using phosphoramidite chemistry. Molecules 28:5380

120. Herdewijn P (2000) Heterocyclic modifications of oligonucleotides and antisense technology. Antisense Nucleic Acid Drug Dev 10:297–310
121. Epple S, El-Sagheer AH, Brown T (2021) Artificial nucleic acid backbones and their applications in therapeutics, synthetic biology and biotechnology. Emerg Top Life Sci 5:691–697
122. Mangos MM, Damha MJ (2002) Flexible and frozen sugar-modified nucleic acids – modulation of biological activity through furanose ring dynamics in the antisense strand. Curr Top Med Chem 2:1147–1171
123. Yoshida T, Hagihara T, Uchida Y et al (2024) Introduction of sugar-modified nucleotides into CpG-containing antisense oligonucleotides inhibits TLR9 activation. Sci Rep 14:11540
124. Ji D, Feng H, Liew SW et al (2023) Modified nucleic acid aptamers: development, characterization, and biological applications. Trends Biotechnol 41:1360–1384
125. Zhao XC, Dong HL, Li XL et al (2022) 5-methyl-cytosine stabilizes DNA but hinders DNA hybridization revealed by magnetic tweezers and simulations. Nucleic Acids Res 50:12344–12354
126. Kropp HM, Diederichs K, Marx A (2019) The structure of an archaeal B-family DNA polymerase in complex with a chemically modified nucleotide. Angew Chem Int Ed 58: 5457–5461
127. Hottin A, Marx A (2016) Structural insights into the processing of nucleobase-modified nucleotides by DNA polymerases. Acc Chem Res 49:418–427
128. Wagner RW, Matteucci MD, Lewis JG et al (1993) Antisense gene inhibition by oligonucleotides containing C-5 propyne pyrimidines. Science 260:1510–1513
129. Gutierrez AJ, Terhorst TJ, Matteucci MD et al (1994) 5-Heteroaryl-2′-deoxyuridine analogs. Synthesis and incorporation into high-affinity oligonucleotides. J Am Chem Soc 116:5540–5544
130. Terrazas M, Kool ET (2009) RNA major groove modifications improve siRNA stability and biological activity. Nucleic Acids Res 37:346–353
131. Egli M, Minasov G, Tereshko V et al (2005) Probing the influence of stereoelectronic effects on the biophysical properties of oligonucleotides: comprehensive analysis of the RNA affinity, nuclease resistance, and crystal structure of ten 2’-O-ribonucleic acid modifications. Biochemistry 44:9045–9057
132. Evich M, Spring-Connell AM, Germann MW (2017) Impact of modified ribose sugars on nucleic acid conformation and function. Heterocycl Commun 23:155–165
133. Manoharan M (1999) 2′-carbohydrate modifications in antisense oligonucleotide therapy: importance of conformation, configuration and conjugation. Biochim Biophys Acta 1489: 117–130
134. Monia BP, Johnston JF, Sasmor H et al (1996) Nuclease resistance and antisense activity of modified oligonucleotides targeted to Ha-ras. J Biol Chem 271:14533–14540
135. Shaer DA, Musaimi OA, Albericio F et al (2022) 2021 FDA TIDES (peptides and oligonucleotides) harvest. Pharmaceuticals 15:222
136. Martin P (1995) Ein neuer zugang zu 2′-O-alkylribonucleosiden und eigenschaften deren oligonucleotide. Helv Chim Acta 78:486–504
137. Manoharan M, Akinc A, Pandey RK et al (2011) Unique gene-silencing and structural properties of 2′-F modified siRNAs. Angew Chem Int Ed 50:2284–2288
138. Kawasaki AM, Casper MD, Freier SM et al (1993) Uniformly modified 2′-deoxy-2′-fluorophosphorothioate oligonucleotides as nuclease-resistant antisense compounds with high affinity and specificity for RNA targets. J Med Chem 36:831–841
139. Jepsen JS, Sørensen MD, Wengel J (2004) Locked nucleic acid: a potent nucleic acid analog in therapeutics and biotechnology. Oligonucleotides 14:130–146
140. Kamali MJ, Salehi M, Fatemi S et al (2023) Locked nucleic acid (LNA): a modern approach to cancer diagnosis and treatment. Exp Cell Res 423:113442
141. Discovering the first microRNA-targeted drug | Journal of Cell Biology | Rockefeller University Press. https://rupress.org/jcb/article/199/3/407/37008/Discovering-the-first-microRNA-targeted. Accessed 6 June 2024

142. Sabat N, Katkevica D, Pajuste K et al (2023) Towards the controlled enzymatic synthesis of LNA containing oligonucleotides. Front Chem 11:1–10
143. Pichon M, Hollenstein M (2024) Controlled enzymatic synthesis of oligonucleotides. Commun Chem 7:1–11
144. Bonora GM (2002) Polymer-conjugated bioactive oligonucleotides. J Bioact Compat Pol 17: 375–389
145. Shokrzadeh N, Winkler AM, Dirin M et al (2014) Oligonucleotides conjugated with short chemically defined polyethylene glycol chains are efficient antisense agents. Bioorg Med Chem Lett 24:5758–5761
146. Nair JK, Willoughby JLS, Chan A et al (2014) Multivalent N-acetylgalactosamine-conjugated siRNA localizes in hepatocytes and elicits robust RNAi-mediated gene silencing. J Am Chem Soc 136:16958–16961
147. Wolfrum C, Shi S, Jayaprakash KN et al (2007) Mechanisms and optimization of in vivo delivery of lipophilic siRNAs. Nat Biotechnol 25:1149–1157
148. Eguchi A, Meade BR, Chang YC et al (2009) Efficient siRNA delivery into primary cells by a peptide transduction domain–dsRNA binding domain fusion protein. Nat Biotechnol 27:567–571
149. McNamara JO, Andrechek ER, Wang Y et al (2006) Cell type-specific delivery of siRNAs with aptamer-siRNA chimeras. Nat Biotechnol 24:1005–1015
150. Song E, Zhu P, Lee SK et al (2005) Antibody mediated in vivo delivery of small interfering RNAs via cell-surface receptors. Nat Biotechnol 23:709–717
151. Benizri S, Gissot A, Martin A et al (2019) Bioconjugated oligonucleotides: recent developments and therapeutic applications. Bioconjug Chem 30:366–383
152. Freund N, Taylor AI, Arangundy-Franklin S et al (2023) A two-residue nascent-strand steric gate controls synthesis of 2′-O-methyl- and 2′-O-(2-methoxyethyl)-RNA. Nat Chem 15:91–100

Part IV
Other Biotherapeutics

Top Med Chem
https://doi.org/10.1007/7355_2024_175

Published online: 5 January 2025

Conventional and Emerging Cell Therapy Strategies

Mailín Casadei, Mariel Fusco, María José Cantero, Juan Bayo, Mariana Malvicini, and Esteban Fiore

Contents

Abstract Cell therapy is the transfer of cells into a patient that has emerged as a promising approach to improve or even cure a wide range of diseases. This strategy

M. Casadei, M. J. Cantero, J. Bayo, and E. Fiore (✉)
Programa de Hepatología Experimental y Terapia Génica, Instituto de Investigaciones en Medicina Traslacional, Universidad Austral-Consejo Nacional de Investigaciones Científicas y Técnicas, Pilar, Argentina
e-mail: mcasadei@austral.edu.ar; mcantero-conicet@austral.edu.ar; jbayofina@austral.edu.ar; efiore@austral.edu.ar

M. Fusco and M. Malvicini
Cancer Immunobiology Laboratory, Instituto de Investigaciones en Medicina Traslacional, Universidad Austral-Consejo Nacional de Investigaciones Científicas y Técnicas, Pilar, Argentina
e-mail: mfusco-conicet@austral.edu.ar; mmalvicini@austral.edu.ar

takes advantage of the use of living cells that interact actively with the host organism to overcome biological barriers to target specific cell types and tissues. Cell therapy could be based on the use of either stem cells, unspecialized and self-renewing cells that can differentiate into several cell types, or non-stem cells/somatic cells highly specialized that can be further manipulated or treated before reintroduction into humans. This chapter outlines the general concept and types of cell therapies, with a focus on two of the most advanced types used in clinical practice: mesenchymal stem cells and chimeric antigen receptor (CAR)-T cells.

Keywords Cell therapy, Chimeric antigen (CAR)-T cells, Mesenchymal stem cells

1 An Overview of Cell Therapy

Cell therapy is the transfer of cells into a patient for medical purposes [1, 2]. These types of emerging therapies are a promising approach to improving or even curing a broad spectrum of diseases. In general, cell therapy offers some clinical and therapeutic benefits. On one hand, living cells can not only respond simultaneously to systemic and local chemical, physical, and biological signals but also overcome biological barriers to target specific cell types and tissues. In addition, cell therapies provide support for further therapeutic applications, being used as delivery systems for gene therapy[1] or factories of biological active compounds [1, 2]. Therefore, cell therapies are currently one of the most extensively researched therapeutic modalities in both preclinical and clinical settings, with many products already approved and numerous others undergoing clinical investigation.

Cell therapy could be separated into unicellular therapies, where only one type of cell is administered, and multicellular therapies, where a mixture of cell types is administered. In addition, the cells could be obtained from the same person (autologous use) or from another person (allogeneic use). Furthermore, cell therapy might involve genetic engineering or manipulations in formulation. Regarding the administration route, cells can be administered topically or as injectables, infusions, bioscaffolds, or scaffold-free systems [3]. Finally, cell therapy could be based on the use of either stem cells or non-stem cells.

1.1 Stem Cells Based Therapy

Stem cells are unspecialized and self-renewing cells that can differentiate into several cell types. Their potential to differentiate into a specific cell type is

[1] See Chapter "Gene Therapy".

influenced by both internal signals and the microenvironment guiding these cells to switch from proliferation to specialization. Stem cell therapies can be categorized into two groups: pluripotent stem cells (PSCs) and adult stem cells (ASCs). PSCs can differentiate into almost any type of specialized cell. The first source of PSCs was embryonic, but ethical controversies have limited the use of these cells. However, the development of technology to de-differentiate somatic cells has allowed the creation of induced pluripotent stem cells (iPSCs), which do not involve ethical issues [4]. On the other hand, ASCs can specialize in multiple cell types within a specific organ or tissue, including skin, neural, mesenchymal, and hematopoietic stem cells that could be obtained either from neonatal or adult tissues [2, 5]. Currently, iPSCs are under intensive study, but there are no approved therapies using them. Actually, all the approved stem cell therapies involve ASCs, such as hematopoietic stem cells (HSCs), mesenchymal stem cells (MSCs), and limbal stem cells (LSCs) [1, 3].

HSCs are multipotent stem cells mainly obtained from bone marrow, peripheral blood, and umbilical cord blood. They can self-renew and differentiate into the myeloid and lymphoid blood cell lineages. As a result, HSCs can be used to reconstitute the hematopoietic and immune systems for treating inherited and acquired blood disorders [6, 7]. In this line, HSCs obtained through bone marrow aspiration or peripheral blood mobilization are frequently indicated to reconstitute the patient's hematopoietic capability after the elimination of cancerous cells through myeloablative treatment [8]. Autologous HSCs or matched-donor HSCs are the most ideal candidates for transplantation since they have a lower probability to result in graft-versus-host disease (GVHD), graft rejection, and engraftment syndrome [7]. Allogeneic HSCs have an additional advantage in cancer treatment because they can elicit graft-versus-tumor effects [9]. Additionally, in the global cell therapy market, there are currently 10 approved treatments involving HSC products registered for the pharmaceutical industry, most of which are derived from cord blood and applied as allogeneic transplants [1, 3].

MSCs are multipotent stem cells capable of differentiating primarily into mesodermal cell types (e.g., chondrocytes) and, to a lesser extent, into ectodermal (e.g., neurocytes) and endodermal lineages (e.g., hepatocytes) [10]. Most of the therapeutic benefits of MSC treatment are based on their modes of action[2] including tissue repair or immunomodulation [8, 11]. Therefore, they have broad applications in clinical use for the treatment of degenerative diseases, autoimmune diseases, inflammatory diseases, and trauma, among others. In the global cell therapy market, there are 12 MSC products that have been approved worldwide, although none have received approval from the U.S. Food & Drug Administration (FDA) [1].

[2] See Chapter "Medicinal Chemistry and Biotherapeutics: Overview and Definitions".

1.2 Non-stem Cell Based Therapy

Non-stem cell therapies use somatic cells obtained from the human body, which are isolated, expanded, and administered to patients as curative or preventative treatments [12, 13]. The techniques for isolation and propagation depend on the cell type, location, and specific use. Somatic cells are highly specialized and can be further manipulated or treated before reintroduction into humans. These cells serve as in vivo sources of enzymes, cytokines, and growth factors. They are also used in adoptive cell therapy (ACT) for cancers, in the transplantation of cells as scaffolds for treating open wound injuries such as burns or lesions, or to treat metabolic disorders [12]. Non-stem cell therapeutics can be categorized into non-immune and immune cell types. Non-immune cells are mainly used for regenerative medicine and tissue engineering applications. For example, autologous skin cells (i.e., keratinocytes, fibroblasts, and melanocytes) were used for the treatment of extensive burns, skin substitutes for venous leg ulcers and diabetic foot ulcers, or autologous chondrocyte scaffolds for the repair of cartilage defects [14]. On the other hand, non-stem cell therapy with immune cells includes T cells, dendritic cells (DCs), natural killer (NK) cells, and macrophages [15]. However, only T cells and DCs have been approved as therapeutic products on the market to date. Among them, most approved T cell products are chimeric antigen receptor (CAR)-T therapies for hematologic malignancies, whereas DC products are used as vaccines for solid tumors. In addition, the cell source of these approved products can be either autologous, from the patients themselves, or allogenic, from the other donors [1, 3].

Considering that the majority of currently approved cell therapies in the global market are based on T cells and MSCs, the next section will provide an in-depth explanation of the concepts and characteristics of these cell types.

2 Cell Therapy with Mesenchymal Stem Cells: General Concepts, Mechanisms of Action, and Clinical Applications

2.1 Definition and Characteristic of MSCs

Mesenchymal stromal/stem cells (MSCs) were described as a heterogeneous population of multipotent cells found in many tissues of the body. The first isolation and culture of MSCs was reported by Friedenstein in 1970 using bone marrow mononuclear cells from guinea pigs [16]. When these cells were seeded in tissue culture dishes, they developed colonies of spindle-shaped plastic-adherent cells, known as colony-forming unit fibroblastic (CFU-F). These results confirmed that bone marrow is also composed of non-hematopoietic stromal cells, with a clonal characteristic. In 1991, Caplan adopted the term "mesenchymal stem cell" to describe stem cells in the marrow involved in bone and cartilage formation [17]. Following these reports, the

MSC name became the subject of debate because not all cells adhered to plastic, isolated from sources such as bone marrow and fat, have stem-like properties. Later, the International Society for Cell and Gene Therapy (ISCT) released the minimal criteria for cells to be referred to as MSCs. Since then, to be defined as a MSC, a culture should: (1) be plastic-adherent; (2) express (≥95%) mesenchymal superficial markers (CD29, CD73, CD90, and CD105) and lack the expression (≤2% positive) of hematopoietic markers (CD11b, CD14, CD19, CD34, CD45, CD79, and HLA-DR); (3) be able to differentiate to adipocytes, chondroblasts, and osteoblasts [18]. Even more, the ISCT MSC committee recommends performing supplement studies with information on the tissue-source origins of MSCs and restricts the use of the term "mesenchymal stem cells" to instances where in vitro and in vivo evidence proves their self-renewal and differentiation properties [19]. Lastly, MSCs are considered "immunoprivileged" cells because they lack molecules of the major complex of histocompatibility (MHC) class II and costimulatory molecules (CD40, CD80, and CD86) and have low levels of MHC class I, all proteins involved in immunological detection [20]. Nevertheless, it is important to consider that although they offer an interpretation quite close to their biological nature, MSCs cultured in vitro differ from their natural physiological state [21].

2.2 Sources of MSCs

Since the first report obtaining MSCs from bone marrow, these cells have been isolated from both perinatal tissues (amniotic fluid, umbilical cord, and placenta) and adult tissues (lung, heart, dermis, liver, pancreas, dental pulp, brain, spleen, thymus, skeletal muscle, adipose tissue, dermal tissue, and dental tissue) [22]. It is important to highlight that depending on the tissue origin, MSCs exert different phenotypic and functional properties [23]. The principal sources of MSCs for clinical use are bone marrow, adipose tissue, the placenta and the umbilical cord [24, 25]. For bone marrow-derived MSCs, cells are usually extracted from the iliac crest, femur, or tibia. Then, MSCs are isolated using a density gradient centrifugation with low viscosity and low osmotic pressure solutions to separate the mononuclear cells of bone marrow based on their density. Last, cells are seeded into a culture dish/flask at a low density of 10^3 to 10^6 cells/cm^2 for expansion. The bone marrow MSCs expanded in culture retain their differentiation potential, although their viability is limited to seven culture passages [26]. Similarly, adipose tissue is an attractive source of MSCs for cell therapy because it is relatively easy to obtain and associated with minimal risks. However, like bone marrow MSCs, their regenerative capability is dependent on the donor's age. MSCs from adipose tissue are obtained via liposuction of visceral or subcutaneous fat from the abdomen or gluteal areas. The isolation process involves enzymatic digestion with collagenases, removal of red blood cells, and subsequent seeding of the cells onto plastic surfaces [26]. Compared to adult tissues, MSCs from the placenta and umbilical cord are ideal candidates for regenerative medicine. These tissues, considered medical waste, pose no ethical issues and are obtained through no invasive procedures. Additionally, their immature cells exhibit better proliferative activity, lower immunogenicity, and fewer mutations, thereby reducing the risk of initiating tumorigenesis [27]. The protocol to isolate these cells from the umbilical cord is similar to that used for those obtained from adipose tissue [28].

Recently, induced pluripotent stem cells (iPSCs) have become a novel source with clinical potential for MSCs. iPSCs are obtained from the reprogramming of somatic cells through the expression of the pluripotency genes (Oct4, Sox2, c-Myc, and Klf4) and can differentiate into tissue cells belonging to the three embryonic germ layers [29, 30]. This technology allows to obtain pluripotent cells that are histoidentical with the initial donor cell used for reprogramming [31]. Many studies have reported the successful generation of functional MSCs from iPSCs (iMSCs) [31–33]. iMSCs can be obtained from adult accessible tissue, like skin or blood, for example, by first reprogramming adult somatic cells to iPSCs and then differentiated to MSCs [34]. Likewise, it has been shown that iMSCs not only maintain characteristics and properties of MSCs from other sources, but also exhibit greater proliferation potential than MSCs from conventional adult tissues [35].

2.3 Biological Properties of MSCs

Mesenchymal stromal/stem cells play a critical role in the bone marrow microenvironment; they are involved in creating the adequate environment for renewal, proliferation, and differentiation of hematopoietic stem cells (Fig. 1). Additionally, MSCs have the capacity to migrate toward injured tissues, where they act as biological "pumps," releasing cytokines, chemokines, and trophic factors that could induce tissue regeneration. In addition, MSCs can modulate the inflammation through their interaction with both the innate and adaptive immune systems.

2.3.1 Migration

In response to specific signaling molecules secreted by injured tissues, MSCs are recruited to exert their pro-regenerative and immunomodulatory properties [26, 36]. Research over the last few decades has uncovered the role of several key molecules involved in cell trafficking that influence the migration and homing behavior of MSCs. These include chemokines like stromal cell-derived factor 1 (SDF-1 or CXCL12) and its receptor CXCR4, as well as adhesion molecules and matrix metalloproteinases (MMPs) [37–39]. To initiate the migration process, MSCs first roll along and adhere to vascular endothelial cells, a process mediated by the very late antigen-4/vascular cell adhesion molecule-1 (VLA-4/VCAM-1), before subsequently crossing the endothelial barrier [40, 41].

A vast number of studies using different animal models support the idea that MSCs have enormous therapeutic potential based on this unique property. These

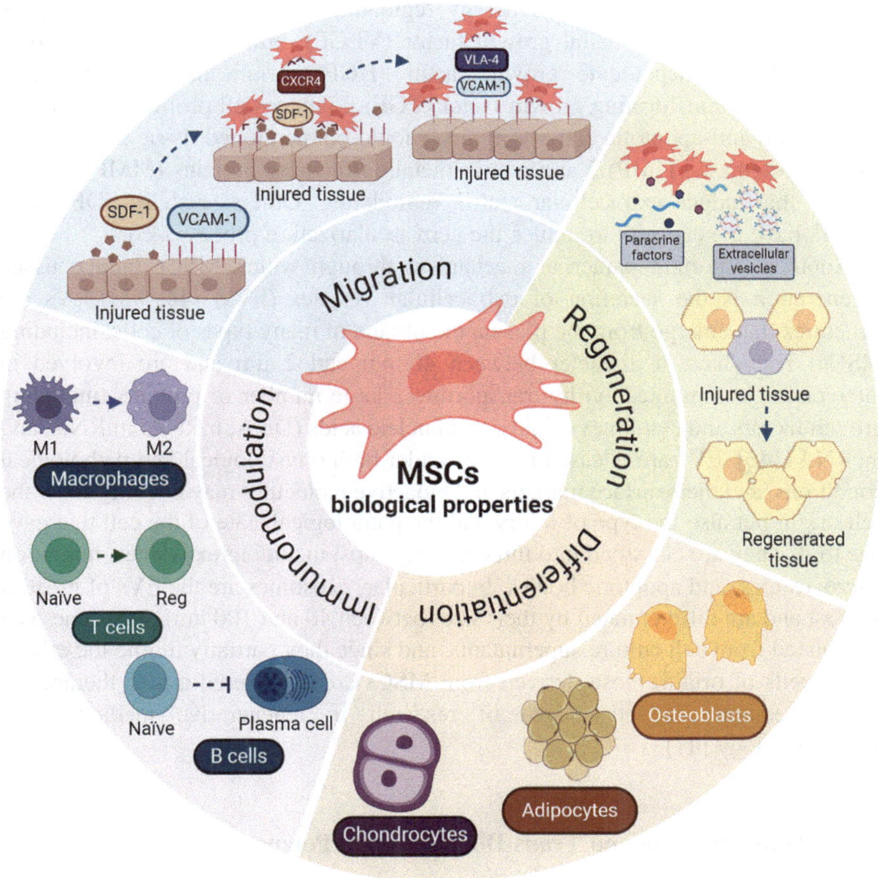

Fig. 1 Biological properties of MSCs

studies have shown encouraging results, highlighting the potential for clinical use of MSCs to regenerate damaged tissues or organs [42]. Likewise, easy ex vivo manipulation also allows them to be used as therapeutic gene vehicles, such as for the overexpression of antitumor genes in clinical anticancer therapy.

2.3.2 Tissue Regeneration by Paracrine Factors Secretion

Constitutively or upon stimulation, MSCs secrete a wide array of soluble factors that influence various cells within the damaged tissue. MSCs secrete interleukin (IL)-1, IL-6, IL-8, IL-7, IL-10, IL-11, IL-12, IL-14, IL-15, prostaglandin E2 (PGE2), leukemia inhibitory factor (LIF), granulocyte colony-stimulating factor (G-CSF), granulocyte macrophage colony-stimulating factor (GM-CSF), stromal cell-derived factor 1 (SDF-1 or CXCL2), CCL2, CXCL1, CXCL2, CXCL6, macrophage colony-stimulating factor (M-CSF) that act by regulating the immune system [43]. In addition, vascular endothelial growth factor (VEGF), brain-derived neurotrophic factor (BDNF), hepatocyte growth factor (HGF), insulin-like growth factor 1 (IGF-1) and transforming growth factor (TGF)-β induce cell proliferation, regeneration, and anti-apoptotic activity. It also should be mentioned tissue inhibitors of metalloproteinase (TIMP) 2 and matrix-metalloproteinase proteins (MMP)-1, 2, 9, and 13 that induce extracellular matrix remodeling. Others, such as SDF-1 and VEGF, have the capacity to induce the neovascularization process [43].

Another important paracrine mechanism through which MSCs induce tissue regeneration is the secretion of extracellular vesicles (EVs) [44, 45]. EVs are structures that emerge from the plasma membrane of many types of cells, including MSCs. They have a diameter between 40 nm and 2 μm and are involved in intercellular communication by transporting a large number of proteins (including growth factors and cytokines), lipids, and nucleic acids (DNA, mRNA, miRNA, and lncRNA) [46]. EVs are released by cells under both physiological and pathological conditions, and their surface markers and bioactive molecules may not only show the cell origin but also the type of injury and the pathological state of the cell that gave rise to it. They are classified into three main groups, including exosomes, migration microvesicles, and apoptotic bodies. In particular, exosomes are the EVs of greatest interest and are differentiated by their size (between 40 and 100 nm). Exosomes can be isolated from cell culture supernatants, and since they partially mimic the effects of the cells of origin, those derived from MSCs are considered a new therapeutic alternative and an active area of research in regenerative medicine and nanotechnology [47].

2.3.3 Differentiation and Trans-Differentiation Potential

As multipotent stem cells, MSCs have the capability to differentiate into cells derived from mesodermal germ layers, including osteoblasts, adipocytes, and chondrocytes [17]. Additionally, under the right conditions, it could transdifferentiate in vitro into different lineages, such as epithelial, endothelial, and neural cells. Altogether, these abilities enable them to form and regenerate mature tissues. It should be noted that to differentiate MSCs to different cell types, specific key molecular regulators and appropriate activation conditions in vivo are required.

Under physiological conditions, the microenvironment can support MSCs and determine their fate. For instance, the osteogenic differentiation of bone marrow MSCs can be enhanced by IGF-1, an important growth factor in the bone matrix, via the mTOR pathway [48]. Additionally, this differentiation can be stimulated by estrogens when they bind to their receptors and activate the p38 MAPK/NF-κB and BMP/WNT/β-catenin pathways [49].

A recent study identified a transcriptional pathway crucial for the in vivo differentiation fate of MSCs. The MSCs genome exhibits a high degree of overlap in binding sites for master transcriptional regulators [50]. These regions are epigenetically reduced in size following differentiation, displaying high plasticity that allows

MSCs to trans-differentiate between adipocytes and osteoblasts. Altogether, these findings emphasize the importance of in vivo and in vitro conditions in establishing the molecular basis for MSCs multipotency and may lead to enhanced therapies. Nevertheless, this evidence suggests that the therapeutic benefits of MSCs in treating diseases extend beyond their differentiation potential, as this is only one aspect of the overall mechanism underlying their effectiveness.

2.3.4 Immunomodulation

MSCs are involved in both sensing and modulating immune responses of both innate and adaptive immunity. In this line, MSCs are able to stimulate inflammation when the immune system is underactive and restrain inflammation when the immune system is overactive [51]. The polarization process of MSCs toward either an anti-inflammatory or a pro-inflammatory phenotype is mediated by Toll-like receptors (TLRs). Activation of TLR3 leads to an anti-inflammatory phenotype in MSCs, whereas activation of TLR4 results in a pro-inflammatory phenotype [52]. These immunomodulatory functions are achieved through interactions with immune cells via direct cell-to-cell contact and through the secretion of soluble factors and EVs.

Interestingly, the modulation of macrophages/monocytes by MSCs appears to be a critical step in their immunomodulatory mechanisms. They regulate macrophages mainly through the secretion of soluble factors and extracellular vesicles. MSCs can induce macrophage transformation from proinflammatory type M1 to anti-inflammatory type M2 through PGE2, TGF-β, and CCL2 pathways [53]. In addition, MSCs have also been shown to inhibit differentiation of monocytes into dendritic cells and induce a more tolerant dendritic cell phenotype, which promotes the formation of regulatory T (Treg) cells [54]. As for macrophages, MSCs shift the activated neutrophil phenotype into an aged neutrophil phenotype, reducing chemotaxis, reactive oxygen species (ROS) production, and the secretion of granules. In addition, it has been demonstrated that MSCs inhibit B cell differentiation, proliferation, activation, and IgG secretion [54, 55].

2.4 *Clinical Application of MSCs*

Since their discovery and isolation in the 1970s, MSCs have been investigated for both autologous and allogeneic transplantation, demonstrating success without the need for immunosuppressants. This is attributed to their low expression of HLA class I molecules, as well as the absence of HLA class II and co-stimulatory molecules. Furthermore, they are easily isolated and cultured in vitro in sufficient quantities for the doses typically administered to patients [56].

Initially, MSCs were mostly addressed in their potential to reconstruct damaged or diseased tissues due to their selective migration to injured tissues and differentiation competence. However, recent studies have shown that MSCs can influence various physiological and pathological processes, by releasing biologically active molecules.

Regarding the administration route, different clinical indications and pathologies require specific routes for MSC administration to achieve optimal therapeutic efficacy [56]. The most common delivery method involves the topical application or injection of MSCs directly into a specific site or tissue, ensuring high precision and enhancing the engraftment of therapeutic cells at the desired location. This approach is the least invasive and has shown significant potential for treating burns and wounds related to diabetes.

There is considerable evidence indicating that MSCs primarily exert their effects through direct cell-to-cell interactions and localized paracrine actions, rather than by engrafting and differentiating within the target organ [42]. Moreover, the presence of immune cells and tissue repair signals in the bloodstream likely influences the distribution to injured tissue and the response of MSCs. While intravenous administration is the most commonly used method for delivering MSCs, its use carries some risks, including the potential accumulation of MSCs in the lungs and spleen. Several additional routes of MSC administration are notable for consideration for some specific applications. For instance, administering MSCs during ex vivo perfusion of solid organs by normothermic machine perfusion has emerged as a notable approach. Recent studies have demonstrated that machine perfusion supports MSC function and survival, facilitates direct interaction with the target tissue, and prevents cellular entrapment in the lungs and spleen [57, 58].

Currently, numerous studies are exploring the application of MSCs in cell therapy, where they are utilized to treat hematological diseases, graft-versus-host disease (GVHD), liver, kidney, and lung disorders, as well as cardiovascular, bone and cartilage, neurological, and autoimmune diseases.

2.4.1 Graft-Versus-Host Disease

Graft-versus-host disease is a significant complication associated with allogeneic bone transplantation, as well as the transplantation of solid organs rich in lymphoid cells and the transfusion of unirradiated blood. The pathobiology of acute GVHD is attributed to the activation of cytotoxic CD8+ T cells within the graft, which recognize the recipient's MHC proteins as "non-self" and initiate an immune response between the donor and the recipient. This adverse reaction occurs shortly after transplant, usually within the first 100 days. The pathophysiology of chronic GVHD is unclear and is believed to involve both cellular and humoral immunity [59]. Currently, the methods for preventing GVHD involve the removal of T lymphocytes and the administration of immunosuppressives. However, eliminating T lymphocytes diminishes the grafts' antitumor efficacy and can result in tumor recurrence. Moreover, prolonged use of immunosuppressive drugs and hormones frequently causes severe side effects [60]. Given the immunological nature of

GVHD, there is substantial interest in utilizing MSCs for both the prevention and treatment of this serious condition. This interest is based on their strong immunosuppressive capacities and their ability to produce cytokines that support hematopoiesis. MSCs used to treat this pathology are almost exclusively allogeneic and can be derived from either the allogeneic HSC donor or from a third-party donor, whether related or unrelated. In pre-clinical studies, MSCs were co-administered with HSCs either on the same day/week or months later of the HSC transplant. The first observation of the clinical efficacy of MSCs in acute GVHD occurred in a nine-year-old boy who was unresponsive to other therapies and received bone marrow MSCs from a third donor 73 days following HSC transplantation [61]. Following this study, in 2006, six out of eight patients with steroid-resistant grade III to IV GVHD demonstrated complete remission after receiving MSC treatment. Although studies on the use of MSCs for treating acute GVHD have yielded promising results, the therapeutic efficacy of MSCs in chronic GVHD remains uncertain due to the limited number of investigations.

Today, there are two agency-approved MSC products for the treatment of GVHD. The product available in Japan is ***Temcell HS***, an allogeneic cell therapy derived from human BM-MSCs used for treating acute and refractory GVHD in both children and adult patients undergoing bone marrow transplants for hematologic diseases. The efficacy of ***Temcell HS*** was evaluated in 381 consecutive patients who received up to eight intravenous infusions of 2.00×10^6 cells/kg. Of the 306 evaluable patients, 56% achieved overall response, defined as complete resolution of a GVHD symptoms, on day 28 after treatment initiation. Of the 153 patients who received ***Temcell HS*** as a second-line therapy following first-line steroid therapy, the overall response was 61%. Similarly, the MSC therapy ***Remestemcel-L*** was approved in Canada for the treatment of steroid-refractory acute GVHD in pediatric patients. A total of 260 patients were enrolled in a multicenter randomized study to evaluate the efficacy of ***Remestemcel-L*** in addition with second-line therapy for the treatment of steroid-refractory acute GVHD. Patients received eight intravenous infusions of 2.00×10^6 BM-MSC sc/kg or placebo, given over 4 weeks, in addition to second-line therapy according to institutional standards. In post hoc analyses, 64% of pediatric patients had a higher overall response with MSCs compared with placebo, although similar rates of adverse events were observed between treatment groups [62]. Based on that, the FDA recommended more randomized controlled trials to provide additional information about the therapeutic mechanism and efficacy.

2.4.2 Autoimmune Diseases

Autoimmune diseases encompass a wide array of related conditions in which the body's immune system erroneously identifies its own healthy tissues as foreign and subsequently targets them for attack. . Based on their ability to modulate immune responses, MSCs have also been proposed as a treatment for these kinds of diseases

[63]. Clinical studies have demonstrated that MSCs are effective in the treatment of refractory and severe systemic lupus erythematosus patients, resulting in the induction of clinical remission, improvements in serological markers of organ dysfunction, and stabilized renal function [64].

The role of MSCs has also been documented in Crohn's disease, a chronic inflammatory disorder of the gastrointestinal tract. Three pilot studies involving patients with Crohn's disease treated with autologous bone marrow or adipose tissue-derived MSCs showed that MSCs were well tolerated and potentially beneficial. In 2012, the South Korean Ministry of Food and Drug Safety approved ***Cupistem***, an allogeneic product of human adipose-MSCs, for the treatment of fistulas in Crohn's disease. ***Cupistem*** demonstrated that 82% of patients with complex Crohn's fistulas experienced complete healing by week 8 post-treatment, with 81% maintaining their response at week 96 [65]. Six years later, the European Medicines Agency approved a similar product, ***Alofisel***, for the treatment of complex anal fistulas in adult patients with Crohn's disease. Its mode of action appears to involve MSCs' capacity to suppress the proliferation of activated lymphocytes, which in turn reduces the production of pro-inflammatory cytokines.

2.4.3 Cardiovascular Disease

MSC therapy is an attractive candidate for cardiovascular repair due to its regenerative and immunomodulatory properties. MSCs can protect the myocardium by reducing inflammation, facilitating the differentiation of myocardial cells in regions surrounding infarcts, promoting angiogenesis, enhancing resistance to apoptosis, and inhibiting fibrosis [66].

In 2011, the Republic of Korea developed ***Cellgram***, a method aimed at enhancing left ventricular ejection fraction in patients who have experienced acute myocardial infarction and undergone reperfusion via coronary angioplasty within 72 h of chest pain. . The manufacturer has proven that human BM-MSCs can differentiate into cardiac muscle and induce angiogenesis through the secretion of VEGF, IL-6, and MCP-1, which may be a significant benefit for ischemic tissue. Subsequently, another clinical study was performed with 69 patients with acute myocardial infarction, who were treated with either autologous MSCs or standard saline as controls. The MSC-treated group exhibited significant improvements in cardiac function compared to the control group, with no adverse events [67]. In 2016, ***Stempeucel***, a human BM-MSC allogeneic product obtained from the bone marrow aspirates of several consenting donors, was approved in India for the treatment of critical limb ischemia due to Buerger's disease and peripheral arterial disease. They found that the intramuscular administration of ***Stempeucel*** alleviated limb necrosis and enhanced limb function within 28 days by releasing a variety of pro-angiogenic growth factors, including VEGF.

2.4.4 Bone and Cartilage Defects

Skeletal disorders rank among the most prevalent debilitating conditions impacting millions of individuals globally. The application of MSCs may facilitate the repair and regeneration of skeletal tissues through various mechanisms, including homing, induction of angiogenesis, differentiation, and modulation of inflammatory responses. One of the most common disorders of the skeletal system is osteoarthritis, characterized by the degeneration of articular cartilage, thickening of subchondral bone, and the formation of osteophytes [68]. One approved MSC therapy, ***Cartistem***, is available in the Republic of Korea. This biotherapeutic is based on allogeneic umbilical cord blood-derived mesenchymal stem cells and is indicated for the treatment of knee cartilage defects associated with severe osteoarthritis, where no visible cartilage remains. A recent study evaluated 15 randomized controlled clinical trials (RCTs) and 11 nonrandomized controlled clinical trials using culture-expanded MSCs in the treatment of knee osteoarthritis. They observed overall positive effects of MSCs in reducing pain and symptoms of the disease as well as in protecting and/or repairing cartilage [69].

2.4.5 Neurological Diseases

One of the most active clinical trials using MSCs is for neurological disorders treatment, however, without sufficient efficacy data [70].

Spinal cord injuries lead to sensory and motor impairments below the site of injury. In the chronic phase, complications may arise, including dysfunction in the respiratory, cardiovascular, urogenital, and gastrointestinal systems, along with motor nerve spasms. The pathological mechanisms underlying this condition involve inflammation in the injured region, resulting from the infiltration of macrophages, T cells, microglial cells, and neutrophils through the blood-spinal cord barrier. This leads to the release of inflammatory factors, fibrosis, and vascular rupture, all of which contribute to neuronal tissue degradation. Astrocytes help separate necrotic tissue from healthy tissue, but their excessive proliferation results in scars that affect axonal regeneration [71].

One MSC product, called ***Stemirac***, has been approved in Japan for the treatment of adult patients with spinal cord injuries. Reports from the phase 2 clinical study with ***Stemirac*** showed neurological improvement in 12 of the 13 patients who received an infusion of 50–200 × 10^6 BM-MSCs. Although the study was unblinded and did not exclude placebo effects [72].

Another significant neurological disease is amyotrophic lateral sclerosis (ALS), which primarily affects the upper and lower motor neurons [73]. Muscle weakening and gradual atrophy are the first signs of the disease, impacting eating, speaking, moving, breathing, and shortening life expectancy to about 5 years from onset. Unfortunately, Riluzole, a glutamatergic antagonist, is the only drug approved for managing this disease, and there is still no definitive cure [74]. In this context, the use of MSCs for the treatment or management of ALS has been common in preclinical and clinical studies for many years. ***NeuroNata-R***, an autologous BM-MSCs product, was approved by the Republic of Korea in 2013 for the treatment of ALS. Phase 1 and 2 clinical trials demonstrated that ***NeuroNata-R*** has a neuroprotective effect, slowing the progression of ALS by extending motor nerve cell survival. Additionally, a recent clinical study with 157 participants who received ***NeuroNata-R*** showed that the treatment increased survival probability compared to the external control group and highlighted the importance of multiple injections for greater effectiveness. Moreover, no adverse drug reactions were observed during the year-long assessment after the first administration [75].

2.4.6 Cancer

Another use of MSCs is as a vehicle for therapeutic and anti-tumoral genes. Recent studies in animal models have demonstrated promising results utilizing MSCs as vehicles for targeted therapies [76, 77]. Currently, clinical trials are evaluating MSCs engineered to express antitumor genes for at least six different types of tumors. The use of MSCs as biological vehicles in cell therapy has involved numerous advantages, including immune evasion, tropism for tumor tissue, facile and fast isolation, ex vivo expansion, multilineage differentiation, and the capacity to deliver a variety of therapeutic agents. In addition, several reports have indicated that the tumor microenvironment provides a privileged niche for the homing and survival of MSCs. However, the clinical application of MSC-based therapy in cancer treatment remains fraught with numerous challenges [78]. For instance, elucidating the interactions between MSCs and tumor cells to enhance the clinical safety of MSC-based therapeutic interventions is essential. Future research should prioritize long-term follow-up studies of MSC-treated tumor-bearing animals to thoroughly report all safety concerns associated with the plasticity of MSCs and their potential pro-tumorigenic effects, particularly in solid tumors.

Regarding malignancies of T and B lymphocytes, allogeneic hematopoietic stem cell transplantation (allo-HSCT) has seen significant advancements over the past few decades and is increasingly being utilized worldwide. The success of haploidentical HSCT has effectively ensured a donor for nearly every patient. Outcomes for patients who receive transplantation while in remission are generally favorable, whereas those transplanted at advanced stages of the disease face a poorer prognosis. In this scenario, CAR-T cell therapy has developed as a pivotal advancement in the therapy of relapsed or refractory (R/R) B cell acute lymphoblastic leukemia (B-ALL) [79], demonstrating high remission rates in patients previously treated with multiple lines. The relatively high relapse rate remains a significant obstacle to its establishment as a definitive cure. Combining allo-HSCT and CAR-T cell therapy presents an attractive research avenue to enhance the prognosis of R/R B-ALL [80]. Thus, it next will describe adoptive cell therapy with T lymphocytes as a novel strategy for cancer treatment.

3 Cell Therapy with Chimeric Antigen Receptor (CAR)-T Cells: Concepts, Clinical Application, and Manufacturing Procedure

3.1 Adoptive Cell Transfer with Lymphocytes as Cancer Immunotherapy

The concept driving cancer immunotherapy appears from gathering evidence displaying the vital role of the immune system in malignant cells' recognition and restricting their progression. Adoptive cell therapy (ACT) represents a strategy of cancer immunotherapy methodologies that exploit the intrinsic capabilities of host immune cells to target and eliminate tumor cells. Reinfused autologous cells, particularly T lymphocytes, have successfully controlled advanced progression in patients with many types of cancer. In addition, donor-derived immune cells may be utilized under specific circumstances.

3.2 CAR-T Cells Concept

T-cells are crucial in the identification and eliminating of abnormal cells, including tumor cells. Chimeric antigen receptor (CAR)-T cell therapy is a treatment in which T cells are collected from the patient and reengineered in vitro in the laboratory to express a CAR that recognizes and binds to specific antigens on the surface of cancer cells [81].

The CAR consists of three main components (Fig. 2a) [82, 83]:

The extracellular domain: It has the function of antigen binding and recognition, and is a single-chain fragment composed of the variable light and heavy chain regions of an immunoglobulin.

The transmembrane region: This region is a hydrophobic alpha helix and transmits the receptor-binding signal.

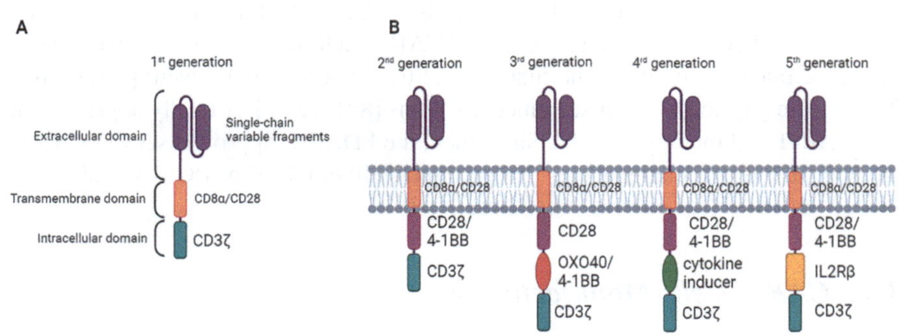

Fig. 2 Different generations of CARs

Table 1 FDA-approved CAR-T cell therapies

Name	Target	Disease	FDA reference
Tisagenlecleucel	CD19	B-cell acute lymphoblastic leukemia (ALL)	[89]
		B-cell non-Hodgkin lymphoma (NHL)	
Axicabtagene ciloleucel	CD19	B-cell non-Hodgkin lymphoma (NHL)	[90]
		Follicular lymphoma	
Brexucabtagene autoleucel	CD19	Mantle cell lymphoma (MCL)	[91]
		B-cell acute lymphoblastic leukemia (ALL)	
Lisocabtagene maraleucel	CD19	B-cell non-Hodgkin lymphoma (NHL)	[92]
		Follicular lymphoma	
Idecabtagene vicleucel	BCMA	Multiple myeloma	[93]
Ciltacabtagene autoleucel	BCMA	Multiple myeloma	[94]

The intracellular domain: It is involved in the trigger of the T-cell effector functions. This intracellular compound can be used to classify CARs into different generations based on their structure and composition (Fig. 2b).

To the second and third generations have been added a CD28/4-1BB domain, a co-stimulatory receptor, to enhance proliferation and activation. The fourth generation contains a cytokine release domain. Finally, the fifth generation has a cytoplasmic IL-2 receptor.

The concept of CAR-T cell therapy emerged in the 1980s with the pioneering work of Dr. Zelig Eshhar at the Weizmann Institute in Israel [84, 85]. He developed the first-generation CARs, essentially combining an antibody fragment that recognizes cancer cells with a T cell activation signal. Besides the great development, these early CARs did not persist in the body and were not clinically effective. Later, in 1998, Dr. Sadelain and collaborators showed that introducing a co-stimulatory molecule (CD28) into engineered T cells allows them to persist and remain active, making the second generation of CARs [86]. In 2002, a second generation of CAR-T cells was able to proliferate, survive, and in vitro eliminate prostate cancer cells [87]. Further studies directing CAR-T cells to CD19 to kill leukemia cells in a murine model confirmed the feasibility of CAR-T cell therapy for tumor treatment. This preclinical study led to the first clinical trial, whose reports were published in 2013 by the physician-scientist Renier Brentjens [88]. The FDA finally approved the first CAR-T cell therapy in 2017. Since then, the FDA has approved six CAR-T cell therapies for B cell malignancies that recognize either CD19 or BCMA (Table 1).

3.3 CAR-T Cells Manufacturing

To obtain patient immune cells, T cells are extracted using a procedure called leukapheresis. In this procedure (Fig. 3), the T cells are separated magnetically

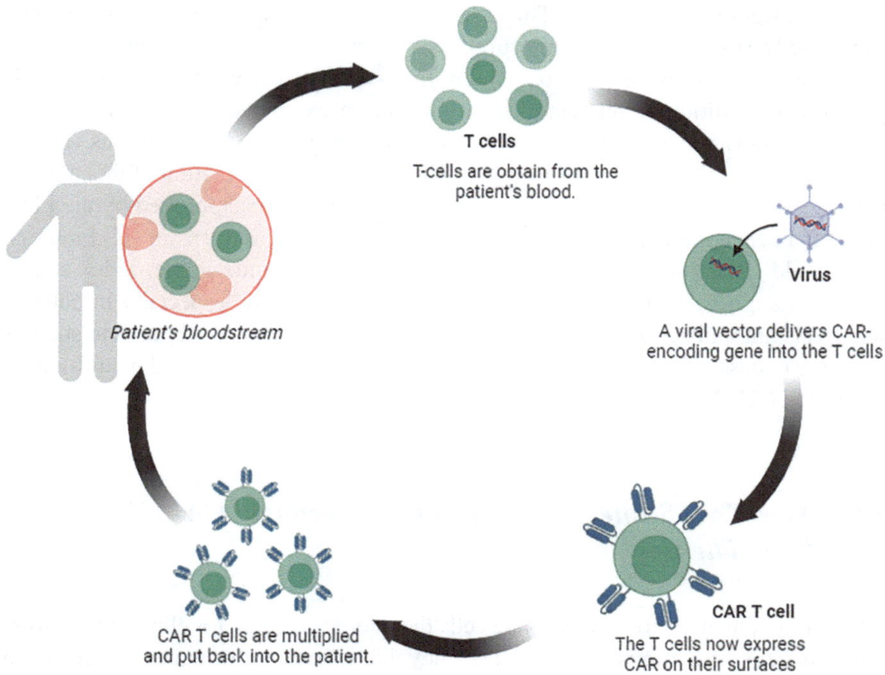

Fig. 3 The ex vivo CAR-T cells generation process

using a phenotypic marker such as CD3 from the rest of the blood, which is returned to the patient. Subsequently, the T cells need to be activated and expanded in the laboratory. This is achieved by exposing them to interleukin-2 (IL-2), a cytokine with the ability to induce T cell proliferation. Then, T cells are genetically modified using a virus or a plasmid to express an antigen-specific CAR [95, 96]. This process involves several steps. First, the lentiviral vector is constructed by inserting the desired gene into the viral genome. Then, this vector is packaged into functional lentiviral particles using specialized packaging cells. The lentiviral particles are incubated with the target cells, in this case, the patient's T cells, allowing them to enter the cells. Once inside the cell, the viral RNA undergoes reverse transcription to DNA, which is then integrated into the host cell's genome. At this point, the integrated gene is transcribed into mRNA and translated into the desired protein product [97]. These cells are further expanded and cultured in vitro and tested to ensure that they are functioning properly and are free of contamination. Finally, the CAR-T cells are ready to be infused back into the patient [95, 96].

3.4 CAR-T for Solid Tumors

Despite FDA-approved therapies for several blood cancers (Table 1, FDA-approved CAR-T cell therapies) [3], the development of effective therapies for solid tumors remains a significant challenge. This is, at least in part, due to the effects triggered by cancer cells carrying to an immunosuppressive tumor microenvironment enriched in regulatory and suppressor cells, which limits the efficacy of effector T cells [98, 99]. In addition, solid tumors may present an excessive extracellular matrix, which leads to an increased intratumoral pressure, the generation of tortuous vessels, and therefore limits the access of reinfunded cells. Another limitation of this therapy is the selection of targets for the CAR design. Most therapies approved by the FDA target B lineage markers. In the absence of tumor-specific antigens in solid tumors, this could lead to off-target CAR-T activity and undesirable side effects [99]. Although CAR-T cell therapy has a high response rate, especially in patients with resistant and refractory disease, it is also associated with side effects such as cytokine release syndrome (CRS) and immune effector cell-associated neurotoxicity syndrome (ICANS) [100].

3.5 New CAR Strategies and Future Directions: CAR-NK Cells Therapy

NK cells are innate cytotoxic immune cells that are responsible for eliminating virus-infected cells and tumor cells [101]. They have some advantages, such as no restrictions in MHC recognition; they are able to infiltrate the tumor; have cytolytic capacity; and they have shown minimal side effects [101]. Therefore, it would be advantageous to use these cells as CAR-NK cells. These CAR-NK cells have intrinsic cytolytic activity and do not need to recognize a target antigen like CAR-T cells.

Although few functional NK cells can be isolated from patients to generate CAR-NK cells because their function and number are usually compromised by curative treatments, there are several sources from which they can be obtained. NK derived from human embryonic stem cells (hESCs) and induced pluripotent stem cells (iPSCs) can be obtained in large quantities for use in immunotherapy [102].

As with CAR-T cells, there are similar limitations to using CAR-NK cells as antitumor therapies [103]. Similar to the limitations of CAR-T cells, these include both manufacturing and post-administration challenges, such as migration of NK cells to tumor sites and the tumor milieu. Today, active research is currently ongoing to optimize the structure of NK CAR, with the idea of enhancing their tumor infiltration, maintaining their cytotoxicity capability in the immunosuppressive tumor microenvironment, and minimizing on-target and off-tumor toxicity. Another important consideration is that NK cells can be alive for a short time (<10 days) [104], which is a double-edged sword because, in the event of severe toxicity, this is an advantage but also poses the challenge that repeated administrations may be necessary to reach a durable response.

4 Concluding Remarks

Cell therapy has emerged as a promising approach to improving or even curing a wide range of diseases. Both stem cell and non-stem cell-based therapies broaden the possibilities in medical practice.

The multiple properties of MSCs underline their therapeutic potential in tissue regeneration and the treatment of inflammatory diseases. However, it is necessary to generate more information about the mechanisms involved in tissue regeneration, the optimal source to isolate them to avoid mixed cell populations, and the effects that different culture conditions have on therapeutic potential. Despite the open questions that remain to be answered, we are observing a significant increase in clinical research of these cells for tissue regeneration or to contribute to the resolution of inflammatory processes.

Regarding cell therapy opportunities for cancer, the CAR generation for improving adoptive cell transfer represents an innovative and promising approach to cancer treatment, which entails genetically modifying a patient's T-cells to target and eliminate cancer cells. Here, it provided a comprehensive examination of the most recent advancements and clinical trials of CAR-T cell therapy, elucidating both the fundamental concept and diverse applications of this therapeutic strategy. Additionally, the chapter addresses the inherent limitations and potential adverse effects associated with CAR-T cell therapy, such as the substantial financial cost, the risk of cytokine release syndrome, and neurotoxicity. CAR-T cell therapy has demonstrated significant efficacy in the management of hematologic malignancies; however, further research is imperative to enhance the therapeutic efficacy and safety, as well as to extend its applicability to solid tumors. With sustained research and development efforts, CAR-T cell therapy holds the potential to revolutionize cancer treatment and substantially improve patient outcomes.

Compliance with Ethical Standards This chapter is a review and no humans or animals were involved in the preparation of the chapter.

References

1. Wang LL-W, Janes ME, Kumbhojkar N et al (2021) Cell therapies in the clinic. Bioeng Transl Med 6:e10214
2. El-Kadiry AE-H, Rafei M, Shammaa R (2021) Cell therapy: types, regulation, and clinical benefits. Front Med 8:756029
3. Bashor CJ, Hilton IB, Bandukwala H et al (2022) Engineering the next generation of cell-based therapeutics. Nat Rev Drug Discov 21:655–675
4. Yamanaka S (2020) Pluripotent stem cell-based cell therapy—Promise and challenges. Cell Stem Cell 27:523–531
5. Gurusamy N, Alsayari A, Rajasingh S et al (2018) Adult stem cells for regenerative therapy. Prog Mol Biol Transl Sci 160:1–22
6. Amouzegar A, Dey BR, Spitzer TR (2019) Peripheral blood or bone marrow stem cells? Practical considerations in hematopoietic stem cell transplantation. Transfus Med Rev 33:43–50

7. Mosaad YM (2014) Hematopoietic stem cells: an overview. Transfus Apher Sci 51:68–82
8. Attia N, Mashal M (2021) Mesenchymal stem cells: the past present and future. Adv Exp Med Biol 1312:107–129
9. Karadurmus N, Sahin U, Bahadir Basgoz B et al (2018) A review of allogeneic hematopoietic stem cell transplantation in metastatic breast cancer. Int J Hematol Oncol Stem Cell Res 12: 111–116
10. Pittenger MF, Mackay AM, Beck SC et al (1999) Multilineage potential of adult human mesenchymal stem cells. Science 284:143–147
11. Schuppan D, Kim YO (2013) Evolving therapies for liver fibrosis. J Clin Invest 123:1887–1901
12. Mount NM, Ward SJ, Kefalas P et al (2015) Cell-based therapy technology classifications and translational challenges. Philos Trans R Soc Lond B Biol Sci 370:20150017
13. Golchin A, Farahany TZ (2019) Biological products: cellular therapy and FDA approved products. Stem Cell Rev Rep 15:166–175
14. De Pieri A, Rochev Y, Zeugolis DI (2021) Scaffold-free cell-based tissue engineering therapies: advances, shortfalls and forecast. NPJ Regen Med 6:18
15. Weber EW, Maus MV, Mackall CL (2020) The emerging landscape of immune cell therapies. Cell 181:46–62
16. Friedenstein AJ, Chailakhjan RK, Lalykina KS (1970) The development of fibroblast colonies in monolayer cultures of guinea-pig bone marrow and spleen cells. Cell Tissue Kinet 3:393–403
17. Caplan AI (1991) Mesenchymal stem cells. J Orthop Res 9:641–650
18. Dominici M, Le Blanc K, Mueller I et al (2006) Minimal criteria for defining multipotent mesenchymal stromal cells. The International Society for Cellular Therapy position statement. Cytotherapy 8:315–317
19. Boland LK, Burand AJ, Boyt DT et al (2019) Nature vs. nurture: defining the effects of mesenchymal stromal cell isolation and culture conditions on resiliency to palmitate challenge. Front Immunol 10:1080
20. Wang Y, Chen X, Cao W et al (2014) Plasticity of mesenchymal stem cells in immunomodulation: pathological and therapeutic implications. Nat Immunol 15:1009–1016
21. Drela K, Stanaszek L, Nowakowski A et al (2019) Experimental strategies of mesenchymal stem cell propagation: adverse events and potential risk of functional changes. Stem Cells Int 2019:7012692
22. da Silva ML, Chagastelles PC, Nardi NB (2006) Mesenchymal stem cells reside in virtually all post-natal organs and tissues. J Cell Sci 119:2204–2213
23. Costa LA, Eiro N, Fraile M et al (2021) Functional heterogeneity of mesenchymal stem cells from natural niches to culture conditions: implications for further clinical uses. Cell Mol Life Sci 78:447–467
24. Pittenger MF, Discher DE, Peault BM et al (2019) Mesenchymal stem cell perspective: cell biology to clinical progress. NPJ Regen Med 4:22
25. Shammaa R, El-Kadiry AE-H, Abusarah J et al (2020) Mesenchymal stem cells beyond regenerative medicine. Front Cell Dev Biol 8:72
26. Bayo J, Fiore E, Aquino JB et al (2014) Human umbilical cord perivascular cells exhibited enhanced migration capacity towards hepatocellular carcinoma in comparison with bone marrow mesenchymal stromal cells: a role for autocrine motility factor receptor. Biomed Res Int 2014:837420
27. Hass R, Kasper C, Bohm S et al (2011) Different populations and sources of human mesenchymal stem cells (MSC): a comparison of adult and neonatal tissue-derived MSC. Cell Commun Signal 9:12
28. Sarugaser R, Lickorish D, Baksh D et al (2005) Human umbilical cord perivascular (HUCPV) cells: a source of mesenchymal progenitors. Stem Cells 23:220–229
29. Takahashi K, Yamanaka S (2006) Induction of pluripotent stem cells from mouse embryonic and adult fibroblast cultures by defined factors. Cell 126:663–676

30. Wernig M, Meissner A, Foreman R et al (2007) In vitro reprogramming of fibroblasts into a pluripotent ES-cell-like state. Nature 448:318–324
31. Villa-Diaz LG, Brown SE, Liu Y et al (2012) Derivation of mesenchymal stem cells from human induced pluripotent stem cells cultured on synthetic substrates. Stem Cells 30:1174–1181
32. Liu Y, Goldberg AJ, Dennis JE et al (2012) One-step derivation of mesenchymal stem cell (MSC)-like cells from human pluripotent stem cells on a fibrillar collagen coating. PLoS One 7:e33225
33. Guzzo RM, Gibson J, Xu R-H et al (2013) Efficient differentiation of human iPSC-derived mesenchymal stem cells to chondroprogenitor cells. J Cell Biochem 114:480–490
34. Eberle I, Moslem M, Henschler R et al (2013) Engineered MSCs from patient-specific iPS cells. Adv Biochem Eng Biotechnol 130:1–17
35. Carpenter MK, Rao MS (2015) Concise review: making and using clinically compliant pluripotent stem cell lines. Stem Cells Transl Med 4:381–388
36. Chinnadurai R, Moll G, Viswanathan S (2023) Next generation MSC therapy manufacturing, potency and mechanism of action analysis. Front Immunol 14:1192636
37. Nitzsche F, Müller C, Lukomska B et al (2017) Concise review: MSC adhesion cascade-insights into homing and transendothelial migration. Stem Cells 35:1446–1460
38. Cuesta-Gomez N, Graham GJ, Campbell JDM (2021) Chemokines and their receptors: predictors of the therapeutic potential of mesenchymal stromal cells. J Transl Med 19:156
39. Guan SP, Lam ATL, Newman JP et al (2018) Matrix metalloproteinase-1 facilitates MSC migration via cleavage of IGF-2/IGFBP2 complex. FEBS Open Bio 8:15–26
40. Ullah M, Liu DD, Thakor AS (2019) Mesenchymal stromal cell homing: mechanisms and strategies for improvement. iScience 15:421–438
41. Han Y, Yang J, Fang J et al (2022) The secretion profile of mesenchymal stem cells and potential applications in treating human diseases. Signal Transduct Target Ther 7:92
42. Merimi M, El-Majzoub R, Lagneaux L et al (2021) The therapeutic potential of mesenchymal stromal cells for regenerative medicine: current knowledge and future understandings. Front Cell Dev Biol 9:661532
43. Liang X, Ding Y, Zhang Y et al (2014) Paracrine mechanisms of mesenchymal stem cell-based therapy: current status and perspectives. Cell Transplant 23:1045–1059
44. Tang J, Wang X, Lin X et al (2024) Mesenchymal stem cell-derived extracellular vesicles: a regulator and carrier for targeting bone-related diseases. Cell Death Dis 10:212
45. Zhang B, Tian X, Hao J et al (2020) Mesenchymal stem cell-derived extracellular vesicles in tissue regeneration. Cell Transplant 29:963689720908500
46. Gowen A, Shahjin F, Chand S et al (2020) Mesenchymal stem cell-derived extracellular vesicles: challenges in clinical applications. Front Cell Dev Biol 8:149
47. Fuloria S, Subramaniyan V, Dahiya R et al (2021) Mesenchymal stem cell-derived extracellular vesicles: regenerative potential and challenges. Biology 10:172
48. Feng X, Huang D, Lu X et al (2014) Insulin-like growth factor 1 can promote proliferation and osteogenic differentiation of human dental pulp stem cells via mTOR pathway. Dev Growth Differ 56:615–624
49. Zhu S, Chen W, Masson A et al (2024) Cell signaling and transcriptional regulation of osteoblast lineage commitment, differentiation, bone formation, and homeostasis. Cell Discov 10:71
50. Wu H, Gordon JAR, Whitfield TW et al (2017) Chromatin dynamics regulate mesenchymal stem cell lineage specification and differentiation to osteogenesis. Biochim Biophys Acta Gene Regul Mech 1860:438–449
51. Song N, Scholtemeijer M, Shah K (2020) Mesenchymal stem cell immunomodulation: mechanisms and therapeutic potential. Trends Pharmacol Sci 41:653–664
52. Giallongo C, Tibullo D, Camiolo G et al (2019) TLR4 signaling drives mesenchymal stromal cells commitment to promote tumor microenvironment transformation in multiple myeloma. Cell Death Dis 10:704

53. Lu D, Xu Y, Liu Q et al (2021) Mesenchymal stem cell-macrophage crosstalk and maintenance of inflammatory microenvironment homeostasis. Front Cell Dev Biol 9:681171
54. Negi N, Griffin MD (2020) Effects of mesenchymal stromal cells on regulatory T cells: current understanding and clinical relevance. Stem Cells 38:596–605
55. Feng B, Feng X, Yu Y et al (2023) Mesenchymal stem cells shift the pro-inflammatory phenotype of neutrophils to ameliorate acute lung injury. Stem Cell Res Ther 14:197
56. Caplan H, Olson SD, Kumar A et al (2019) Mesenchymal stromal cell therapeutic delivery: translational challenges to clinical application. Front Immunol 10:1645
57. Foguenne M, MacMillan S, Kron P et al (2023) Current evidence and future perspectives to implement continuous and end-ischemic use of normothermic and oxygenated hypothermic machine perfusion in clinical practice. J Clin Med Res 12:3207
58. Sierra Parraga JM, Rozenberg K, Eijken M et al (2019) Effects of normothermic machine perfusion conditions on mesenchymal stromal cells. Front Immunol 10:765
59. Justiz Vaillant AA, Modi P, Mohammadi O (2024) Graft-versus-host disease. In: StatPearls. StatPearls Publishing, Treasure Island, FL
60. Nachmias B, Zimran E, Avni B (2022) Mesenchymal stroma/stem cells: haematologists' friend or foe? Br J Haematol 199:175–189
61. Le Blanc K, Rasmusson I, Sundberg B et al (2004) Treatment of severe acute graft-versus-host disease with third party haploidentical mesenchymal stem cells. Lancet 363:1439–1441
62. Kebriaei P, Hayes J, Daly A et al (2020) A phase 3 randomized study of Remestemcel-L versus placebo added to second-line therapy in patients with steroid-refractory acute graft-versus-host disease. Biol Blood Marrow Transplant 26:835–844
63. Shandil RK, Dhup S, Narayanan S (2022) Evaluation of the therapeutic potential of mesenchymal stem cells (MSCs) in preclinical models of autoimmune diseases. Stem Cells Int 2022: 6379161
64. Liang J, Zhang H, Hua B et al (2010) Allogenic mesenchymal stem cells transplantation in refractory systemic lupus erythematosus: a pilot clinical study. Ann Rheum Dis 69:1423–1429
65. Vosough M, Nikfam S, Torabi SH et al (2022) Mesenchymal stromal cell therapy improves refractory perianal fistula in Crohn's disease: case series clinical interventional study. Cell J 24:62–68
66. Guo Y, Yu Y, Hu S et al (2020) The therapeutic potential of mesenchymal stem cells for cardiovascular diseases. Cell Death Dis 11:349
67. Chen S-L, Fang W-W, Qian J et al (2004) Improvement of cardiac function after transplantation of autologous bone marrow mesenchymal stem cells in patients with acute myocardial infarction. Chin Med J (Engl) 117:1443–1448
68. Kangari P, Talaei-Khozani T, Razeghian-Jahromi I et al (2020) Mesenchymal stem cells: amazing remedies for bone and cartilage defects. Stem Cell Res Ther 11:492
69. Copp G, Robb KP, Viswanathan S (2023) Culture-expanded mesenchymal stromal cell therapy: does it work in knee osteoarthritis? A pathway to clinical success. Cell Mol Immunol 20:626–650
70. Zhang X, Kuang Q, Xu J et al (2024) MSC-based cell therapy in neurological diseases: a concise review of the literature in pre-clinical and clinical research. Biomolecules 14:538
71. Inoue M, Yamaguchi R, He CCJ et al (2023) Current status and prospects of regenerative medicine for spinal cord injury using human induced pluripotent stem cells: a review. Stem Cell Investig 10:6
72. Honmou O, Yamashita T, Morita T et al (2021) Intravenous infusion of auto serum-expanded autologous mesenchymal stem cells in spinal cord injury patients: 13 case series. Clin Neurol Neurosurg 203:106565
73. Mead RJ, Shan N, Reiser HJ et al (2023) Amyotrophic lateral sclerosis: a neurodegenerative disorder poised for successful therapeutic translation. Nat Rev Drug Discov 22:185–212
74. Fang T, Al Khleifat A, Meurgey J-H et al (2018) Stage at which riluzole treatment prolongs survival in patients with amyotrophic lateral sclerosis: a retrospective analysis of data from a dose-ranging study. Lancet Neurol 17:416–422

75. Nam J-Y, Chun S, Lee TY et al (2023) Long-term survival benefits of intrathecal autologous bone marrow-derived mesenchymal stem cells (Neuronata-R®: lenzumestrocel) treatment in ALS: propensity-score-matched control, surveillance study. Front Aging Neurosci 15: 1148444
76. Mohammadi M, Jaafari MR, Mirzaei HR et al (2016) Mesenchymal stem cell: a new horizon in cancer gene therapy. Cancer Gene Ther 23:285–286
77. Cantero MJ, Bueloni B, Gonzalez Llamazares L et al (2024) Modified mesenchymal stromal cells by in vitro transcribed mRNA: a therapeutic strategy for hepatocellular carcinoma. Stem Cell Res Ther 15:208
78. Harrell CR, Volarevic A, Djonov VG et al (2021) Mesenchymal stem cell: a friend or foe in anti-tumor immunity. Int J Mol Sci 22:12429
79. Feins S, Kong W, Williams EF et al (2019) An introduction to chimeric antigen receptor (CAR) T-cell immunotherapy for human cancer. Am J Hematol 94:S3–S9
80. Zhang M, Huang H (2020) How to combine the two landmark treatment methods-allogeneic hematopoietic stem cell transplantation and chimeric antigen receptor T cell therapy together to cure high-risk B cell acute lymphoblastic leukemia? Front Immunol 11:611710
81. Labanieh L, Mackall CL (2023) CAR immune cells: design principles, resistance and the next generation. Nature 614:635–648
82. Tokarew N, Ogonek J, Endres S et al (2019) Teaching an old dog new tricks: next-generation CAR T cells. Br J Cancer 120:26–37
83. Amorós-Pérez B, Rivas-Pardo B, Gómez Del Moral M et al (2024) State of the art in CAR-T cell therapy for solid tumors: is there a sweeter future? Cells 13:725
84. Gross G, Waks T, Eshhar Z (1989) Expression of immunoglobulin-T-cell receptor chimeric molecules as functional receptors with antibody-type specificity. Proc Natl Acad Sci U S A 86: 10024–10028
85. Eshhar Z, Waks T, Gross G et al (1993) Specific activation and targeting of cytotoxic lymphocytes through chimeric single chains consisting of antibody-binding domains and the gamma or zeta subunits of the immunoglobulin and T-cell receptors. Proc Natl Acad Sci U S A 90:720–724
86. Krause A, Guo HF, Latouche JB et al (1998) Antigen-dependent CD28 signaling selectively enhances survival and proliferation in genetically modified activated human primary T lymphocytes. J Exp Med 188:619–626
87. Maher J, Brentjens RJ, Gunset G et al (2002) Human T-lymphocyte cytotoxicity and proliferation directed by a single chimeric TCRzeta /CD28 receptor. Nat Biotechnol 20:70–75
88. Brentjens RJ, Davila ML, Riviere I et al (2013) CD19-targeted T cells rapidly induce molecular remissions in adults with chemotherapy-refractory acute lymphoblastic leukemia. Sci Transl Med 5:177ra38
89. US FDA, KYMRIAH. In: U.S. Food and Drug Administration. https://www.fda.gov/vaccines-blood-biologics/cellular-gene-therapy-products/kymriah. Accessed 4 Aug 2024
90. US FDA, YESCARTA. In: U.S. Food and Drug Administration. https://www.fda.gov/vaccines-blood-biologics/cellular-gene-therapy-products/yescarta. Accessed 4 Aug 2024
91. US FDA, TECARTUS. In: U.S. Food and Drug Administration. https://www.fda.gov/vaccines-blood-biologics/cellular-gene-therapy-products/tecartus. Accessed 4 Aug 2024
92. US FDA, BREYANZI. In: U.S. Food and Drug Administration. https://www.fda.gov/vaccines-blood-biologics/cellular-gene-therapy-products/breyanzi-lisocabtagene-maraleucel. Accessed 4 Aug 2024
93. US FDA, ABECMA. In: U.S. Food and Drug Administration. https://www.fda.gov/vaccines-blood-biologics/abecma-idecabtagene-vicleucel. Accessed 4 Aug 2024
94. US. FDA, CARVYKTI. In: U.S. Food and Drug Administration. https://www.fda.gov/vaccines-blood-biologics/cellular-gene-therapy-products/carvykti. Accessed 4 Aug 2024
95. De Marco RC, Monzo HJ, Ojala PM (2023) CAR T cell therapy: a versatile living drug. Int J Mol Sci 24:6300

96. Watanabe N, Mo F, McKenna MK (2022) Impact of manufacturing procedures on CAR T cell functionality. Front Immunol 13:876339
97. Lana MG, Strauss BE (2020) Production of lentivirus for the establishment of CAR-T cells. Methods Mol Biol 2086:61–67
98. Hou AJ, Chen LC, Chen YY (2021) Navigating CAR-T cells through the solid-tumour microenvironment. Nat Rev Drug Discov 20:531–550
99. Albelda SM (2024) CAR T cell therapy for patients with solid tumours: key lessons to learn and unlearn. Nat Rev Clin Oncol 21:47–66
100. Lee DW, Santomasso BD, Locke FL et al (2019) ASTCT consensus grading for cytokine release syndrome and neurologic toxicity associated with immune effector cells. Biol Blood Marrow Transplant 25:625–638
101. Maalej KM, Merhi M, Inchakalody VP et al (2023) CAR-cell therapy in the era of solid tumor treatment: current challenges and emerging therapeutic advances. Mol Cancer 22:20
102. Knorr DA, Ni Z, Hermanson D et al (2013) Clinical-scale derivation of natural killer cells from human pluripotent stem cells for cancer therapy. Stem Cells Transl Med 2:274–283
103. Pan K, Farrukh H, Chittepu VCSR et al (2022) CAR race to cancer immunotherapy: from CAR T, CAR NK to CAR macrophage therapy. J Exp Clin Cancer Res 41:119
104. Kruschinski A, Moosmann A, Poschke I et al (2008) Engineering antigen-specific primary human NK cells against HER-2 positive carcinomas. Proc Natl Acad Sci U S A 105:17481–17486

Top Med Chem
https://doi.org/10.1007/7355_2024_176

Published online: 11 January 2025

Low Molecular-Mass Hormones: Peptidic and Steroidal Hormones

Luis Gabriel Valdvieso Gelves, Angela Cristina Leal Badaro Trindade, and Lidia Moreira Lima

Contents

Abstract Hormones are chemical messengers released by endocrine glands into the bloodstream to regulate various functions like metabolism, growth, reproduction, and responses to stress. These signaling molecules, which can be lipid, amide, or peptide-based, act slowly over time to influence many bodily processes. The endocrine system, including glands such as the hypothalamus and pituitary gland, coordinates these functions by sending signals to target cells. In contrast, exocrine glands release substances through ducts, such as those in the digestive system. Proper hormonal balance is crucial for health, as imbalances can lead to a range of diseases affecting overall well-being.

Keywords Hormones, Peptide hormones, Steroid hormones

L. G. V. Gelves, A. C. L. B. Trindade, and L. M. Lima (✉)
Laboratório de Avaliação e Síntese de Substâncias Bioativas (LASSBio), Federal University of Rio de Janeiro, Rio de Janeiro, Brazil

1 Introduction

Hormones are signaling molecules essential for transmitting messages to target tissues and regulating numerous bodily functions, including metabolism, growth, sexual characteristics, and other functions [1, 2]. Some endogenous molecules that play this role are peptide hormones, such as *growth hormone*, *oxytocin*, *vasopressin*, *thyroxine*, and steroidal hormones, *cortisol*, *estradiol*, and *testosterone*, among many others. These hormones vary significantly in size, structure, and function. In the context of biotherapeutics with hormonal function (or hormonal antagonists), which have small structures, three groups stand out: hormones derived from amino acids, peptide hormones, comprising from 2 to 60 amino acids, and steroid hormones.

2 Amino Acid-Derived Hormones

Considering the concept that hormones are biomolecules produced by glands capable of transmitting information to other tissues, it can consider different molecules derived from amino acids capable of reaching the bloodstream and reaching different organs. In this context, some biological molecules are recognized as "hormones," but they can also exert more local actions, such as certain neurotransmitters, such as *norepinephrine*, *epinephrine*, *dopamine*, *levodopa*, *serotonin*, *melatonin*, among others, which are preferably produced by chemical synthesis, but which can eventually be produced and isolated from living organisms.

Norepinephrine and *epinephrine* are biosynthesized in the adrenal and adrenergic neurons from the amino acid L-tyrosine, with other important molecules such as dopamine and *levodopa* (*L-DOPA*) as synthesis intermediates (Fig. 1). As hormones, *norepinephrine* and *epinephrine* are synthesized and released into the circulation, binding to alpha 1 receptor in blood vessels, causing vasoconstriction, to beta

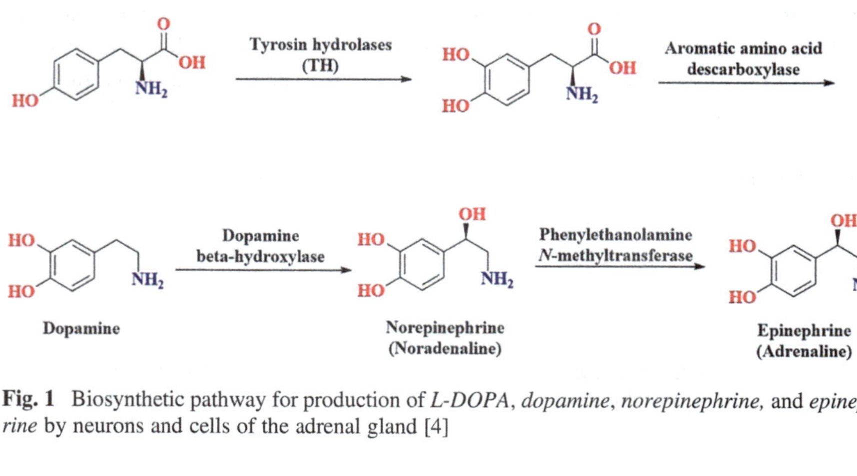

Fig. 1 Biosynthetic pathway for production of *L-DOPA*, *dopamine*, *norepinephrine*, and *epinephrine* by neurons and cells of the adrenal gland [4]

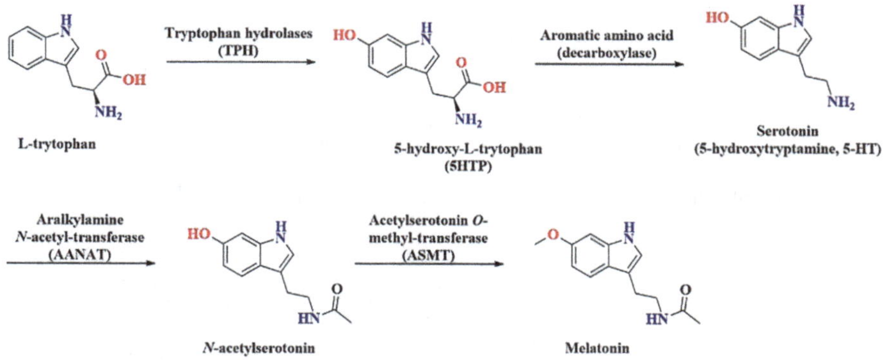

Fig. 2 Biosynthetic pathway of L-tryptophan derivatives: *serotonin* and *melatonin* [4]

1 receptors in the heart, leading to an increase in heart rate and strength, and to beta 2 receptors in the bronchi, resulting in bronchodilator effect [3, 4].

Due to the rapid vasoconstrictor effect, *norepinephrine*, *epinephrine,* and *dopamine* are used in therapy, especially in patients in an acute hypotensive state, such as in anaphylactic, cardiogenic, and septic shock. *Epinephrine* is also used to stop asthma attacks. *Levodopa* is used therapeutically in the treatment of Parkinson's disease, as it can cross the blood–brain barrier, reach the central nervous system (CNS), and be converted into *dopamine*, an central neurotransmitter involved in the pathophysiology of Parkinson's disease [4].

Industrially, *epinephrine* can be obtained from adrenal gland of livestock or by chemical synthesis. *Norepinephrine* and *dopamine* are obtained essentially by chemical synthesis. Because of the asymmetric center, the synthetic production of these catecholamines may involve a stereoselective reaction step or the separation of the enantiomers by resolution with tartaric acid. *Levodopa* can be obtained from L-tyrosine and a microbial hydroxylation step, or by total synthesis from L-tyrosine, vanillin, or piperonal [4]. Some derivatives of the amino acid tryptophan also have an interesting pharmacological action for their therapeutic use, such as *serotonin* and *melatonin* (Fig. 2).

Serotonin (or 5-HT) binds to receptors (5-HT1 to 5-HT7) to trigger different responses. It has been used in homeopathic products for temporary relief of nervousness, anxiety, mood swings, joint pain, weakness, drowsiness, itching, and lethargy. The classic method of obtaining serotonin is by chemical synthesis. Alternatively, the precursor amino acid L-tryptophan is commercially available as a food supplement.

Melatonin is a hormone produced by the pineal gland that by binding to specific receptors (MT1 and MT2, both G protein-coupled receptors), regulates sleep–wake cycles and when provided has beneficial effects on sleep-onset latency. It also has antioxidant and anti-inflammatory actions, among others. It is commercially available as an over-the-counter supplement and used for jet lag, shift-work disorder, insomnia, circadian rhythm disorders in the blind, and benzodiazepine and nicotine

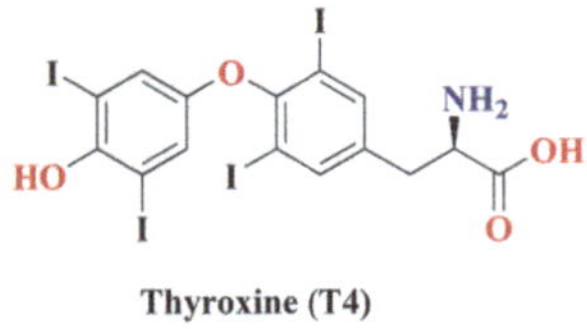

Fig. 3 Chemical structures of *thyroxine* (*T4*) and *triiodothyronine* (*T3*)

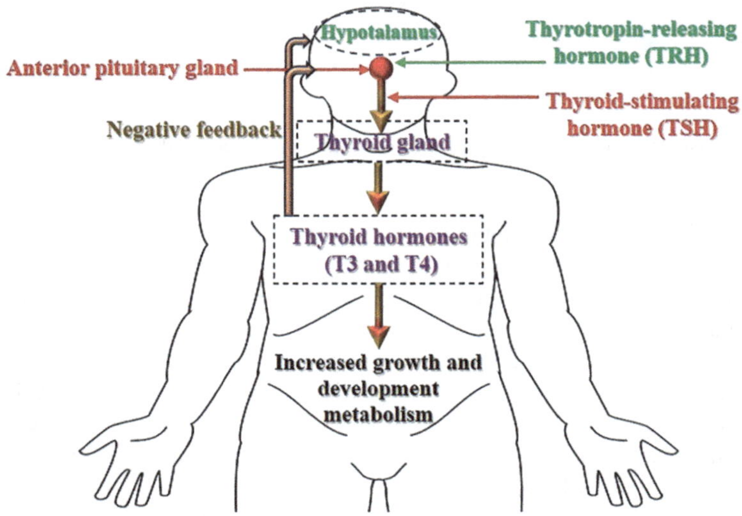

Fig. 4 Regulation of the release of thyroid hormones T3 (triiodothyronine) and T4 (thyroxine); TRH, thyrotropin-releasing hormone; TSH, thyroid-stimulating hormone

withdrawal. Commercial *melatonin*, like other hormones derived from amino acids mentioned above, is obtained mainly by chemical synthesis. However, good results have been achieved in the preparation of this hormone by biotechnology, from the culture of bacteria or fungi, with appreciable yields [5].

Thyroid hormones are small molecules produced by thyroid gland and released into the blood circulation to reach target organs and tissues. The most important hormones are *thyroxine*, or *T4* (3,5,3′,5′-tetraiodothyronine), and *triiodothyronine*, or *T3* (3,5,3′-triiodothyronine) (Fig. 3). These compounds are derived from the amino acid tyrosine on thyroglobulin and are released on plasma to reach target organs and tissues. Their release is regulated by the peptide hormones *TRH* (*thyrotropin-releasing hormone*) on hypothalamus, and *TSH* (*thyroid-stimulating hormone*) on pituitary gland (Fig. 4). *T4* is mostly converted into its more active metabolite *T3* through deiodination (by type I or type II 5′-deiodinases) [6]. They are transported in plasma, in inactive form, bound to plasma proteins. Because they are lipophilic, they cross the plasma membrane passively or via energy-dependent transporters. They act on the cell nucleus, binding to thyroid hormone receptors in DNA regions called thyroid hormone response elements (TREs) near genes. This complex of *T4*, *T3*, DNA and other coregulatory proteins causes a conformational change and a resulting shift in transcriptional regulation of nearby genes, synthesis of messenger RNA, and cytoplasmic protein production [7].

While *T4* is the major product secreted by the thyroid gland, *T3* exerts most of the physiological effects of the thyroid hormones; *T4* and *T3* have a relative potency of ~1:4 (*T4*:*T3*). *Thyroxine* and *triiodothyronine* act to increase basal metabolism, allow the correct fetal development and the proper functioning of the cardiovascular [8], hematopoietic, and musculoskeletal systems. They also play a role in the body's responses to heat production, oxygen consumption, and regulation of other hormones.

Levothyroxine, or synthetic *T4*, is indicated as replacement therapy in primary (thyroidal), secondary (pituitary), and tertiary (hypothalamic) congenital or acquired hypothyroidism, resulting in diminished downstream effects of these hormones, such as fatigue, an increased heart rate, depression, dry skin and hair, muscle cramps, constipation, weight gain, memory impairment, and poor tolerance to cold temperatures. It is also indicated as an adjunct to surgery and radioiodine therapy in the management of thyrotropin-dependent well-differentiated thyroid cancer. The exogenous *T3* (or *liothyronine*) product is indicated to treat hypothyroidism, to suppress *TSH*, and to help in the diagnosis of hyperthyroidism.

Thyroid hormone drugs bind more than 99% to plasma proteins, including thyroxine-binding globulin (TBG), thyroxine-binding prealbumin (TBPA) and albumin (TBA). The half-life is 6 to 7 days for *T4* and 1 to 2 days for *T3*.

3 Oligopeptide and Polypeptide Hormones

Hormones containing small or polypeptides are most common in living organisms, being essential for most functions, like regulation of other hormones releasing, metabolic and cardiovascular regulation, growth, sexual behaviors, and reproductive process control. These biomolecules are elaborated and released mainly by hypothalamus (CNS), pituitary gland (or hypophysis), thyroid gland, pancreas, gastrointestinal (GI) tract, liver, and kidneys. The biotherapeutic hormones can be the native peptides or modified peptides to achieve "better" properties, including less immunological reactions, more stability, and new biological activity or use. The medicinal chemistry of some representative oligo and polypeptide hormones is discussed here, focusing on those with up to 60 amino acids in their constitution. Insulin and other important peptide hormones, due to their importance in the group of biotherapeutics, are discussed in other chapter of this book.[1]

Due to their peptide nature, these substances have hydrophilic characteristics and are usually administered parenterally due to their low absorption from the GI tract, good plasma distribution, and low CNS penetration. In general, they interact with

[1] See Chapter "Therapeutic Proteins".

membrane receptors to trigger an intracellular change and trigger the corresponding biological response.

3.1 Hypothalamic and Pituitary Hormones

There is a complex relationship between hypothalamus, pituitary gland, and other important glands, such as thyroid. For example, the hypothalamus releases *thyrotropin-releasing hormone* (*TRH*), stimulating the release of *thyrotropin* (or *thyroid-stimulating hormone*, *TSH*) from the pituitary gland. *Thyrotropin* will act on the thyroid gland and stimulate the production of *thyroxine* and *triiodothyronine* (Fig. 5). In turn, the increased plasma levels of these two thyroid hormones results in the inhibition of the action of *TRH* on the hypophysis, by feedback regulation. Other hormones are also released from hypothalamus and pituitary gland and are depicted in Fig. 5.

Hormones released from the anterior pituitary stimulate the production of hormones by a peripheral endocrine gland, or other tissues, or act directly on target tissues. *ADH* and *oxytocin* are produced by neurons of hypothalamic supraoptic and paraventricular nucleus and stored in the posterior pituitary. *Prolactin* and the hormones released from the posterior pituitary act directly on target tissues. Hypothalamic factors regulate the release of anterior pituitary hormones [9].

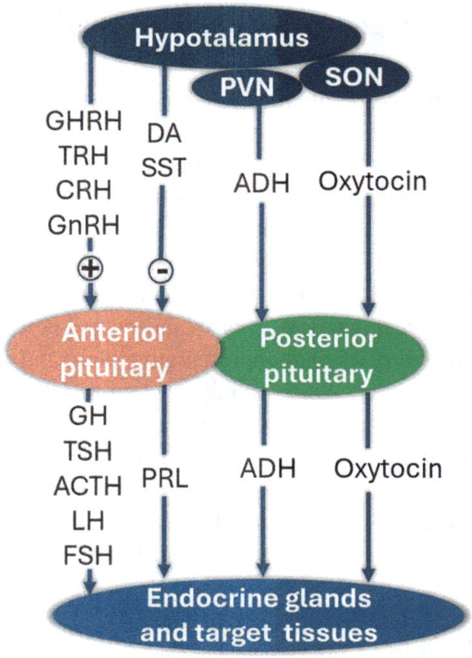

Fig. 5 The hypothalamic–pituitary endocrine system. ACTH, adrenocorticotropin; ADH, antidiuretic hormone [vasopressin]; CRH, corticotropin-releasing hormone; DA, dopamine; FSH, follicle-stimulating hormone; GH, growth hormone; GHRH, growth hormone-releasing hormone; GnRH, gonadotropin-releasing hormone; LH, luteinizing hormone; PRL, prolactin; PVN, paraventricular nucleus; SON, supraoptic nucleus; SST, somatostatin; TRH, thyrotropin-releasing hormone; TSH, thyroid-stimulating hormone (Adapted from [9])

Although hypothalamic and pituitary hormones are available for use in humans, only a few are clinically important. Among them are *somatostatin* (*SST*) and *gonadotropin releasing hormone* (*GnRH*), both produced and released by hypothalamus, the *growth hormone* (*GH*) and the *gonadotropins* (*follicle-stimulating hormone*, *FSH*, and *luteinizing hormone*, *LH*), released by the adenohypophysis (or anterior pituitary), and *vasopressin* (*antidiuretic hormone*, *ADH*) and *oxytocin* released by the neurohypophysis (or posterior pituitary) (Fig. 5).

3.1.1 Somatostatin

Somatostatin, also known as *somatotropin release-inhibiting factor* (*SRIF*) or *growth hormone inhibiting hormone* (*GHIH*), is a cyclic peptide originally discovered as a hypothalamic neuropeptide [10]. It is produced in GI tract, pancreas, and different locations in CNS. It has broad antisecretory activity, including growth hormone, thyroid-stimulating hormone, insulin, *glucagon*, *gastrin*, *cholecystokinin* (*CCK*), and *ghrelin* [11]. In human and animal models, somatostatin prevents angiogenesis and displays anti-proliferative effects on healthy and cancerous cells.

Two active forms of the somatostatin hormone are known, differing in the size of the amino acid sequences (14 amino acids versus 28 amino acids) (Fig. 6). They have considerable overlaps in activity and differ mainly in their site of effect. The shorter peptide acts mainly in the brain, while the longer form acts in the gastrointestinal tract.

Somatostatin receptors (SST_1 to SST_5) are G protein coupling receptors (GPCRs) that belong to the rhodopsin-like family. All of the receptors are coupling to inhibitory G protein ($G\alpha_{i/0}$) and causes a decrease in cAMP levels and adenylate cyclase activity. It also inhibits calcium influx into the pituitary cells and suppresses glucose-induced pancreatic insulin secretion by activating and deactivating potassium and calcium ions permeability, respectively [11]. *Somatostatin* suppresses the

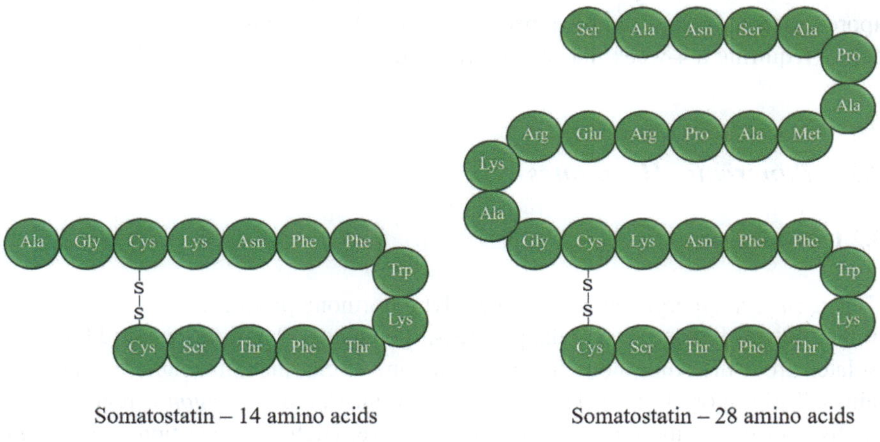

Fig. 6 Two forms of *somatostatin* peptide hormone

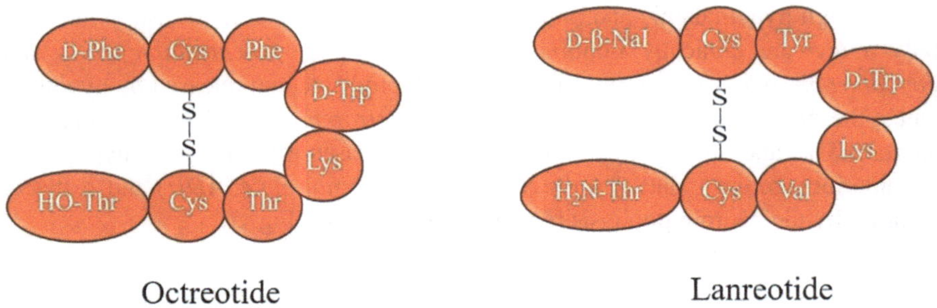

Fig. 7 Somatostatin analogs: octreotide and lanreotide

release of *GH* from the pituitary gland and, originally, could be used for treatment of acromegaly. However, due to its short half-life ($t_{1/2}$), between 1 and 3 min, the therapeutic potential of natural hormone is limited. Consequently, highly potent and metabolically stable analogs were developed. Nonetheless, *somatostatin* has been used in the clinical setting for the diagnosis of acromegaly and GI tract tumors. Other uses of *somatostatin* are to treat acute bleeding from esophageal varices, GI ulcers, and gastritis, to prevent pancreatic complications after surgery and to restrict secretions of the upper intestine, pancreas, and biliary tract.

Octapeptide octreotide is the first *somatostatin* analog approved for clinical use (Fig. 7) [12]. These peptide biotherapeutic, in acetate salt form, are used to treat acromegaly as well as diarrhea associated with metastatic carcinoid tumors and vasoactive intestinal peptide secreting tumors [13].

Other long-acting *somatostatin* analog, *lanreotide* (Fig. 7), was introduced for the long-term treatment of acromegaly. *Lanreotide* also indicated in the treatment of adult patients with unresectable, well- or moderately differentiated, locally advanced or metastatic gastroenteropancreatic neuroendocrine tumors (GEP-NETs) [14]. It is an inhibitor of various endocrine functions, but its main effects are associated with inhibiting insulin-like growth factor 1 (IGF1) [15]. Half-life of *lanreotide acetate* is approximately 22 days and has been used by subcutaneous or intramuscular drug depot, requiring a 4-week interval between doses.

3.2 *Pancreatic Hormones*

3.2.1 Glucagon

Glucagon is a hyperglycemic-glycogenolytic hormone produced by the alpha cells of the islets of Langerhans, in the pancreas. It contains 29 amino acids and has been isolated from the amorphous fraction of a commercial insulin sample, which contains 4% glucagon (Fig. 8). The physiological function of *glucagon* is regulation of the liver-mediated mobilization of stored glucose. High concentration of *glucagon* results in abnormally high concentration of glucose into the blood, thereby causing

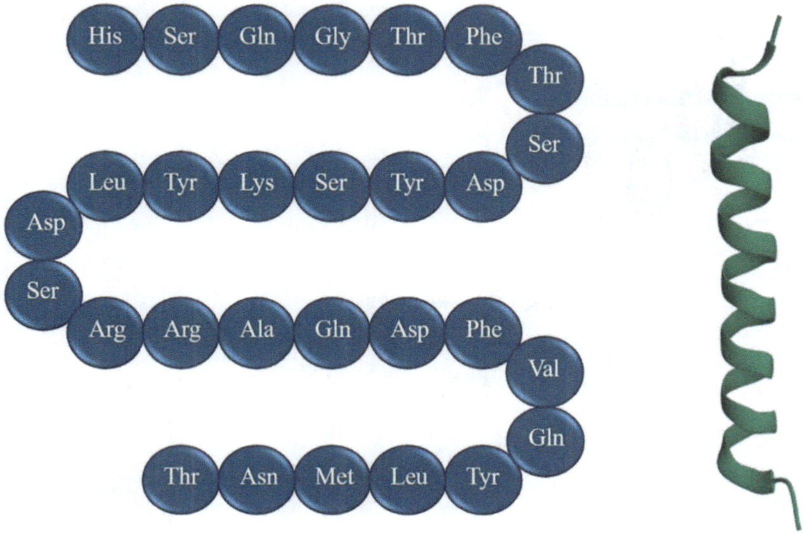

Fig. 8 *Glucagon* structure and their crystal structure (RCSB PDB 6PHI) [15, 16]

hyperglycemia, and therefore, indicating that a relative excess of *glucagon* is an essential factor in the development of diabetes [15, 16]. *Glucagon* binds to membrane-sensitive receptors in target cells. These receptors, called glucagon receptors, are a seven-pass transmembrane G-protein-coupled receptor (G-stimulatory proteins, G_s and G_q) that are abundant in many tissues, such as the liver and kidney. They are found in less amount in the heart, adipocytes, lymphoblasts, spleen, pancreas, brain, retina, adrenal gland, and the GI tract [17, 18]. The interaction of glucagon with its biological receptor stimulates adenylate cyclase, which catalyzes cAMP formation and activates the protein kinase A (PKA) pathway (Fig. 9). PKA catalyzes phosphorylation of a single serine residue of phosphorylase kinase (PPK), which becomes active and forms the active glycogen phosphorylase a (PYGa). Finally, glycogen phosphorylase catalyzes glycogenolysis, which releases glucose 1-phosphate from glycogen polymers, resulting in the hyperglycemic action of *glucagon* [19]. When the *glucagon* receptor is associated with the G_q protein, ligand–receptor interaction activates phospholipase C, increases production of inositol 1,4,5-triphosphate, and releases intracellular calcium [20].

Glucagon exerts other biochemical effects. Gluconeogenesis in the liver is stimulated by *glucagon*, inhibits the incorporation of amino acids into liver proteins and decreases the synthesis of fatty acids, including cholesterol formation. *Glucagon* activates liver lipases and stimulates ketogenesis resulting from increased oxidation of free fatty acids. It also increases the release of glycerol and fatty acids from adipose tissue [15].

As a result of *glucagon*'s effects on carbohydrate and fatty acid metabolism, *glucagon* has important therapeutic uses. It is recommended for the treatment of severe hypoglycemic reactions caused by insulin administration in diabetic or

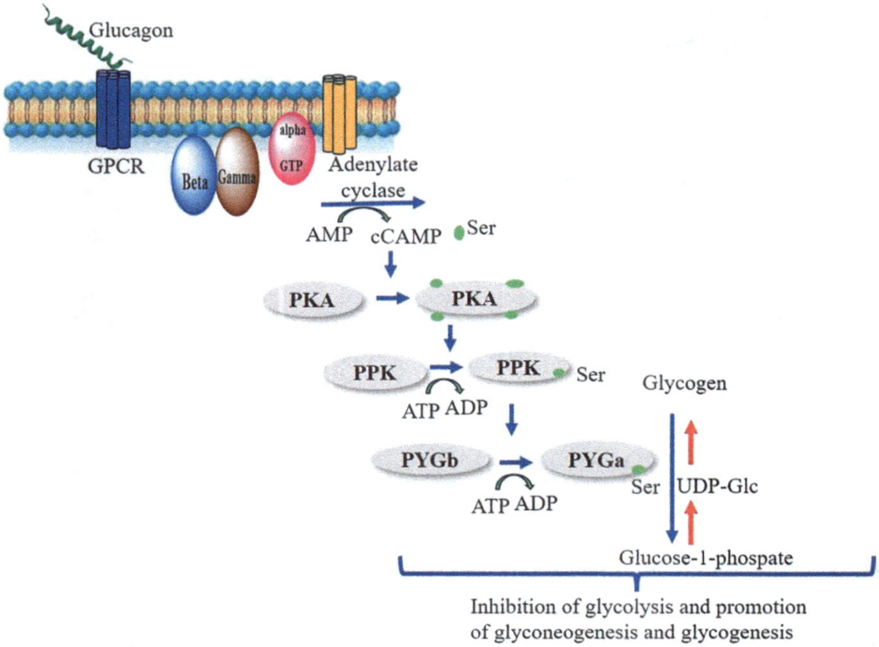

Fig. 9 *Glucagon* signaling pathway by activation of adenylate cyclase. Adapted from [19]

psychiatric patients, being effective only when hepatic glycogen is available. The most frequently reactions to *glucagon* are nausea and vomiting. Due to its short half-life (about 26 min, after intramuscular injection), it has been used in emergencies and dosing is repeated in 20 min, if necessary. Also, it is used as a part of GI imaging procedures by inhibiting the GI tract movement [21]. Currently, *glucagon* has been prepared by bioengineering, using recombinant techniques, but it can also be isolated from commercial insulin, with stability properties comparable to those of insulin [17–20, 22].

3.3 *Gastrointestinal Hormones*

3.3.1 Incretins

Incretin hormones are gut peptides that are secreted after food intake and stimulate insulin secretion and other biological functions. The known incretin hormones are *GIP* (*glucose-dependent insulinotropic polypeptide*) secreted by K cells in the upper small intestine [23, 24], and *GLP-1* (*glucagon-like peptide-1*) secreted by L cells in the lower intestine and colon [23]. Together, they are responsible for the incretin effect: a two- to threefold higher insulin secretory response to oral as compared to intravenous glucose administration [25]. *GIP* and *GLP-1* also present, in addition to

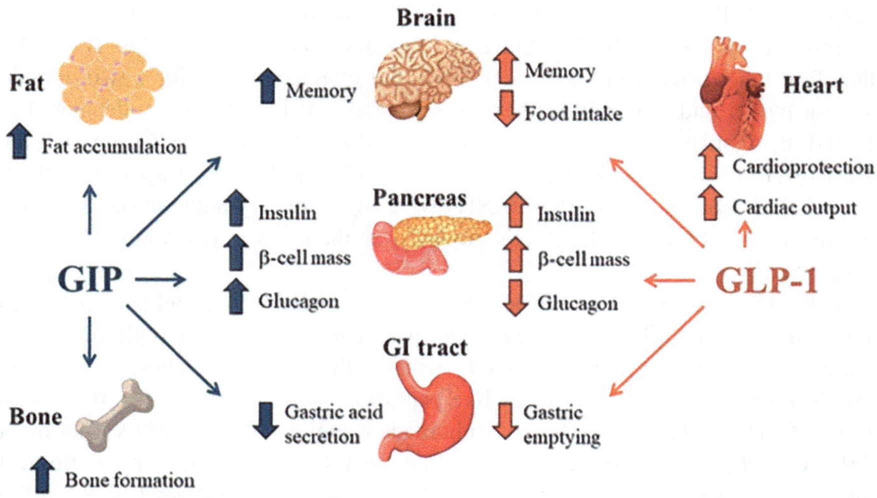

Fig. 10 Biological functions of *GIP* and *GLP-1*. *GIP* acts directly on the endocrine pancreas, bone, fat, GI tract, and brain. *GLP-1* acts directly on the endocrine pancreas, GI tract, heart, and brain [26]. The image contains fragments taken from the free collection of the Freepik website (https://br.freepik.com/), accessed on 4 August 2024

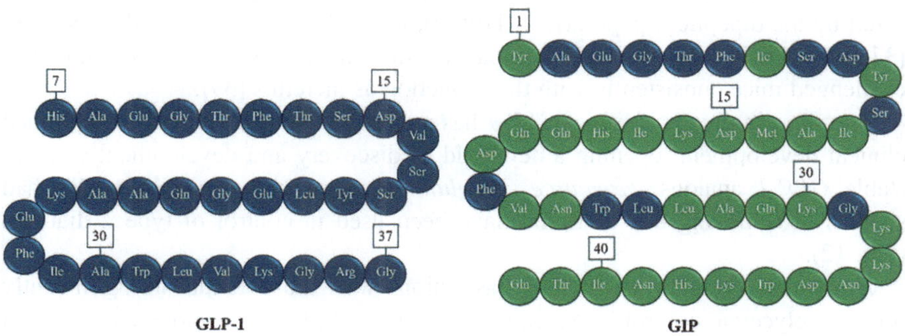

Fig. 11 *GLP-1* and *GIP* amino acid sequences. Blue circles: similar amino acid sequences. Green circles: different amino acid sequences of *GIP* compared to *GLP-1*

stimulating insulin secretion, other important biological functions in pancreas, bone, fat, GI tract, heart, and brain (Fig. 10) [26].

GIP is a peptide containing 42 amino acids (Fig. 11), produced from the precursor protein preproGIP (153 amino acids), which plays a key role in glycemia regulation and energy balance [27]. This hormone was first isolated from porcine intestine based on its ability to inhibit gastric acid secretion [28]. Its ability to stimulate insulin secretion in healthy volunteers was later described [29]. *GIP* acts directly on pancreatic islets to stimulate insulin secretion [30] and it is currently under investigation in clinical trial in patients with hepatic nuclear factor 1 alpha gene mutation (HNF1A) diabetes [27]. The *GIP* receptor (GIPR) belongs to the G protein coupled receptor (GPCR) superfamily and more specifically, the class B1 (or secretin-like) receptor family [18, 26, 29–33]. The exact activation mechanism of class B GPCRs (therefore the GIPR) is generally unknown, but evidence exists for a two-domain binding model and is currently the leading paradigm. Florent and colleagues has studied the agonist and antagonist peptides based on *GIP* structure. Comparing the effect of GIP(1-42), the naturally occurring GIP(1-30)NH_2 was a agonist peptide, while GIP(3-30)NH_2 and GIP(5-30)NH_2 were high potency antagonists. Receptor residues R183, R190, and R300 are shown to be the key for activation of the *GIP* receptor [27].

GLP-1 (Fig. 11) is a 31 amino acid hormone produced from proglucagon protein (160 amino acids) [34]. Two *glucagon*-like sequences with a high degree of sequence homology were detected within the same precursor molecule (proglucagon) and named *glucagon*-like peptides 1 and 2. Importantly, more than 70% of *GLP-1* secreted from L cells in humans is amidated at its carboxyl-terminal [35]. At that time neither a corresponding peptide product nor any biological function were known. Large amounts of *GLP-1* are secreted from L cells in the lower gut (ileum, colon, and rectum) after meals containing carbohydrate and fat components [36]. *GLP-1* receptor (GLP-1R), like GIPR, belongs to the G-protein coupled receptor family, activating adenylate cyclase and increasing levels of intracellular cAMP in pancreatic beta-cells, thereby stimulating insulin secretion glucose dependently [26]. *GLP-1* is inactivated by the enzymatic hydrolysis catalyzed by the dipeptidyl peptidase-4 (DPP-4), in the *N*-terminal dipeptide His7Ala8 [37, 38]. Deficiency of DPP-4 enhances insulin secretion in orally glucose-challenged mice consistently with their function as incretins [39].

The short in vivo half-life of *GLP-1* has proven a significant barrier to continued clinical development, opening a new field of discovery and development of more stable *GLP-1* analogs. *Exenatide*, *liraglutide*, *semaglutide,* and *tirzepatide* had success in achieving this goal and have been used in control of type 2 diabetes (Fig. 12).

Exenatide (synthetic *exendin-4*) has actions with the potential to significantly improve glycemic control in patients with diabetes [40]. Its development occurred from *exendin-4*, the naturally occurring form of *exenatide*, originally isolated from the *Heloderma suspectum* venom [41]. *Exenatide* have 39 amino acids, consisting of a single chain and have some amino acid alterations (purple circles in Fig. 12) to avoid degradation by DPP-4 and to extend its plasma half-life, especially by modification on position 2 (alanine to glycine substitution) [41]. *Exenatide* presents a combination of mode of actions which may include glucose-dependent stimulation of insulin secretion, suppression of glucagon secretion, enhancement of beta-cell mass, slowing of gastric emptying, inhibition of food intake, and modulation of glucose trafficking in peripheral tissues. One of the most provocative areas of recent research is the finding that *exenatide* enhances beta-cell mass, thereby impeding or even reversing disease progression. *Exenatide* was approved by the U.S. Food & Drug Administration (FDA) in 2005, for subcutaneous administration, twice daily.

Currently approved *GLP-1* analogs generally are based on the scaffold of either naturally occurring *exendin-4* or human endogenous *GLP-1*. Starting with the

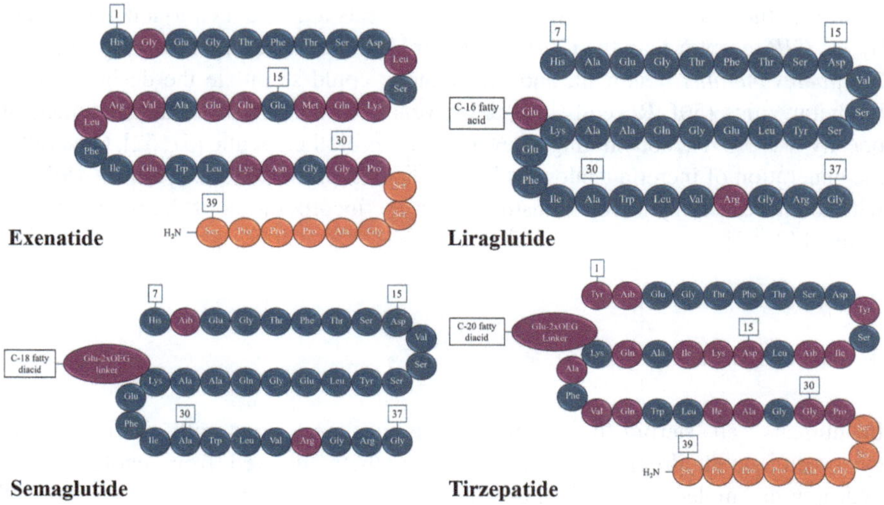

Fig. 12 Incretin analogs *exenatide*, *liraglutide*, *semaglutide,* and *tirzepatide*. Blue circles: amino acid sequences like *GLP-1*. Purple circles: alterations in the amino acid sequence or added side chains. Orange circles: additional peptide fragments. Aib: -NHC$(CH_3)_2$CO-; Glu-2xOEG linker: $[COCH_2O(CH_2)_2O\text{-}(CH_2)_2NH]_2CO(CH_2)_2CH(CO_2H)NH$-; C-16 fatty acid: $CO(CH_2)_{14}CH_3$; Fatty diacid: $CO(CH_2)_nCO_2H$ (*semaglutide*: $n = 16$; *tirzepatide*: $n = 18$)

human *GLP-1* amino acid sequence was developed *liraglutide* (FDA approval: 2010), a peptide with approximately 97% of homology to natural incretin (blue circle in Fig. 12). *Liraglutide* is made by attaching a C-16 fatty acid (palmitic acid) with a glutamic acid spacer on the remaining lysine residue at position 26 of the peptide precursor, extending its half-life to approximately 13 h [41]. This made it possible to administer it subcutaneously, once a day.

Some new ideas came up in designing *GLP-1* analogs, in which synthetic amino acids are used in improving pharmacokinetics and pharmacodynamics [41]. *Semaglutide* [42], approved by FDA in 2017 for type 2 diabetes, has 94% of homology to natural *GLP-1*. It contains two amino acid substitutions compared with human *GLP-1*, with alanine changed to alpha-amino isobutyric acid at position 8 and lysine to arginine at position 34 [43]. The lysine at position 26 was acylated and linked to a glutamic acid moiety combined with a C-18 fatty diacid side chain [44]. Owing to these structural modifications, it has a long half-life (168 h), attributed to its albumin binding, making it possible to have only one subcutaneous administration per week. It has also been reported that *semaglutide* has better efficiency in controlling weight, helping with obesity treatment [45].

Tirzepatide is a dual *GIP* and *GLP-1* analog [46, 47], being approved for type 2 diabetes by FDA in 2022. Their half-life is approximately 5 days, also allowing its subcutaneous administration once a week. *Tirzepatide* is a peptide whose structure was modified from *GIP*, with a C-20 fatty diacid moiety linked to the lysine at position 20 [48], pharmacology study indicated that this biotherapeutic has higher affinity to the *GIP* receptor, meaning that its effects are mediated to a higher extent by the *GIP* receptor than the *GLP-1* receptor [49]. The dual target can improve type 2 *diabetes mellitus* treatment, and this strategy could stimulate the design of new biotherapeutics [50]. Recently, the effectiveness of *tirzepatide* in the treatment of obesity was proven, constituting a new use for this biotherapeutic [51, 52]. Generally, the generation of incretin analogs is based on solid-phase peptide synthesis (SPPS), with the necessary synthetic adjustments for the incorporation of the modified side chains [42, 53].

4 Steroidal Hormones

In humans, all steroid or steroidal hormones are structurally derived from cyclopentanoperhydrophenantrene and biosynthetic derived from cholesterol, a hydrophobic molecule that is essentially insoluble in water. The production of steroid hormones involves several precise modifications to the cholesterol structure, with different series of modifications occurring in different pathways. Vitamin D3 and its metabolites are also derived from cholesterol. The cholesterol and steroid backbones are represented in Fig. 13. The 3-hydroxyl, the 18- and 19-methyl groups and the sidechain attached to the 17-carbon all project above the plane of the fused ring system. Substituents located above the ring plane are denoted β. Substituents located below the ring plane (such as the hydrogens at the C3 and C17 positions) are denoted α. The difference between α and β is often the difference between an active and an inactive compound.

The chemical nomenclature for each steroid is based on describing the modifications to one of four possible backbones: cholestane, pregnane, androstane, and estrane. In fact, all physiological steroids are derivatives of one of these four backbone structures. The cholestane (which has 27 carbons), pregnane (which has 21 carbons), and androstane (which has 19 carbons) differ from one another according to the length of the sidechain at the carbon 17. In mammalian systems, there are six families of steroid hormones that can be classified on both a structural and biological (hormonal) basis (Fig. 14). They are the estrogens and progestins (female sex steroids), androgens (male sex steroids), mineralocorticoids (aldosterone), glucocorticoids (cortisol), and vitamin D [$1\alpha,25(OH)_2D_3$]. It is also worth mentioning that bile acids are structurally related to cholesterol, particularly to cholestane backbone and thus constitute a seventh member of the steroid family.

4.1 *Biosynthetic Production of Steroid Hormones*

The biosynthesis of the six essential steroid hormones classes starts by the effect of specific stimulatory or tropic hormone on endocrine glands, which are the site of production of a different specific steroid hormone, as depicted in Fig. 14. For

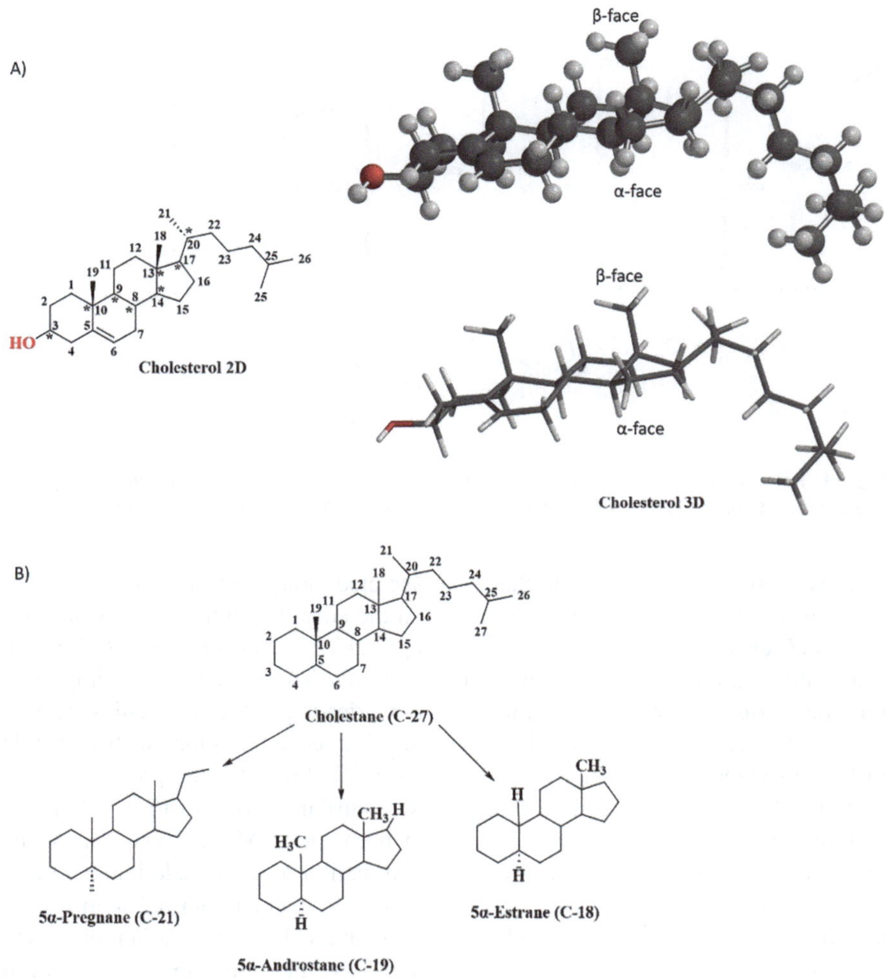

Fig. 13 2D and 3D structure of cholesterol (**a**) and general steroid backbones (**b**)

example, the *adrenocorticotropic hormone* (*ACTH*) – a 39 amino acid peptide secreted from the anterior pituitary gland under the control of *corticotropin-releasing hormone* – moves through the circulatory system to the zona fasciculata region of the adrenal cortex where it binds to its partner receptor, to trigger the metabolism of cholesterol to produce the glucocorticoid steroid hormones. The juxtaglomerular cells of the kidneys, in response to perceived low blood pressure and renal perfusion, released the aspartyl protease *Renin* that also acts as a hormone. It binds to pro-*renin* receptors catalyzing the conversion of *angiotensinogen* to *angiotensin I*. The latter is cleaved into *angiotensin II* by the angiotensin-converting enzyme (ACE), located in the endothelium of the lungs and kidneys. *Angiotensin II* triggers the release of *aldosterone* from the zona glomerulosa of the adrenal cortex. *Luteinizing hormone*

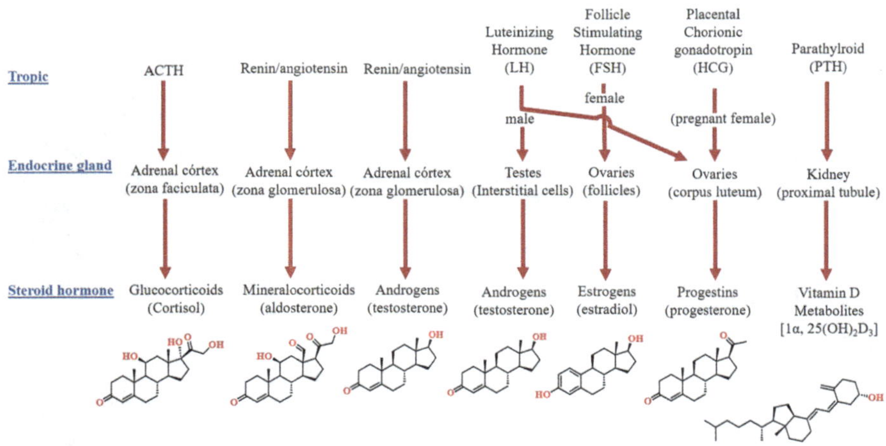

Fig. 14 Relationships between stimulatory peptides or protein hormones and the endocrine glands that are the site of production of six steroid hormones and vitamin D. Adapted from [54]

(*LH*) is a glycoprotein hormone that is co-secreted along with *follicle-stimulating hormone* (*FSH*) by the gonadotrophin cells in the adenohypophysis (anterior pituitary). *LH* release is stimulated by *gonadotropin-releasing hormone* (*GnRH*) and it causes the testes' Leydig cells to produce *testosterone* in men. LH also has important hormonal effects in women. It stimulates the ovaries to produce progestins. FSH, a glycoprotein dimer with alpha and beta subunits, is a second hormone with a key role in the biosynthesis of sex hormones. In men, FSH stimulates the proliferation of Sertoli cells, which is the factor that most contributes to testicular volume in children. These cells produce anti-mullerian hormone (AMH), which causes the involution of Muller's ducts, preventing the formation of the female internal genitalia. Together with testosterone, FSH is necessary to maintain normal sperm count and function. In women, FSH stimulates the granulosa cells in the ovarian follicles to synthesize aromatase, which is responsible for converting the androgens produced by the thecal cells into estradiol. Another important tropic hormone is human chorionic gonadotrophin (HCG), a hormone produced mainly by the syncytiotrophoblastic cells of the placenta during pregnancy. *HCG* is a glycoprotein composed of two subunits, the alpha and beta subunits that stimulate the corpus luteum to produce *progesterone* to maintain the pregnancy. Finally, the *parathyroid hormone* (*PTH*), a polypeptide hormone with 84 amino acids responsible for maintaining serum calcium homeostasis, stimulates the activity of the 1-alpha hydroxylase enzyme in the renal proximal tubule, enhancing the synthesis of *1.25 dihydroxy-vitamin D* (*1.25OH2D*).

4.2 Historical Aspects

The word "hormone" was first used in 1945 by Ernest Starling, a professor of physiology at University College London, UK, in one of four Croonian Lectures delivered at the Royal College of Physicians in London [55]. He defined the word as "the chemical messengers which, passing from cell to cell along the bloodstream, can coordinate the activities and growth of different parts of the body" [56]. Since then, elucidating the biological properties and functions of these chemical messengers has been the subject of a great deal of research. Many elucidated hormones were named after the gland or organ from which they were secreted, and they had their chemical nature deciphered. Less than 20 years after Starling coined the word "hormone," Edward Calvin Kendall, from the Mayo Clinic in Rochester, NY, USA, purified and determined the structures of *cortisone* (a glucocorticoid) and *thyroxine*.

In the 1920s and 1930s, Adolf Butenandt, Tadeus Reichstein, and Edward Adelbert Doisy discovered and characterized various steroid hormones, including *estrogen*, *testosterone,* and *progesterone*, being all of them later awarded Nobel Prizes. Still in the 1930s, Russell Marker, then a chemistry professor at the Pennsylvania State College, proposed a remarkable new molecular structure for sarsasapogenin, a plant steroid isolated from sarsaparilla that could be used to prepare steroid hormones, since he had demonstrated that under "Marker Degradation" diosgenin was converted to *progesterone*. However, due to the high cost of sarsasapogenin and the major expense involved in these final manipulations, Marker began a search for a plant steroid of the sapogenin class with a ring structure more like *progesterone*. He found it in Beth root, a plant of the lily family, which already was being used in a patent medicine called "Lydia Pinkham's Compound." Japanese chemists previously had isolated this steroid from a yam of the *Dioscorea* family and named it diosgenin. Nevertheless, neither Beth root nor the Japanese Dioscorea was an economical source of diosgenin. In Marker's search for a better source of diosgenin, he was struck by the root of a *Dioscorea* that grows in the Mexican state of Veracruz, near Orizaba, drew his attention, called *cabeza de negro* by the natives. From Mexican Dioscorea, Marker converted the syrup of *cabeza de negro* roots to progesterone. Later, the *cabeza de negro* was replaced by another yam called barbasco, which contained five times as much diosgenin. Following the pioneering work of Marker and colleagues, other steroid hormones were prepared, such as testosterone. Encouraged by their success in converting diosgenin into male and female sex hormones, Carl Djerassi and colleagues studied the conversion of diosgenin to *cortisone*. In 1951, scientists at the Upjohn Co. introduced a microbiological process, which specifically oxidized progesterone to a product that was easily converted to *cortisone*. Further, focused on the discovery of an orally active *progesterone* mimic, Djerassi's group designed and synthesized (in 1951) *norethindrone*, the active ingredient in the first practical oral contraceptive, that has had an enormous impact on society [55, 56]. Ever since, several microbial bioconversions of steroids and sterols have been reported ever since, focusing mainly on steroid

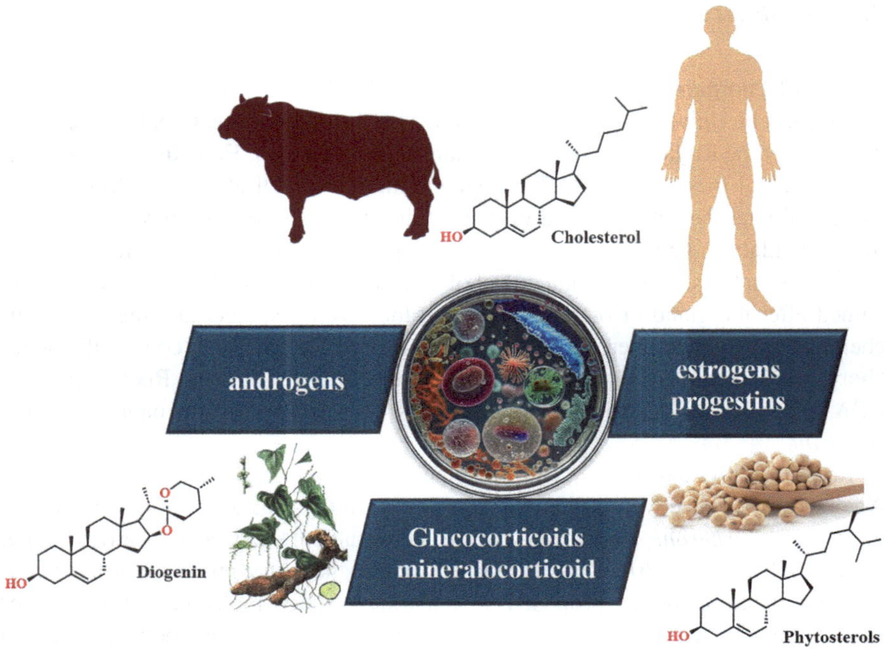

Fig. 15 Steroid hormones production from natural sterols using biotechnological approach. The image contains fragments taken from the free collection of the Freepik websites (https://br.freepik.com/) and diArk (https://www.diark.org/diark)

hydroxylations, 1-dehydrogenation, and sterol side-chain cleavage (Fig. 15). Nowadays steroid manufacturing occupies a prominent place in the pharmaceutical industry with an annual global market over $10 billion. The synthesis of steroidal active pharmaceutical ingredients (APIs) such as sex hormones (estrogens, androgens, and progestogens) and corticosteroids is currently performed by a combination of microbiological and chemical processes [57, 58].

4.3 Sex Steroid Hormones

There are three major classes of sex steroid hormones: androgens, estrogens, and progestogens. Androgens are known as primary "male sex hormones" because of their masculinizing effects and estrogens, together with progestins, as primary "female sex hormones." However, all these hormones are synthesized in both females and males in different concentrations. They are called sex steroids because they were first found to be synthesized in the male and female gonads, respectively. Now we know they can also be synthesized in other peripheral as well as nervous tissues. Although sex steroid hormones are best known for their role as regulators of sexual differentiation, secondary sexual characteristics, sexual behavior and reproduction, they also exert hormonal actions on various other systems, such as the skeletal, immune, muscular, and cardiovascular. More recently, their pleiotropic effects on the CNS have been demonstrated, promoting neurogenesis and neuroprotection, as well as learning and memory [59, 60].

Historically, it was believed that the actions of androgens and estrogens were slow nuclear processes, mediated by hormone receptors located in the cytoplasm, complexed to chaperons, or in the nucleus (i.e. intracellular ARs and ERs). The interaction of sex hormones with their respective ARs and ERs receptors results in the transcription of specific genes, known as genomic actions, which can take several hours to take place. For such transcriptional responses, the ligand–receptor complex must be localized in the nucleus. More recent studies have suggested that the biological actions mediated by the interaction of sex hormones with ARs and ERs could also be mediated by ligand–intracellular receptor complexes associated with the plasma membrane, and can stimulate fast, non-classical processes in the cytoplasm occurring within seconds or minutes. These actions include several extranuclear downstream cascades regulating different cellular responses, such as DNA synthesis, cell proliferation, migration, or survival. Therefore, it is possible to infer that the modulation of ARs and ERs receptors by sex hormones results in both genomic and non-genomic responses. The shape and conformational flexibility of the ligands and ligand-binding domains (LBD) determines whether the accommodated ligand will be agonist/antagonist of the genomic or non-genomic action [61].

The two naturally occurring androgens are *testosterone* and the more potent *5α-dihydrotestosterone* (*DHT*). *Testosterone* is mainly produced by the Leydig cells of the testis and can also be produced in most peripheral tissues from the adrenal-produced inactive steroid precursors *dehydroepiandrosterone*, its sulfate, and *androstenedione*. It is converted into *DHT* by the 5α-reductase enzyme expressed in the urogenital tract. From biotechnological point of view, *testosterone* could be produced by a single-step microbial transformation of *Mycobacterium sp* using cholesterol as a starting material [62] or by several fermentative productions of 17-ketosteroids by different microorganisms [57, 63–65].

Estrogens, the so-called female sex hormones, regulate the development and function of the female reproductive organs, and play an important role in the male reproductive system. Estrogen biosynthesis takes place in the ovaries, particularly in the theca cells, beginning with the biosynthesis of androgen and its subsequent conversion into estrogens in the granulosa cells, by the action of the aromatase enzyme. In the male gonad, i.e. the testicles, estrogens are synthesized in Leydig cells, Sertoli cells, and mature spermatozoa. Several biological functions are attributed to estrogens, such as their key role in the immune system, in the development of cancer and in other critical biological processes related to human well-being. Although the gonads are the main sites of estrogen biosynthesis, they are also synthesized at extra-gonadal sites [66]. Circulating estrogen levels are modulated by their ability to be metabolized and clearance. Studies show that the intestinal microbiota influences plasma estrogen levels due to its ability to secrete β-glucuronidase and, therefore, promote deglucuronidation, increasing the concentration of estrogens in their active forms. When this process is impaired by intestinal microbiota dysbiosis, characterized by reduced microbial diversity, decreased deconjugation results in a reduction in circulating estrogens [67].

Progesterone is another steroid hormone linked to female fertility and pregnancy. It is biosynthesized from cholesterol in the corpus luteum of the ovaries and in the placenta during pregnancy. To a lesser extent, it is also biosynthesized in female non-gonadal sites such as the adrenal cortex and adipose tissue. In men, it is biosynthesized in the Leydig cells of the testes [68]. Formation of *progesterone* from cholesterol in fermentation cultures of *Mycobacterium aurum* was described in 1990 by Hörhold and Böhme [69]. More recently, Karpov and colleagues described the *progesterone* production from natural sterols using recombinant strain of *Mycolicibacterium smegmatis* [60–63, 70–80].

4.4 Mineralocorticoids

Mineralocorticoid is a term used to describe those actions of adrenal corticoste-roids that the classical role is to control the Na^+/K^+ balance in the *aldosterone*-sensitive distal nephron producing sodium and fluid retention and potassium excretion. The most potent mineralocorticoid known in male is *aldosterone* that acts on a specific mineralocorticoid receptor, a transcription factor of the nuclear receptor family, present in the kidney and in various other non-epithelial cells including the heart, vasculature, endothelial cells, smooth muscle cells, and inflammatory cells (T cells, macrophages, dendritic cells). *Aldosterone* is synthesized from cholesterol primarily in the zona glomerulosa of the adrenal cortex, which, unlike other regions of this gland, expresses aldosterone synthase (CYP11B2), but can also be synthetized in cardiac and vascular tissues. Renin-angiotensin, which responds to changes in extracellular volume and sodium balance, is the major system that regulates *aldosterone* biosynthesis. *Aldosterone* overproduction leads to primary aldosteronism, the most common defined cause of secondary hypertension [81, 82].

4.5 Glucocorticoids

Glucocorticoids are steroid hormones derived from the metabolic conversion of cholesterol in the adrenal cortex. They are required for many different physiological processes, including the development of organs, carbohydrate homeostasis, protein and fat metabolism, stress adaptation, and immune surveillance. These diverse actions have made glucocorticoids attractive for therapeutic development, and, as a class of compounds, they are among the most prescribed drugs in the world. They are the major hormones of the stress response and regulators of circadian rhythm and are in frequent pharmacological use to combat inflammation with several metabolic side effects. The essential glucocorticoid in the body is *cortisol*. When used as biotherapeutic, it is known as *hydrocortisone*. This steroid hormone could be prepared from *progesterone* using engineered *Escherichia coli* strains with assembled P450s in one pot artificial cascade reactions [83], or even by recombinant *Saccharomyces cerevisiae* strains [84]. *Hydrocortisone* has been manufactured over the past 50 years using fungi such as *Rhizopus arthopus* or *Aspergillus niger* [85–88].

Compliance with Ethical Standards The authors declare that they have no conflict of interest. This chapter does not contain any studies with human participants or animals performed by any of the authors.

References

1. Stárka L, Dušková M (2020) What is a hormone? Physiol Res:S183–S185
2. Ghoshal K (2023) Recent advances in biosensing technologies for detecting hormones. In: Advanced sensor technology. Elsevier, pp 261–295
3. National Center for Biotechnology Information (2024) PubChem annotation record for norepinephrine, source: hazardous substances Data Bank (HSDB). https://pubchem.ncbi.nlm.nih.go
4. Sneader W (2001) The discovery and synthesis of epinephrine. Drug News Perspect 14:491
5. Yang L, Palsson BO (2023) Biomanufacturing of melatonin is now possible. Melatonin Res 6: 72–78
6. Kaplan MM, Breitbart R (1984) Conversion of thyroxine to triiodothyronine in the anterior pituitary gland and the influence of this process on thyroid status. Horm Metab Res Suppl 14: 79–85
7. Cheng S-Y, Leonard JL, Davis PJ (2010) Molecular aspects of thyroid hormone actions. Endocr Rev 31:139–170
8. Klein I, Ojamaa K (2001) Thyroid hormone and the cardiovascular system. New Eng J Med 344:501–509
9. Long RK, Cakmak H (2021) Chapter 37: hypothalamic & pituitary hormones. In: Katzung BG (ed) Basic & clinical pharmacology15th edn. McGraw-Hill
10. Brazeau P, Vale W, Burgus R et al (1973) Hypothalamic polypeptide that inhibits the secretion of immunoreactive pituitary growth hormone. Science 179:77–79
11. Günther T, Tulipano G, Dournaud P et al (2018) International union of basic and clinical pharmacology. CV. Somatostatin receptors: structure, function, ligands, and new nomenclature. Pharmacol Rev 70:763–835
12. Lamberts SWJ, van der Lely AJ, de Herder WW et al (1996) Octreotide. New Eng J Med 334: 246–254
13. Octreotide. In: https://go.drugbank.com/drugs/DB00104
14. Lanreotide. In: https://go.drugbank.com/drugs/DB06791
15. Wilson CO, Beale JM, Block JH, Gisvold O (eds) (2004) Wilson & Gisvold's textbook of organic medicinal and pharmaceutical chemistry. Lippincott Williams & Wilkins, Philadelphia
16. Cherrington A, Vranic M (1971) Role of glucagon and insulin in control of glucose turnover. Metabolism 20:625–628
17. Habegger KM, Heppner KM, Geary N et al (2010) The metabolic actions of glucagon revisited. Nat Rev Endocrinol 6:689–697
18. Meier JJ, Kjems LL, Veldhuis JD et al (2006) Postprandial suppression of glucagon secretion depends on intact pulsatile insulin secretion. Diabetes 55:1051–1056
19. La Sala L, Pontiroli AE (2021) New fast acting glucagon for recovery from hypoglycemia, a life-threatening situation: Nasal powder and injected stable solutions. Int J Mol Sci 22:10643
20. Jiang G, Zhang BB (2003) Glucagon and regulation of glucose metabolism. Am J Physiol-Endocrin Metabol 284:E671–E678
21. Glucagon. In: https://go.drugbank.com/drugs/DB00040. https://go.drugbank.com/drugs/DB00040. Accessed 2 Aug 2024

22. Cherrington A, Vranic M (1971) Role of glucagon and insulin in control of glucose turnover. Metabolism 20:625–628
23. Inagaki N, Seino Y, Takeda J et al (1989) Gastric inhibitory polypeptide: structure and chromosomal localization of the human gene. Mol Endocrinol 3:1014–1021
24. Takeda J, Seino Y, Tanaka K et al (1987) Sequence of an intestinal cDNA encoding human gastric inhibitory polypeptide precursor. Proc Natl Acad Sci U S A 84:7005–7008
25. Nauck MA, Meier JJ (2018) Incretin hormones: their role in health and disease. Diabetes Obes Metab 20:5–21
26. Yabe D, Seino Y (2011) Two incretin hormones GLP-1 and GIP: comparison of their actions in insulin secretion and β cell preservation. Prog Biophys Mol Biol 107:248–256
27. Smit FX, van der Velden WJC, Kizilkaya HS et al (2021) Investigating GIPR (ant)agonism: a structural analysis of GIP and its receptor. Structure 29:679–693.e6
28. Brown JC, Pederson RA (1970) A multiparameter study on the action of preparations containing cholecystokinin-pancreozymin. Scand J Gastroenterol 5:537–541
29. Dupre J, Ross SA, Watson D et al (1973) Stimulation of insulin secretion by gastric inhibitory polypeptide in man. J Clin Endocrinol Metab 37:826–828
30. Yabe D, Seino Y (2011) Two incretin hormones GLP-1 and GIP: comparison of their actions in insulin secretion and β cell preservation. Prog Biophys Mol Biol 107:248–256
31. Gremlich S, Porret A, Hani EH et al (1995) Cloning, functional expression, and chromosomal localization of the human pancreatic islet glucose-dependent insulinotropic polypeptide receptor. Diabetes 44:1202–1208
32. Meier JJ, Kjems LL, Veldhuis JD et al (2006) Postprandial suppression of glucagon secretion depends on intact pulsatile insulin secretion. Diabetes 55:1051–1056
33. Yabe D, Seino Y (2011) Two incretin hormones GLP-1 and GIP: comparison of their actions in insulin secretion and β cell preservation. Prog Biophys Mol Biol 107:248–256
34. Kreymann B, Ghatei MA, Williams G et al (1987) Glucagon-like peptide-1 7-36: a physiological incretin in man. Lancet 330:1300–1304
35. Ørskov C, Rabenhøj L, Wettergren A et al (1994) Tissue and plasma concentrations of amidated and glycine-extended glucagon-like peptide I in humans. Diabetes 43:535–539
36. Nauck MA (1998) Glucagon-like peptide 1 (GLP-1): a potent gut hormone with a possible therapeutic perspective. Acta Diabetol 35:117–129
37. Smith NK, Hackett TA, Galli A et al (2019) GLP-1: molecular mechanisms and outcomes of a complex signaling system. Neurochem Int 128:94–105
38. Mentlein R, Gallwitz B, Schmidt WE (1993) Dipeptidyl-peptidase IV hydrolyses gastric inhibitory polypeptide, glucagon-like peptide-1(7–36)amide, peptide histidine methionine and is responsible for their degradation in human serum. Eur J Biochem 214:829–835
39. Marguet D, Baggio L, Kobayashi T et al (2000) Enhanced insulin secretion and improved glucose tolerance in mice lacking CD26. Proc Natl Acad Sci U S A 97:6874–6879
40. Nielsen LL, Young AA, Parkes DG (2004) Pharmacology of exenatide (synthetic exendin-4): a potential therapeutic for improved glycemic control of type 2 diabetes. Regul Pept 117:77–88
41. Xie Y, Zhou Q, He Q et al (2023) Opportunities and challenges of incretin-based hypoglycemic agents treating type 2 diabetes mellitus from the perspective of physiological disposition. Acta Pharm Sin B 13:2383–2402
42. U.S. Food & Drug Administration (2024) FDA approves first treatment to reduce risk of serious heart problems specifically in adults with obesity or overweight. https://www.fda.gov/news-events/press-announcements/fda-approves-first-treatment-reduce-risk-serious-heart-problems-specifically-adults-obesity-or. Accessed 2 Aug 2024
43. Dhillon S (2018) Semaglutide: first global approval. Drugs 78:275–284
44. Lau J, Bloch P, Schäffer L et al (2015) Discovery of the once-weekly glucagon-like peptide-1 (GLP-1) analogue Semaglutide. J Med Chem 58:7370–7380
45. Christou GA, Katsiki N, Blundell J et al (2019) Semaglutide as a promising antiobesity drug. Obes Rev 20:805–815

46. Frías JP (2020) Tirzepatide: a glucose-dependent insulinotropic polypeptide (GIP) and glucagon-like peptide-1 (GLP-1) dual agonist in development for the treatment of type 2 diabetes. Expert Rev Endocrinol Metab 15:379–394
47. Ludvik B, Giorgino F, Jódar E et al (2021) Once-weekly tirzepatide versus once-daily insulin degludec as add-on to metformin with or without SGLT2 inhibitors in patients with type 2 diabetes (SURPASS-3): a randomised, open-label, parallel-group, phase 3 trial. Lancet 398: 583–598
48. Coskun T, Sloop KW, Loghin C et al (2018) LY3298176, a novel dual GIP and GLP-1 receptor agonist for the treatment of type 2 diabetes mellitus: from discovery to clinical proof of concept. Mol Metab 18:3–14
49. Willard FS, Douros JD, Gabe MBN et al (2020) Tirzepatide is an imbalanced and biased dual GIP and GLP-1 receptor agonist. JCI Insight 5:e140532
50. Wang L (2022) Designing a dual GLP-1R/GIPR agonist from Tirzepatide: comparing residues between Tirzepatide, GLP-1, and GIP. Drug Des Devel Ther 16:1547–1559
51. Jastreboff AM, Aronne LJ, Ahmad NN et al (2022) Tirzepatide once weekly for the treatment of obesity. New Eng J Med 387:205–216
52. Rodriguez PJ, Goodwin Cartwright BM, Gratzl S et al (2024) Semaglutide vs Tirzepatide for weight loss in adults with overweight or obesity. JAMA Intern Med. https://doi.org/10.1001/jamainternmed.2024.2525
53. Liu X, Zhang N, Gu X et al (2020) Total synthesis of Semaglutide based on a soluble hydrophobic-support-assisted liquid-phase synthetic method. ACS Comb Sci 22:821–825
54. Norman AW, Litwack G (1997) Steroid hormones: chemistry, biosynthesis, and metabolism. In: Hormones. Elsevier, pp 49–86
55. Tata JR (2005) One hundred years of hormones. EMBO Rep 6:490–496
56. Starling EH (1905) The Croonian lectures on the chemical correlation of the functions of the body. Lancet 166:339–341
57. Fernández-Cabezón L, Galán B, García JL (2018) New insights on steroid biotechnology. Front Microbiol 9:958
58. Donova M (2021) Microbial steroid production technologies: current trends and prospects. Microorganisms 10:53
59. Sun T, Liu Z, Liu M et al (2019) Hippocampus-specific Rictor knockdown inhibited 17-β-estradiol induced neuronal plasticity and spatial memory improvement in ovariectomized mice. Behav Brain Res 364:50–61
60. Diotel N, Charlier TD, Lefebvre d'Hellencourt C et al (2018) Steroid transport, local synthesis, and signaling within the brain: roles in neurogenesis, neuroprotection, and sexual behaviors. Front Neurosci 12:84
61. Pillerová M, Borbélyová V, Hodosy J et al (2021) On the role of sex steroids in biological functions by classical and non-classical pathways. An update. Front Neuroendocrinol 62: 100926
62. Liu WH, Lo CK (1997) Production of testosterone from cholesterol using a single-step microbial transformation of Mycobacterium sp. J Ind Microbiol Biotechnol 19:269–272
63. Sambyal K, Singh RV (2020) Production aspects of testosterone by microbial biotransformation and future prospects. Steroids 159:108651
64. Wei Y, Mei G, Zhao J et al (2023) Testosterone biosynthesis from 4-androstene-3,17-dione catalyzed via bifunctional ketoreductase. Fermentation 9:998
65. Donova MV, Egorova OV, Nikolayeva VM (2005) Steroid 17β-reduction by microorganisms – a review. Process Biochem 40:2253–2262
66. Barakat R, Oakley O, Kim H et al (2016) Extra-gonadal sites of estrogen biosynthesis and function. BMB Rep 49:488–496
67. Baker JM, Al-Nakkash L, Herbst-Kralovetz MM (2017) Estrogen–gut microbiome axis: physiological and clinical implications. Maturitas 103:45–53
68. Kolatorova L, Vitku J, Suchopar J et al (2022) Progesterone: a steroid with wide range of effects in physiology as well as human medicine. Int J Mol Sci 23:7989

69. Hörhold C, Böhme KH (1990) Formation of progesterone and 1-dehydroprogesterone from cholesterol in fermentation cultures of Mycobacterium aurum. J Steroid Biochem 36:181–183
70. Karpov M, Strizhov N, Novikova L et al (2024) Pregnenolone and progesterone production from natural sterols using recombinant strain of Mycolicibacterium smegmatis mc2 155 expressing mammalian steroidogenesis system. Microb Cell Fact 23:105
71. Leu Y-L, Wang P-H, Shiao M-S et al (2011) A novel testosterone catabolic pathway in bacteria. J Bacteriol 193:4447–4455
72. Heinlein CA, Chang C (2002) Androgen receptor (AR) coregulators: an overview. Endocr Rev 23:175–200
73. Sun T, Liu Z, Liu M et al (2019) Hippocampus-specific Rictor knockdown inhibited 17-β-estradiol induced neuronal plasticity and spatial memory improvement in ovariectomized mice. Behav Brain Res 364:50–61
74. Tan ME, Li J, Xu HE et al (2015) Androgen receptor: structure, role in prostate cancer and drug discovery. Acta Pharmacol Sin 36:3–23
75. Hogg JA (1992) Steroids, the steroid community, and Upjohn in perspective: a profile of innovation. Steroids 57:593–616
76. Mahato SB, Banerjee S (1985) Steroid transformations by microorganisms – II. Phytochemistry 24:1403–1421
77. Mahato SB, Banerjee S, Podder S (1989) Steroid transformations by microorganisms-III. Phytochemistry 28:7–40
78. Ahmad S, Garg SK, Johri BN (1992) Biotransformation of sterols: selective cleavage of the side chain. Biotechnol Adv 10:1–67
79. Mahato SB, Garai S (1997) Advances in microbial steroid biotransformation. Steroids 62:332–345
80. Sedlaczek L, Smith LL (1988) Biotransformations of steroids. Crit Rev Biotechnol 7:187–236
81. Bauersachs J, López-Andrés N (2022) Mineralocorticoid receptor in cardiovascular diseases – clinical trials and mechanistic insights. Br J Pharmacol 179:3119–3134
82. Barrera-Chimal J, Bonnard B, Jaisser F (2022) Roles of mineralocorticoid receptors in cardiovascular and cardiorenal diseases. Annu Rev Physiol 84:585–610
83. Pan H, Chang S, Qu Y et al (2023) Hydrocortisone production using whole-cell biocatalysts in recombinant Escherichia coli. Biochem Eng J 198:109023
84. Szczebara FM, Chandelier C, Villeret C et al (2003) Total biosynthesis of hydrocortisone from a simple carbon source in yeast. Nat Biotechnol 21:143–149
85. Timmermans S, Souffriau J, Libert C (2019) A general introduction to glucocorticoid biology. Front Immunol 10:1545
86. Vettorazzi S, Nalbantoglu D, Gebhardt JCM et al (2022) A guide to changing paradigms of glucocorticoid receptor function – a model system for genome regulation and physiology. FEBS J 289:5718–5743
87. Hardy RS, Raza K, Cooper MS (2020) Therapeutic glucocorticoids: mechanisms of actions in rheumatic diseases. Nat Rev Rheumatol 16:133–144
88. Wiseman A (2004) Double humanized yeast makes hydrocortisone. Trends Biotechnol 22:324–325

Top Med Chem
https://doi.org/10.1007/7355_2024_177

Published online: 1 January 2025

Small Peptides as Biotherapeutics

Diego Vallés, Carolina Villadóniga, and Miriam Barros

Contents

Abstract Peptide-based biotherapies have emerged as a significant class of therapeutics due to their ability to modulate various biological processes through interactions with specific targets, such as receptors, enzymes, or proteins. Peptides are small molecules that can be naturally synthesized in different organisms, derived from in vitro enzymatic hydrolysis of food protein, or synthetically designed to mimic endogenous molecules. These last peptides can be modified to improve their biological functions, which makes them versatile tools as biotherapeutics. In this section, we focus on the use of peptides and peptidomimetic derivatives as biotherapeutics in diseases and pathologies associated with proteolytic enzymes, and the obtaining of bioactive peptides derived from the hydrolysis of food proteins,

D. Vallés (✉) and C. Villadóniga
Laboratorio de Biocatalizadores y sus Aplicaciones, Facultad de Ciencias, Instituto de Química Biológica, Universidad de la República, Montevideo, Uruguay

M. Barros
Departamento de Biociencias, Facultad de Química, Universidad de la República, Montevideo, Uruguay

and on their potential use as antihypertensive, antioxidant, and antibacterial agents. Proteolytic enzymes play an important role in numerous physiological functions; however, their overexpression or dysregulated activity can trigger a wide range of diseases and infections, including cardiovascular disorders; immunological-, inflammatory-, and respiratory-conditions; cancer, as well as neurodegenerative disorders like Alzheimer's disease, among others. Multiple protease inhibitors demonstrated high specific and selective activity in human in vitro and in murine in vivo models, but only a few have been approved as drugs by the U.S. Food & Drug Administration (FDA). In the first part of the section, the major protease inhibitor peptide drugs in the treatment of different pathologies is described. The second part is focused on the bioactive peptides. Bioactive peptides are short sequences of amino acids derived from proteins, which exhibit specific biological activities upon ingestion, absorption, or interaction with target molecules in the body. These peptides can modulate physiological functions such as antimicrobial activity, immunomodulation, antioxidant effects, and regulation of metabolic pathways. Despite bioactive peptides' promising potential as biotherapeutics, several challenges remain, including stability, immunogenicity, regulatory approval, and manufacturing scalability. Addressing these challenges requires a multidisciplinary approach to advance the development and commercialization of bioactive peptide-based therapies.

Keywords Antihypertensive peptides, Antimicrobial peptides, Antioxidant peptides, Bioactive peptides, Biotherapeutic peptide, Peptide-based drug, Protease inhibitor peptide

1 Introduction

Several small peptides serve as therapeutic peptides (drugs), including native, analog (peptidomimetic), and heterologous variants. Native peptides maintain the same sequence as those found in natural peptides isolated from mammalian tissues and other sources. However, most therapeutic peptides on the market arise from chemical synthesis or recombinant expression. Analogs are modified versions of natural peptides with improved properties. In contrast, heterologous peptides are discovered by other strategies such as phage display or synthetic library screening [1].

Furthermore, food proteins serve as reservoirs of peptides that regulate vital physiological processes in organisms, defined as bioactive peptides [2]. This broadens the criteria for assessing protein value, which has traditionally focused on factors like nutritional quality determined by nitrogen availability (including essential and non-essential amino acids), allergenic properties, and the absence of anti-nutrients [3].

The following sections will explore protease inhibitor peptides as biotherapeutics and bioactive peptides derived from food protein hydrolysis in more detail.

2 Protease Inhibitor Peptides as Biotherapeutics

Proteases, also known as peptidases, are crucial enzymes that facilitate the cleavage of peptide bonds within proteins or peptides. They play indispensable roles in cellular processes, from cell division to programmed cell death [4]. Peptidases are not limited to a specific organism or process. They are found in all living organisms and profoundly influence nearly every biological process. Whether it is apoptosis, blood coagulation, or cell signaling cascades, the precise action of proteases in cleaving proteins or pro-enzymes is crucial [5].

Peptidases are a diverse group of enzymes, each with its unique characteristics. They can be categorized into two main groups: (i) exopeptidases, which hydrolyze peptide bonds from either the *N*- or *C*-terminal ends of substrates; and (ii) endopeptidases, which cleave peptide bonds within the interior regions of polypeptide chains. Thus, they can be defined as serine-, cysteine-, threonine-, or aspartic-proteases if the nucleophiles are serine, cysteine, threonine, or aspartic acid, respectively. If the mechanism involves metal ions in catalysis, they are called metallopeptidases [6]. This diversity is a testament to the complexity of these enzymes. Another classification system, proposed by Rawlings and Barret and utilized by the MEROPS database, organizes proteases systematically based on structural and sequence characteristics. This hierarchical classification groups homologous protease sequences into species, species into families, and families into clans, where families contain related sequences and clans contain related structures [7]. Schechter and Berger [8] established a comprehensive naming convention that aids in describing the specificity of peptidases. This convention organizes binding pockets along the active site cleft linearly to accommodate the substrate peptide's specific amino acid side chains (Fig. 1). The letter P indicates the nomenclature for the positions of the substrate residues and that of the subsites where the substrates bind to the surface of a protease by the letter S. Each binding

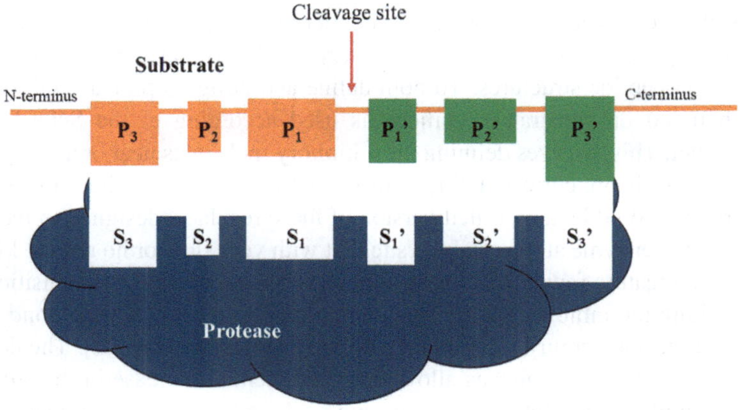

Fig. 1 Representative scheme of the Schechter and Berger nomenclature system for the active site subsites of a peptidase

pocket is numbered according to its position relative to the scissile C-N bond: positions toward the N-terminus of the bond to be cleaved are defined as S1, S2, S3, and so on, whereas positions of the bond to be cleaved toward the C-terminus are defined as S1', S2', S3', etc. For the substrate, the same convention is followed, with P1, P2, and P3 being the positions towards the N-terminal end and P1', P2', P3', towards the C-terminal end, respectively. This naming convention is a powerful tool for understanding the interactions of enzymes and their substrates, and it remains the key to investigating mechanisms of action and specificity.

The activity of proteases is tightly regulated through various mechanisms, including the modulation of peptidases' expression or secretion, activation of pro-peptidases, or specific degradation of mature enzymes. Under normal physiological conditions, protease activity is regulated by protease inhibitors (PIs) [9]. These inhibitors can act through reversible or irreversible mechanisms, forming stable complexes with target proteases and impeding substrate access to the enzyme's active site through various interactions such as disulfide bridges, hydrogen bonds, or hydrophobic interactions [10, 11]. Proteases also play fundamental roles in the infectious processes of pathogens that affect plants and animals, which can lead to important health and economic problems worldwide. Many diseases are also influenced by the deregulation of these enzymes, such as cancer, epilepsy, asthma, malaria, Alzheimer's disease, HIV, and COVID-19, among others [12–20]. Given proteases' critical roles in health and disease, controlling their activity is increasingly important. There is increasing interest in developing inhibitors that can serve as therapeutic agents, effectively targeting proteases.

Protein-based protease inhibitors, also known as peptide protease inhibitors (PPIs), are polypeptides widely distributed in diverse tissues of animals, plants, and microorganisms [21]. Typically comprising small polypeptides ranging from 15 to 60 amino acids, PPIs are often hydrophobic and cationic, with abundant cysteine residues forming disulfide bridges that confer resistance to environmental stresses such as heat, extreme pH, high ionic strength, and proteolysis. It is known that inhibitors simulate the structure of the substrate, interacting with the amino acid residues of the catalytic site in the active site. In proteases, this implies an amino acid sequence (peptide) or chemical structures very similar to these amino acids, which could be non-peptide structures. To both define and design a protease inhibitor, an approach based on substrate specificity is used according to the Schechter and Berger model. This involves defining the inhibitory molecules near or at the position of the cleavable bond, corresponding to residues P1 and/or P′1 of the substrate. This inhibitory unit could be a modified version of these residues, designed to recognize and bind to the enzyme at the catalytic site, but with very slow or no release kinetics. One common feature found in protease inhibitors is the inclusion of a transition state isostere within the inhibitory unit. This component mimics the amide bond hydrolysis transition state, contributing to the inhibitor's effectiveness [22]. The development of bioinformatics tools has allowed an exponential increase in the study and development of new inhibitors using this approach and has been fundamental in the development of new drugs [23, 24]. Although more than 7,000 natural peptides have been identified, the advancement of peptide-based therapies remains a challenge given the high costs and inefficient processes associated with their manufacturing processes [25].

Natural protease inhibitors intended as therapeutic medications are difficult to apply because of their short bioavailability, marked by their short life caused by various organismal actions, such as proteolysis [26]. The short in vivo half-life of peptides poses a significant challenge to the scientific community in developing peptide drug candidates. Many research teams dedicate great efforts to the improvement and design of new inhibitors. Factors determining the in vivo persistence of peptides primarily include the amino acid sequence, terminal modifications, dosage, and administration route [27]. The processes that occur in nature have been the basis on which they have been inspired to establish the mechanistic principles used in the design of small-molecule protease inhibitors [28]. Currently, technological and bioinformatics advances in peptide synthesis have boosted the area of peptide therapy, allowing the exploration of new alternatives to the limitations of conventional therapies due to susceptibility to proteolytic degradation and low oral bioavailability, which many of the peptides used at present [29, 30].

Several studies with protease inhibitors show high therapeutic potential in preclinical trials using animals, and several works have shown initial clinical trials in humans. Most of them are aimed at treating a wide variety of conditions, including viral and parasitic infections, cancer, inflammatory, immunological, and respiratory disorders, as well as cardiovascular and degenerative diseases such as Alzheimer's. However, despite the extensive research, protease inhibitors that have successfully advanced to the clinical trial stage are relatively low. Despite this, those that have reached the market are products that ensure safe and effective treatments. Among them, the most notable are the angiotensin-converting enzyme inhibitors (ACE inhibitors) for high blood pressure, the HIV-1 protease inhibitors for HIV/AIDS, and the thrombin inhibitors for stroke [31]. Next sections describe some examples of the better-studied peptidic or peptide-derived inhibitors of proteases as drugs associated with different pathologies.

2.1 Blood Coagulation

Coagulation is an intricate procedure through which blood coalesces into clots. The primary pathway for triggering blood coagulation involves a sequence of reactions where an inert serine protease precursor, termed a zymogen, along with its cofactor, transitions into active elements. These activated components then catalyze subsequent reactions in the cascade, ultimately leading to the formation of cross-linked fibrin.

Aprotinin, a biotherapeutic polypeptide, serves as a naturally derived competitive inhibitor for a wide range of serine proteases and is obtained from bovine lung tissue [32]. It is a small globular protein (6,512 Da), belonging to the family of Kunitz-type inhibitors, competitive inhibitors of most digestive serine proteinases (trypsin, chymotrypsin, kallikrein). It has a compact structure and is stabilized by 3 disulfide

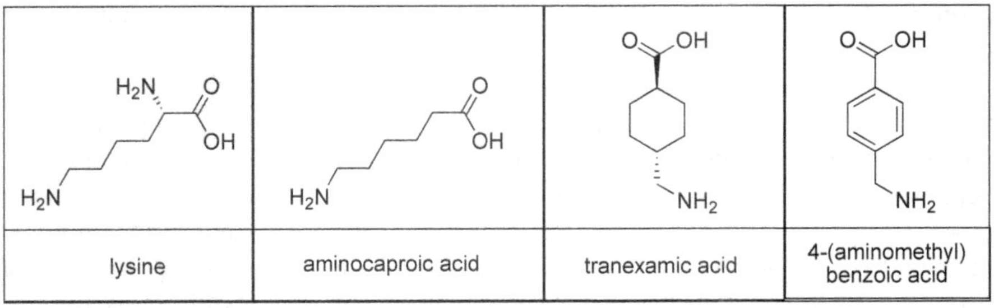

Fig. 2 Structure of lysine and its synthetic analogs to inhibit serine protease

bridges. Located at position 15 of the primary sequence, the side chain of a lysine residue is responsible for the inhibitory activity. There are synthetic analogs of aprotinin, specifically of the lysine residue, such as aminocaproic acid, tranexamic acid. and 4-(aminomethyl) benzoic acid (Fig. 2) [33]. These drugs are ε-amino acid molecules that bind primarily to lysine sites, which are crucial for interactions between plasminogen and the fibrin surface, facilitating tissue-type plasminogen activator (t-PA)-mediated plasmin activation. These agents are used to reduce bleeding in complex surgeries and transfusions [34].

2.2 *Blood Anticoagulants*

Current anticoagulation therapies typically involve the administration of antiplatelet agents such as aspirin, clopidogrel, dipyridamole, and ticlopidine, alongside anticoagulants like heparin and warfarin. While these drugs possess beneficial anticoagulant properties, they are often beset by challenges stemming from drug–drug and drug–disease interactions.

In 1884, John Haycraft demonstrated the anticoagulant prowess of medicinal leeches (*Hirudo medicinalis*), which secrete hirudin, a potent anticoagulant agent. Structurally, hirudin comprises 65 amino acid residues forming a compact *N*-terminal domain with three disulfide bonds and a disordered *C*-terminal domain [35]. However, the extraction of therapeutically viable amounts of natural hirudin from leeches remains challenging due to the presence of various isoforms. Advances in molecular genetics have allowed the development of recombinant hirudin (rH) synthesis, structurally and functionally identical to native hirudin [35]. Several synthetic derivatives have also been developed, using recombinant protein technology, such as lepirudin, desirudin, and bivalirudin, all structurally similar but with differences in their pharmacological action and are often used in different treatments for thrombotic pathologies. This last one, a synthetic 20-amino acid peptide derivative of hirudin, is a notable example. Unlike hirudin, bivalirudin acts as a reversible inhibitor, slowly processed by thrombin between the Pro-Arg and Pro sequences of the drug. This reversibility feature contributes to a reduced risk of severe bleeding compared to standard combination heparin therapy [22].

2.3 Diabetes Mellitus

Diabetes mellitus is a disease caused by a metabolic disorder that causes an increase in blood sugar levels. The cause of this disease is due to insufficient insulin production by the pancreas, the inability of cells to be activated by interaction with it, or the deregulation of other pathways that activate glucose metabolism. The treatment of this disease is based, in most cases, on the endogenous administration of insulin (originally from porcine origin and currently from recombinant production). In addition to insulin, some therapies are based on administering key peptides in managing diabetes mellitus. These include incretin derivatives (hormones secreted by the intestines in response to food intake that stimulate insulin secretion) and dipeptidyl peptidase-4 inhibitors (DPP-IV), which prevent the degradation of incretins [36]. DPP-IV degrades incretins such as glucose-dependent insulinotropic peptide (GIP) and glucagon-like peptide-1 (GLP-1). A significant advantage of DPP-IV inhibitors is their potential for oral bioavailability, thanks to their smaller molecular size. DPP-IV inhibitors can be classified into two groups: those that mimic the DPP-IV molecule, such as vildagliptin and saxagliptin, and those that do not, such as sitagliptin, alogliptin, and linagliptin. These inhibitors act extracellularly, competing with and inhibiting the DPP-IV enzyme [37, 38].

2.4 Blood Pressure

In normal physiology, the renin-angiotensin-aldosterone system is a complex regulator of blood pressure, renal hemodynamics, and volume homeostasis. "Antihypertensive Peptides" section will discuss the specific topic of enzyme-targeting peptide-derived drugs, namely ACE inhibitors.

2.5 Antiviral Peptides and Peptidomimetics

Viruses are tiny microorganisms, obligate intracellular parasites that carry either an RNA or DNA genome enclosed within a protective protein coat encoded by the virus. RNA and DNA viruses infect host organisms by entering and replicating within host cells [39]. Once inside, the virus takes over the cell's machinery to produce multiple copies of itself, eventually releasing these new viral particles to infect additional cells. This cycle of replication and infection is responsible for the progression of the disease [40]. The action of proteolytic enzymes is essential to complete the viral infection cycle. Viruses translate the genomic sequence as a large polyprotein that must be cleaved by these enzymes to obtain mature proteins. Viral proteases have revealed numerous strategies for proteolysis, which include recognition of various combinations of amino acids, and differential speeds in the processing of the polyprotein but always following an ordered sequential processing, essential for obtaining a new viral progeny. This mechanism of action is conserved in different families of viruses, so knowing the specific mechanism will allow the design and development of new peptide inhibitors that can be used as drugs. Proteases of all catalytic mechanisms have been found, generally associated with various families of viruses. For example, serine proteases have been identified in hepatitis C viruses and herpes and flavivirus. Adenoviruses synthesize cysteine proteases, and in human immunodeficiency virus type 1 (HIV1), the essential proteases are aspartic group [41].

2.5.1 HIV Protease

The human immunodeficiency virus (HIV), an RNA virus, is the infectious agent that causes acquired immunodeficiency syndrome (AIDS). In an advanced stage of the disease, patients infected by this virus are susceptible to developing other pathologies caused by different infectious agents or developing cancer that can lead to death. As previously mentioned, in the virus's life cycle, the action of proteolytic enzymes is essential, specifically an aspartic protease. This enzyme, in its active state, is a homodimer, where each 99 kDa monomer has a catalytic aspartic acid residue at position 25 of each polypeptide chain, and together with the threonine, serine, and glycine residues, they make up the active site [42]. Research on this enzyme has allowed the design of several inhibitors of this protease as biotherapeutic agents. However, the high mutation rate of the virus has caused the action of inhibitors to decrease over time, making it necessary to combine several drugs in antiretroviral therapy. This cocktail contains antiviral agents (including several inhibitors), which act on essential agents at different stages of the HIV life cycle.

In the combination of antiviral agents, pepstatin, a general inhibitor of aspartic proteases, is usually included together with specific inhibitors of the HIV protease. These inhibitors have a predominant action in preventing the action of this enzyme on the polyprotein and, therefore, maturation and generation of new virions [43].

Currently, there are nine synthetic PIs approved for clinical use: saquinavir, ritonavir, indinavir, nelfinavir, amprenavir, lopinavir, atazanavir, tipranavir, and darunavir (Fig. 3). All are peptidomimetics, except for tipranavir. Most are prescribed a concomitant low dose of ritonavir as a boosting agent. Apart from tipranavir, all of them are competitive peptidomimetic inhibitors, mimicking the transition state of the natural substrate of the viral protease.

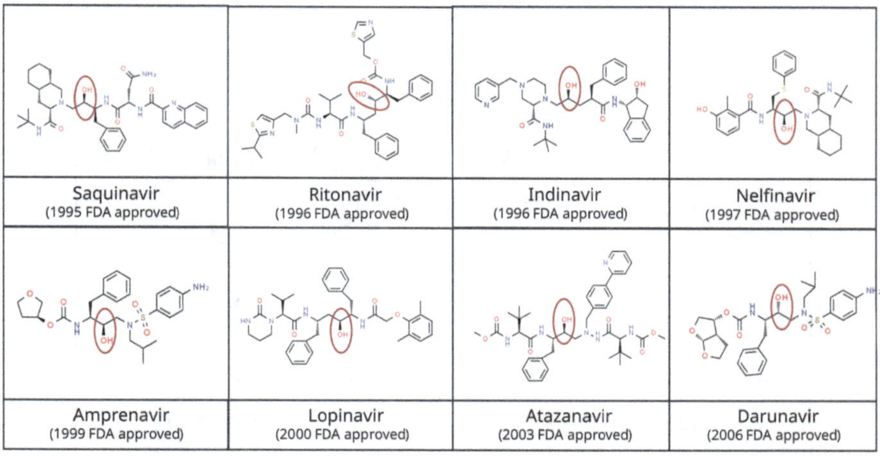

Fig. 3 Chemical structures of the peptidomimetic protease inhibitors, characterized by a hydroxyethylene core (indicated with a red oval) that mimics the transition state, approved for clinical use

2.5.2 COVID Proteases

In December 2019, the first cases of atypical pneumonia were identified in Hubei province in China. Its causal agent was identified as a new coronavirus called SARS-CoV-2, characterized by a high rate of spread and being capable of producing severe respiratory manifestations or death, for which the World Health Organization (WHO) declared the outbreak of this coronavirus as a pandemic [44]. Two-thirds of the SARS-CoV-2 genome comprises two open reading frames (ORFs), translated into two large polyproteins called pp1a and pp1ab. Their cleavage is carried out by two cysteine proteases, one similar to papain ("papain-like protease," PLpro) and the other structurally similar to chymotrypsin ("3-chymotrypsin-like proteases," 3CLpro, or also called "main proteases," Mpro). This processing will give rise to 16 non-structural proteins (NSP), including both proteases, which have a fundamental role in the viral replicative cycle [45]. Due to its intrinsic proteolytic activity and the absence of homologous enzymes in humans, the toxicity of specific inhibitors is expected to be limited [46].

Several existing and some new antivirals have been studied to treat COVID-19. Some of them have been approved, especially in cases of patients with high risks of hospitalization and for different age groups. Most of these antivirals target the virus proteases [47]. Paxlovid is an antiviral drug discovered by Pfizer. It combines nirmatrelvir (fluorinated oral drug containing a CF_3 group that inhibits the viral Mpro protease of SARS-CoV-2 to stop the virus from replicating) with ritonavir (a boosting agent that helps nirmatrelvir work better) (Fig. 4) [48]. According to different studies, the FDA approved the Paxlovid drug in December 2021 for older (age above 65) SARS-CoV-2-infected patients. Others, like Lopinavir/ritonavir

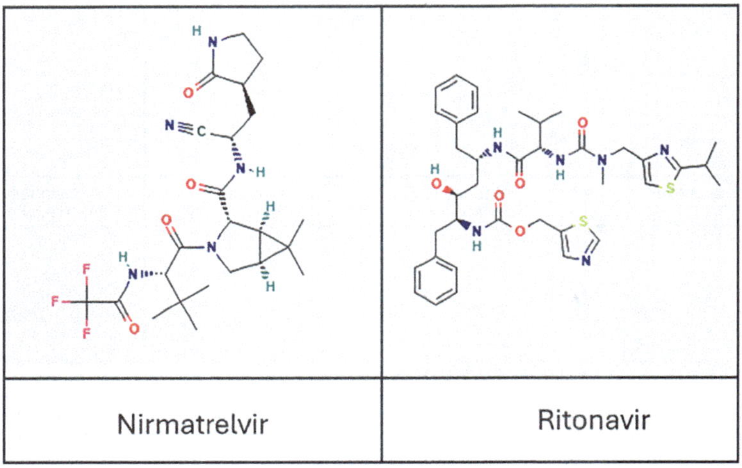

Fig. 4 Nirmatrelvir and ritonavir chemical structure combined in Paxlovid (Pfizer) drugs

(human immunodeficiency virus (HIV) protease inhibitor), showed no efficacy against COVID-19 main proteases.

2.6 Cancer

Much effort has been devoted to elucidating the function of proteases in cancer progression. This has allowed us to define the preponderant role that proteolysis plays in all stages of the disease and, therefore, to join forces to study mechanisms for its regulation through peptide inhibitors that allow the generation of antitumor drugs. Bortezomib, a drug approved by the FDA in 2003, is a dipeptide derived from boronic acid that acts as an inhibitor of the proteasome, a multiprotein complex that plays a vital role in the turnover of cellular proteins essential for cell homeostasis. Is used to treat patients who have developed multiple myeloma (a type of bone marrow cancer) and mantle cell lymphoma (a fast-growing cancer that begins in the cells of the immune system) [49]. Bortezomib binds reversibly to the 26S subunit of the proteasome, similarly to serine proteases, generating the accumulation of pro-apoptotic factors. This activates the apoptotic pathway mediated by caspases, generating the death of neoplastic cells [50]. As with other protease inhibitors, nelfinavir (Fig. 3) was one of the first peptidomimetics synthesized for the treatment of first-generation HIV, which has gradually fallen into disuse for this therapy [51]. However, recent studies have shown an important action as an agent for the treatment of certain types of cancer, repositioning this drug as a strong candidate in certain anticancer therapies. Some of the action pathways of nelfinavir include the induction of apoptosis, autophagy, and oxidative stress, among others, as a response.

These types of effects have been observed for several types of cancer, including pancreatic, breast, and lung cancer, among others [52, 53].

Understanding protease processes in their metabolic context is essential for developing inhibitors that can slow tumor progression and metastasis. However, proteases play key roles at different stages of cancer development, depending on the specific type of cancer. On the other hand, plant-derived products are widely recognized as promising candidates for various applications in biotechnology and medicine. These compounds have been successfully extracted to regulate pathological processes and treat various diseases [54]. Compounds obtained from natural sources are integral to therapeutic approaches in several alternative-based medical systems, including Ayurveda, homeopathy, Unani, and allopathic medicine. Studies have shown that populations with diets rich in seeds, such as beans, corn, and rice, have been found to have lower incidence rates of certain types of cancer, including breast, colon, and prostate cancer. Furthermore, plant-based diets have been found to effectively prevent or mitigate the risks associated with chronic diseases and certain types of cancer [55].

3 Bioactive Peptides

3.1 Definition

Bioactive peptides are short amino acid sequences, typically ranging from 2 to 20 residues in length that elicit a physiologic response within a living organism or cell. They often result from the degradation of endogenous or exogenous proteins, acquiring specific biological functions upon release from their precursor proteins. Also, some bioactive peptides are genetically encoded and produced as peptides, while others are non-ribosomal synthesized [56]. These peptides induce physiological responses by interacting with specific receptors, enzymes, or other targets present in cells [57]. Once they are released from the precursor protein, where they are encrypted, peptide activity is determined by amino acid composition and sequence. These peptides positively impact major body systems, promoting health and decreasing the risk of developing chronic diseases. So, they are recognized as valuable ingredients in nutraceuticals and functional foods [58].

3.2 Bioactive Peptide Sources

Bioactive peptides may be synthesized as natural or released in vivo or in vitro proteolysis of parent proteins.

3.2.1 In Vivo, Proteolytic Activation of Bioactive Peptides

Proteins serve as potential precursors of bioactive peptides, both endogenous compounds within living organisms and exogenous compounds of dietary origin. The release of bioactive peptides through proteolysis has been demonstrated both in vivo and in vitro. Many bioactive peptides have been isolated from bodily fluids and tissues of various organisms (bacteria, fungi, plants, and animals) [59]. Among these are peptides derived from specialized proteins as precursors of hormones [60] or neurotransmitters [61]. In other cases, bioactive peptides are derived from proteins with other functions, such as enzymes [62], structural proteins [63], and transport proteins [64], among others, and it is postulated that these peptides could play a long-term regulatory role in the organism's homeostatic balance. Another endogenous source of bioactive peptides is dietary proteins, which can generate these peptides during gastrointestinal digestion [65]. For example, the presence of opioid and antihypertensive peptides was reported in gastrointestinal digests following the consumption of casein [66]. Furthermore, hydroxyproline-containing dipeptides and tripeptides were detected in human plasma after 1–2 h ingesting collagen hydrolysate [67].

3.2.2 Bioactive Peptide Production by In Vitro Hydrolysis

Enzymatic hydrolysis of food protein has been the most common method to produce bioactive peptides [68]. Bioactive peptides have been obtained through in vitro enzymatic degradation of proteins present in foods such as milk [69], red meat [70], eggs [71], and soybean have been identified [72]. Dietary proteins represent a significant source of bioactive peptides that can be released by in vitro proteolytic degradation. Pancreatic enzymes, mainly trypsin, have been widely used to release bioactive peptides from dietary proteins. Other enzymes and combinations thereof, including chymotrypsin, pancreatin, and pepsin, as well as those from bacterial (alcalase, thermolysin, pronase), fungal, and plant (papain, bromelain) sources have been applied to generate bioactive peptides from various protein sources [73, 74]. Additionally, bioactive peptides produced during food processing (cheeses and cheese whey) or by the action of native proteolytic enzymes (milk), or bacterial enzymes used in food preparation, such as in fermented milk, have been detected [73].

3.3 Classification of Bioactive Peptide Based on Biological Activity

It has been demonstrated that there are bioactive peptides capable of causing beneficial physiological effects in the immune, cardiovascular, nervous, endocrine,

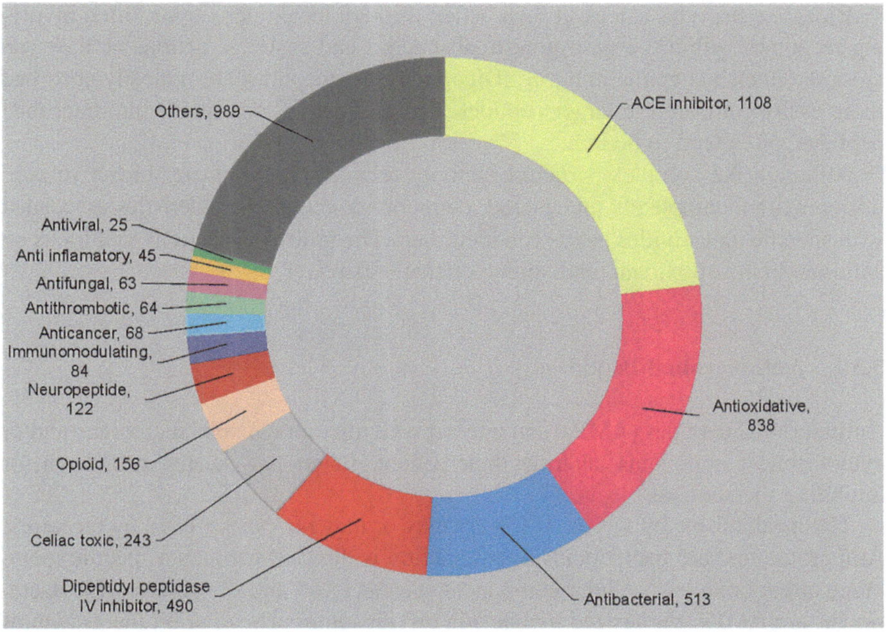

Fig. 5 Bioactive peptides deposited in the BIOPEP database are classified according to the type of bioactivity. The values shown in the graph correspond to the number of peptides for each biological activity

and digestive systems. Some of the activities that have been studied include antihypertensive, antimicrobial, opioid, antitumor, immunomodulatory, calcium transport modulation, antithrombotic, hypocholesterolemic, and antioxidant activities. The BIOPEP database (URL: www.uwm.edu.pl/biochemia), developed by the Chair of Food Biochemistry at the Uniwersytet Warminsko Mazurski [75], is a free repository for numerous reported bioactive peptides. Currently, the database contains 4,808 peptides derived from diverse precursor proteins and exhibiting different activities. Notably, ACE inhibitors, antioxidant peptides, antibacterial peptides, and dipeptidyl peptidase IV inhibitors are the most abundant activities within the BIOPEP database (Fig. 5). This emphasizes the interest in these peptides for their contribution to important health conditions such as cardiovascular diseases, infections, and type-2 diabetes.

There is great diversity within bioactive peptides. Their size generally ranges from two to 20 amino acid residues, although larger peptides have also been identified [76]. These peptides can vary from simple dipeptides or tripeptides to more complex polypeptides. Moreover, their structural diversity encompasses linear or cyclic configurations and glycosylated, phosphorylated, or acylated variants [3]. A single peptide can possess multiple biological functions known as multifunctional peptides. Additionally, overlapping peptide sequences with different functions can also occur, as reported for caseins [77].

Bioactive peptides can exert their action through local interactions, often involving receptors within the gastrointestinal mucosa and systemic actions as they can cross the intestinal epithelium [58]. Dipeptides or tripeptides are typically absorbed more easily compared to larger peptides. Peptides acting locally, like antimicrobial peptides, often tend to be longer.

Although the complete structure–activity relationships of many bioactive peptides remain incompletely understood, common structural characteristics associated with specific bioactivities have been identified. The following sections will focus on antimicrobial, antioxidant, and antihypertensive peptides.

3.3.1 Antimicrobial Peptides

Antimicrobial peptides (AMPs) can interact with microorganisms' membrane and/or cytoplasmic components, altering their cellular functions, causing cell death, or inhibiting metabolism.

The mechanisms by which AMPs act may or may not be mediated by receptors. Among the first are bacteriocins, whose action is directed towards a specific membrane target or an intracellular component such as DNA and ribosomes [78]. Bacteriocins are AMPs synthesized by bacteria on ribosomes. They inhibit the growth of similar bacteria or strains close to those that produce them. Among the most important bacteriocins are nisins, produced by the Gram-positive *Lactococcus lactis*. The FDA permits using nisins in foods. They are widely used as preservatives (code E234), mainly to prevent possible spoilage of cheeses [79].

The cases in which the mechanism is not mediated by receptors are the most abundant and include the most known sequences. These have great diversity in their composition, which is associated with the fact that they have a non-specific and general objective: cell membranes [78].

AMP sizes are quite heterogeneous, generally between 12 and 50 amino acids, although the range can extend, in some cases, to sequences of 10 to 100 amino acid residues. The antimicrobial effect of these peptides is fundamentally based on their insertion into the lipid bilayers of the microbial membrane with the consequent formation of pores and cellular extravasation. In particular, the amino acid composition, amphipathicity, and positive electrical charges at physiological pH are the qualities that facilitate their insertion into the membrane of microorganisms [80]. The net positive charge of AMPs is attracted by the negative charge of the constituent molecules of cell walls and/or membranes, such as lipopolysaccharides present in Gram-negative bacteria or phospholipids and teichoic acid in Gram-positive bacteria. A delicate balance between electrostatic and hydrophobic interactions determines the interaction between peptide and membrane. AMPs are amphipathic molecules and their hydrophobicity allows the cleavage of the lipid bilayer of the target membrane [81]. Due to this type of interaction, in some cases, AMPs additionally exhibit cytolytic activity against tumor cells, which also present a net negative charge on the surface in contrast to normal eukaryotic cells, whose

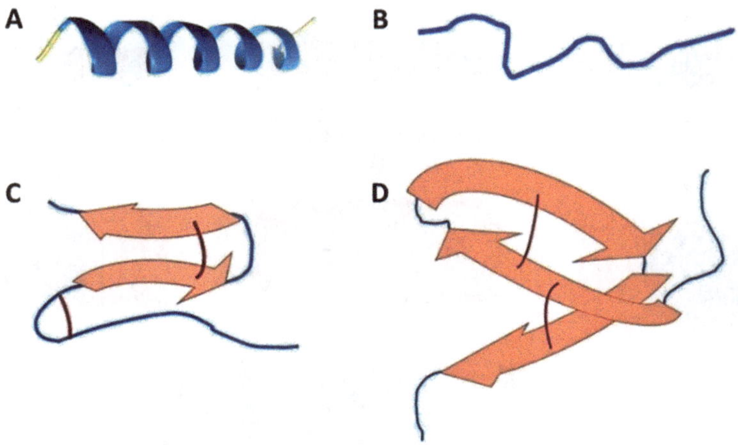

Fig. 6 Structural groups of the AMPs. (**a**) α-Helix structure. (**b**) Extended structure. (**c**) Loop structure stabilized by intramolecular disulfide bonds. (**d**) β-Sheet structure. In (**c**) and (**d**), the disulfide bridges are shown in red

composition consists mainly of zwitterionic phospholipids and, therefore, its net surface charge approaches neutrality [78].

AMPs can also interfere with the biosynthesis of cell wall components, such as peptidoglycans and teichoic acid. They exert this action by binding to their precursor lipids. Some AMPs can inhibit the synthesis of glucans, chitin, and glycoproteins in fungi. Sometimes, AMPs can enter the cell and interact with different organelles or molecules, interrupting replication, transcription, and/or translation processes. The mechanism of action are diverse and include binding to nucleic acids, destroying their conformation, inhibiting enzymes such as topoisomerase I or RNA polymerase, obstructing DNA repair mechanisms, or interacting with ribosomes [82].

Antimicrobial peptides have been classified according to their secondary structure, amino acid composition, and number of disulfide bonds, resulting in four groups as follows (Fig. 6). Certain AMPs may belong to all or none of these groups [80, 83]. The first group comprises peptides that may have a linear or α-helical structure, and do not contain cysteine residues. They are the most widely distributed and the most known. They usually have a broad spectrum of action and few side effects in mammalian cells. They are of great interest at an industrial level. An example of this group is the peptide LL-37, which is the only representative of the cathelicidin family present in the human body. The peptides of the second group do not have a determined secondary structure, usually presenting extended conformations. Like those of first group, they do not present cysteine residues but are usually abundant in proline, arginine, tryptophan, glycine, and histidine. Its antimicrobial activity is mainly due to hydrogen bonds or van der Waals forces with membrane lipids. Most AMPs in this group have intracellular targets. An example of this type of peptide is indolicidin, an AMP especially rich in tryptophan, isolated from bovine neutrophils. The third group of antimicrobial peptides characteristically presents one or more loops stabilized by intramolecular disulfide bonds or by the formation of

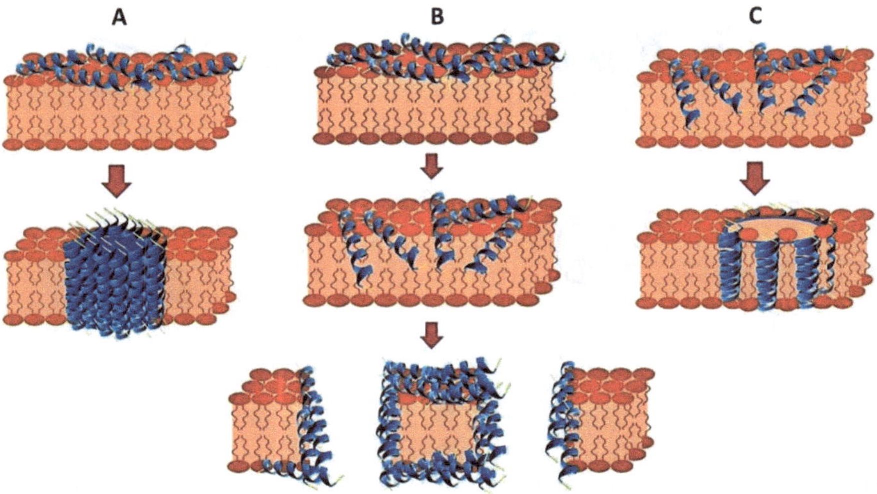

Fig. 7 Main mechanisms postulated for forming transmembrane pores by the action of AMPs. (**a**) Barrel-stave mechanism. (**b**) Carpet mechanism. CWormhole mechanism

cyclic structures in the peptide chain. Examples of cyclic AMPs are polymyxin B and θ defensins. The fourth group of antimicrobial peptides is notable for the presence of antiparallel β-sheets stabilized by 2–4 intramolecular disulfide bonds. Examples of these are α- and β-defensins [82, 83].

The main mode of action of peptides with an α-helical structure is the formation of transmembrane pores, for which several mechanisms have been postulated. Of these, the most accepted by the scientific community are the so-called barrel stave, carpet, and wormhole [82, 83]. The barrel-stave mechanism (Fig. 7A) consists of the formation of channel-shaped transmembrane pores. To achieve this, the hyd rophobic region of the AMP, which is made up of an α helix or a β sheet, interacts with the interior of the membrane while the hydrophilic region is located towards the exterior, thus forming an aqueous pore. The pore size is related to the number of inserted peptides. The first stage of the mechanism consists of positioning a certain number of peptides on the surface of the membrane. These aggregations cause their hydrophilic regions to confront each other, forming the pore lumen and energetically facilitating the complex's insertion into the membrane.

The carpet mechanism (Fig. 7B) is a membrane permeabilization process that acts with a mechanism analogous to the action of detergents. Unlike the barrel-stave mechanism, peptides are not inserted into the membrane but accumulate on the surface, forming a "carpet." Due to the high density of peptides, a very high number of electrostatic interactions occur that finally cause the destabilization of the membrane due to changes in fluidity, collapsing into several micelles.

The wormhole mechanism, also called toroidal pore (Fig. 7C), has the particularity that the AMPs are intercalated in the pore with membrane lipids. As in the barrel and carpet stave mechanisms, peptides initially accumulate on the membrane surface. However, in this new situation, the hydrophobic amino acid residues begin to displace the polar heads of the lipids, causing a tension that leads to the destabilization of the membrane surface with the consequent extravasation of the cellular contents. The assembly of AMPs and LPS has the additional effect of achieving flocculation of the latter, suppressing their systemic dissemination and, therefore, the uncontrolled immune response that can lead to septic shock. Similarly, some AMPs cause bacterial aggregation, they have been reported to play a key role in maintaining pathogens in a commensal state, thus minimizing their dangerousness [84].

In addition to direct bactericidal effects, AMPs can exert indirect actions by inhibiting the formation of biofilms or acting on the immune system by regulating pro- and anti-inflammatory responses and chemotaxis. Approximately 15% of AMPs also have antiviral properties. They can exert this action by altering the lipid bilayer of the viral envelope and preventing infection of host cells. Other mechanisms include blocking the interaction with cell surface receptors, inhibition of gene expression, and immunomodulation [83, 85].

In recent years, due to the growing problem of acquiring microbial resistance to conventional antibiotics, AMPs have been considered as alternative therapeutic agents. Since synergistic effects between these have been demonstrated, therapeutic strategies that combine traditional antibiotics with AMPs are also possible.

3.3.2 Antioxidant Peptides

Antioxidant peptides may minimize the damage caused by free radicals in organisms. Free radicals are chemical species that contain one or more unpaired electrons in their outermost orbital and can maintain an independent existence. They can be unstable, which gives them great power as reactant. Examples of these are reactive oxygen species (ROS) and reactive nitrogen species (RNS) [86]. ROS include superoxide, hydroxyl, peroxyl, hydroperoxyl, and alkoxyl radicals, mainly produced in the biological systems by the mitochondrial respiratory chain. When the imbalance is in favor of free radicals, they could damage various biological molecules such as proteins, DNA, membrane lipids, etc. The action of free radicals on proteins results in oxidation of amino acids, association of protein fragments through cross-linking by disulfide bridges, cleavage of peptide bonds, marked decrease or complete loss of affinity for metals, and increase in hydrophobicity. As a result, proteins lose their biological functions and become more susceptible to proteolytic degradation. Regarding their action on DNA, free radicals can damage various genes, thus affecting the expression of many proteins and, therefore, the functions of the cell. It has been reported that the hydroxyl radical can cause modifications in deoxyribose, which can lead to the release of nitrogenous bases and the breaking of one or both chains. ROS can also induce errors in transcription and translation processes. The donation of electrons from free radicals to polyunsaturated fatty acids leads to lipoperoxidation and the formation of various decomposition products. Consequently, changes are generated in the molecular structure of the cell membrane and loss of permeability due to the formation of disulfide bonds between membrane proteins. The resulting instability leads to cell death [87].

Organisms have defense systems that allow them maintain an adequate balance between production and elimination, either using enzymes such as superoxide dismutase, catalase, and glutathione peroxidase or through the action of non-enzymatic compounds, such as peptides (e.g., melatonin, glutathione, and carnosine), vitamins (tocopherol and ascorbic acid), or amino acids [88]. When the accumulation of ROS exceeds the cell's capacity to eliminate these free radicals, a disruption of redox homeostasis occurs. This imbalance, oxidative stress, is implicated in the development of many chronic diseases, such as cancer, atherosclerosis, and diabetes [89]. With increasing age, the natural antioxidant system becomes less efficient, and the capacity to absorb nutrients, including exogenous antioxidant substances, decreases. Due to this, the risk of oxidative stress increases with aging. This situation can be aggravated by numerous exogenous factors that favor the formation of free radicals, such as exposure to X-rays, smoking, air pollution, certain industrial products, some medicines, and inappropriate eating habits [87].

In general, antioxidant peptides are formed by sequences between 2 and 20 amino acids. However, those with the highest bioactivity reported contain between 5 to 9 amino acids, with molecular mass between 500 to 1,800 Da [90]. Due to their small size, these molecules do not have a structure of higher order than the peptide sequence, so the predominant effect on antioxidant activity is exerted by the nature and position of certain amino acids that constitute the peptide [90]. The most reactive antioxidant peptides contain aromatic residues (tryptophan, tyrosine, histidine, and phenylalanine) and sulfur-containing amino acids such as methionine and cysteine. In the case of hydrophobic amino acids, they promote the solubility of peptides in lipid phases and facilitate interactions with free radicals, thus improving the ability to inhibit peroxidation. For example, the hydroxyl of tyrosine and the indole group of tryptophan can act directly to scavenge free radicals. Histidine also has a great capacity to chelate metal ions due to the influence of its imidazole group. The amino acids tryptophan, tyrosine, and phenylalanine also have metal ion chelation capacity. Regarding the contribution of sulfur amino acids, the thiol group of the side chain of cysteine can provide hydrogen radicals, thus eliminating free radicals. In contrast, the sulfur atom of methionine can be oxidized to sulfoxide by the supply of electrons. Acidic and basic amino acids also have an important role in the antioxidant capacity of peptides since they can act as chelators due to the ability of their charged groups to form complexes with metal ions [88]. Antioxidant peptides may also exhibit a mode of action that involves physical blocking protection. Due to their surfactant properties, peptides could inhibit lipid peroxidation in emulsion systems by forming a barrier at the oil–water interface around the lipid droplets, thus preventing their contact with the oxidizing agents [90]. While synthetic antioxidants, such as butylated hydroxytoluene (BHT), butylated hydroxyanisole (BHA), t-butylhydroquinone (TBHQ), and propyl gallate, could effectively control nutrient oxidation in foods and consequently prevent the emergence of chronic diseases related to oxidative stress, they have the disadvantages of high costs and the risk of toxicity [88]. For this reason, interest in identifying and applying antioxidant peptides in the prevention and treatment of various pathologies has greatly increased in recent decades.

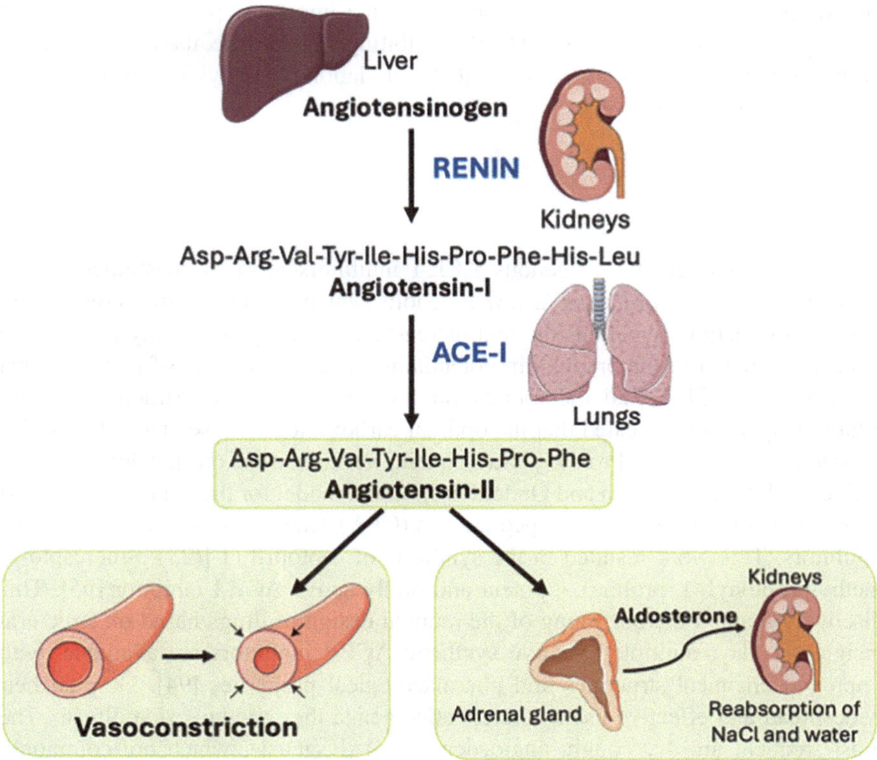

Fig. 8 Renin-Angiotensin-Aldosterone System (RAAS). Parts of this figure were drawn using images from Servier Medical Art (http://smart.servier.com/, accessed on 5 April 2024), licensed under Creative Commons by 4.0

3.3.3 Antihypertensive Peptides

Antihypertensive peptides are of special interest because they effectively lower blood pressure. Hypertension is a key and preventable risk factor for developing cardiovascular diseases and premature death worldwide [91].

Mechanism of Action

Most antihypertensive peptides effectively lower blood pressure by inhibiting angiotensin-I converting enzyme (ACE-I). This enzyme is crucial for regulating blood pressure and maintaining electrolyte balance within the renin-angiotensin-aldosterone system (Fig. 8). ACE-I catalyzes the proteolytic cleavage of angiotensin I (an inactive decapeptide) to produce angiotensin II (a strong vasoconstrictor octapeptide). Additionally, it degrades bradykinin, a potent endogenous vasodilator peptide [92]. ACE-I inhibitors are considered first-line antihypertensive therapy in treating cardiovascular diseases [93]. By inhibiting this enzyme, there is a reduction in the formation of angiotensin II and the degradation of bradykinin, resulting in a decrease in blood pressure.

Synthetic ACE-I Inhibitors

The initial discovery of exogenous ACE-I inhibitors arose from studies on the alcoholic extracts of *Bothrops jararaca* venom, which exhibited effects on smooth muscle contraction, hypertension, and increased capillary permeability [94]. These effects, attributed to the bradykinin potentiating factor (a mixture of peptides) that could inhibit ACE-I, laid the foundation for further research. Structure–activity relationship studies revealed that the optimal carboxy-terminal sequence for ACE-I inhibitors was Phe-Ala-Pro [95]. Subsequent work focused on developing orally active inhibitors. Cushman and Ondetti proposed a model for the structure of ACE-I based on its similarity to carboxypeptidase A (CPA), leading to the rational design of inhibitors. This work resulted in the synthesis of captopril (1-[(2*S*)-3-mercapto-2-methylpropionyl]-L-proline), a potent and orally active ACE-I inhibitor [96]. This discovery marked the beginning of the rational design of drugs based on structural insights. Today, multiple effective synthetic ACE-I inhibitors are available, with improved chemical structures and pharmacological properties [94]. Despite their widespread and effective use as therapeutic agents, they produce side effects. The most frequent are dry cough, angioedema, and dysgeusia, which are commonly associated with bradykinin accumulation [97]. Consequently, modifying these drugs or developing new ones with fewer side effects is an open area of research.

Natural Sources of ACE-I Inhibitory Peptides

A natural (non-pharmacological) and safe alternative for controlling blood pressure and related diseases is using bioactive peptides derived from food proteins. Many peptides with ACE-I inhibitory properties have been identified in the enzymatic hydrolysates of various proteins from foods such as milk, chicken, fish, eggs, soy, wheat [57, 98], and proteins from marine species [99]. Milk proteins serve as a primary source of a wide range of bioactive peptides [100, 101], offering the additional advantage of having high consumer acceptance. Various authors have achieved the release of ACE-I inhibitory peptides from dairy proteins through in vitro enzymatic hydrolysis with digestive enzymes [102, 103], microbial enzymes [104, 105], fermentative processes [106, 107], or through gastrointestinal digestion of dietary proteins [108]. Table 1 summarizes some of these findings. Plant peptidases have also been used in the production of ACE-I inhibitory peptides. Studies have reported the production and identification of ACE-I inhibitory peptides by hydrolysis of whey protein using papain [109], as well as non-commercial plant peptidases such as those from cardoon flowers (*Cynara cardunculus*) [110] and

Table 1 Some examples of peptides from milk proteins with in vitro ACE-I inhibitory activity (expressed as IC_{50}) and in vivo antihypertensive effect assayed on spontaneously hypertensive rats (SHR)

Source	Sequence	IC_{50} (μM)	Preparation	Gastrointestinal stability	In vivo activity on SHR	Reference
αS1-casein	RYLGY	0.71	Pepsin hydrolyzate	Low resistance (93% hydrolyzed); released peptides were active	SBP reduction (−20 mmHg) after 4 h post-oral administration of peptide (5 mg/kg BW)	[103]
	AYFYPEL	6.58	Pepsin hydrolyzate of casein	Partially resistant (53% hydrolyzed); released peptides were less active	nd	[103]
	RPKHPIKHQ	13.4	Water-soluble peptides of Gouda cheese (8-months aged)	nd	SBP reduction (−24,7 mmHg)after 6 h of oral administration of the sample (6.1 to 7.5 mg/kg BW)	[114]
	TTMPLW	16	Tryptic hydrolyzate of casein	nd	Intravenously administration of purified peptide (−31 mg/kg) to the rats antagonizes the pressure effect of injected angiotensin I (30 or 100 ng/kg BW)	[115]
αS2-casein	MKP	0.43	Casein hydro-lysate using a combination of subtilisin, bacillolysin, and trypsin	Intact 14C-MKP orally administered detected in the plasma of SHR	SBP reduction (−14,4 mmHg) after 8 h of single oral administration of the sample (1 mg/kg BW)	[104]
	PYVRYL	2.4	Casein hydrolysate using porcine pepsin A	nd	nd	[102]
	LKKISQ	2.6				
	FALPQYLK	11.9				
	FALPQYLK	4.3	Tryptic hydrolysate of S 2 casein	nd	nd	[116]
	FALPQY	4.3				
	FPQYLQY	14				
	TVY	15				

(continued)

Table 1 (continued)

Source	Sequence	IC_{50} (μM)	Preparation	Gastrointestinal stability	In vivo activity on SHR	Reference
β-Casein	IPP	5	Lactobacillus helveticus and Saccharomyces cerevisiae fermented milk	nd	SBP reduction (– 20.0, –21.8, –17.7 mmHg after 4, 6, and 8 h of single-dose oral administration of fermented milk (5 mg/kg BW)	[106, 117]
	VPP	9		nd		
	KVLVO	5	Synthetic		SBP reduction (–32 mmHg) after 6 h of single oral administration of the sample (1 mg/kg BW)	[118]
α-Lact albumin	KGYGGVSLPEW	0.8	Cynara cardunculus extract whey hydrolysate	Partial hydrolysis; ACE-inhibitory effect did not change to a significant extent	nd	[119]
	DKVGINYW	25.2		Partial hydrolysis; reduced ACE-inhibitory effect	nd	
	WLAHK	77	Tryptic hydrolyzate of α-lactalbumin	nd	nd	[120]
	VGINYWLAHK	327		nd	nd	
β-Lacto globulin	DAQSAPLRVY	13	Cynara cardunculus extract whey hydrolysate	Fully hydrolyzed; the ACE-inhibitory effect moderately increased.	nd	[119]
	ALPMHIR	43	Tryptic hydro-lysate of β-lactoglobulin	Resistant to further hydrolysis with pepsin and very low extent of digestion with chymotrypsin.	nd	[121]
	MKG	71.8	Thermolysin hydrolysate of β-lacto globulin	nd	nd	[122]

BW = body weight; nd = not determined; IC_{50} = half-maximal inhibitory concentration, In vitro activity assay of ACE (of rabbit lung) with synthetic substrate; SBP = systolic blood pressure

Arctium minus [111], from fruits of *Maclura pomifera* [112] or *Bromelia antiacantha* [113], among others.

Structure–Activity Relationship

Many ACE-I inhibitory peptides have a chain-size ranging from 2 to 12 amino acid residues [98], with dipeptides and tripeptides being recognized as particularly efficient in their inhibitory activity [123]. While most ACE-I inhibitors are competitive inhibitors [124, 125], noncompetitive inhibitors have also been identified [125], and in some cases, uncompetitive inhibitors were found [126]. ACE-I is a zinc metallopeptidase, making the interaction between peptides and the Zn^{2+} ion at the enzyme's active site important for mediating inhibition [98].

The relationship between structure and activity is not fully established, but some characteristic structural features have been established. It is believed that the *C*-terminal tripeptide is crucial for activity, and the presence of hydrophobic residues in any position of the *C*-terminal tripeptide is favorable. The presence of residues with negative charge (Glu or Asp) at the *C*-terminal position, as well as the presence of Pro at the penultimate *C*-terminal position, is unfavorable. Whereas, if the tripeptide presents Pro both at the *C*-terminal position and at the third position from the *C*-terminal end, the structure is favorable [110]. A quantitative structure–activity relationship (QSAR) study conducted to establish the relationship between pentapeptide sequence and ACE-I activity predicted that the residue at the *C*-terminal position is the most important in determining the inhibitor's potency, followed by the residue at the fourth position from the *C*-terminal end. The presence of aliphatic and small residues at the *C*-terminal position was favorable, but the presence of Pro at this position did not have a favorable prognosis. For the position, four residues from the *C*-terminal end, residues predicted to be favorable were bulky hydrophobic and those containing sulfur groups [127].

In another QSAR model conducted with dipeptides and tripeptides, it was predicted that for dipeptides, residues with bulky and hydrophobic side chains are favorable. It is suggested that residues with aromatic side chains and Pro are favorable at the *N*-terminal end, and residues with branched aliphatic side chains are favorable at the *C*-terminal position. For tripeptides, the most favorable residues, both at the *C*-terminal and *N*-terminal ends, were aromatic residues, while residues with positive charge were more favorable at the intermediate position [128].

In Vivo Activity

For ACE-I inhibitory peptides to exert an antihypertensive effect in vivo, they must reach the cardiovascular system in an active form. This requires that, after oral administration, they resist degradation by gastrointestinal proteases and brush border peptidases while being absorbed through the intestinal wall without losing their physiological activity. During this oral delivery, peptides can be degraded, which may either activate or deactivate their biological effects. This helps to explain the differences observed between in vitro ACE-I inhibitory activity and the actual antihypertensive effects following the intake of dietary protein hydrolysates [68]. Scientific evidence regarding the hypotensive effect of ACE-I activity peptides in humans is controversial [129, 130]. Evaluating the effect of food on health is complex, among other things, because it is difficult to quantify the active principle in the food, there may be interactions with the food matrix, and consumption of food may lead to dietary changes. In studies with lactotripeptides (IPP and VPP) reporting positive results, the maximum hypotensive effect observed was a reduction of 13 mmHg (systolic pressure) and 8 mmHg (diastolic pressure) after 8–12 weeks of treatment compared to placebo. Effective doses range from 3.07 to 52.5 mg/day and must be consumed for 1–2 weeks to begin seeing effects. Other reported qualities of lactopeptides include no effect on normal blood pressure, effectiveness only at elevated blood pressures, effectiveness across diverse ethnicities, and no interference with other antihypertensive medications (primarily calcium antagonists, β-blockers, and ACE-I inhibitors). Regarding the safety and toxicity of lactopeptides, in vitro and in vivo studies have demonstrated their safety [131].

4 Considerations Regarding the Pharmaceutical Administration of Bioactive Peptides

Although the administration of peptides can present several advantages, such as easy absorption and reduced allergenicity compared to intact proteins, there are several challenges to delivering drugs consisting of bioactive peptides. Besides those specifically designed to remain stable in the tissue environment, proteolytic degradation and rapid excretion through the kidneys will dramatically reduce their biological effect.

Because AMPs have a net positive charge and are amphiphilic, they have a high affinity for negatively charged serum proteins, causing them to be rapidly cleared from the bloodstream and instead accumulate in the reticuloendothelial system. Furthermore, physiological salt concentration may cause loss of their antibacterial properties.

Another significant concern in the therapeutic area is the possibility that the delivered peptides generate unwanted immune responses. This could be addressed by various strategies, such as stimulating the endogenous production of bioactive peptides through inducers delivered systemically or topically. On the other hand, this antigenic capacity can be used to formulate adjuvants (immune response boosters), immune enhancers, or a combination of these [132, 133].

On the other hand, in some cases it has been shown that certain peptides are able to resist these physiological processes and reach the circulation. A small portion can cross the intestinal barrier and, although this proportion is not significant enough to be considered of nutritional importance, it could be capable of causing biological effects at the tissue level. In addition, many of the bioactive peptides exert their effect directly in the intestinal lumen or by interacting with receptors in the intestinal cell wall, thus avoiding the metabolic consequences of their absorption [57].

Despite all these drawbacks, bioactive peptides are of great interest for targeted therapies with minimal side effects because they usually have a wide margin of safety. Drug delivery systems[1] could offer approaches for protecting against proteolytic degradation until peptides have reached their site of action and for reducing serum protein binding and clearance-related activity loss [29, 83, 134].

Compliance with Ethical Standards This chapter is a review and no humans or animals were involved in the preparation of the chapter.

References

1. Lau JL, Dunn MK (2018) Therapeutic peptides: historical perspectives, current development trends, and future directions. Bioorg Med Chem 26:2700–2707
2. Sánchez A, Vázquez A (2017) Bioactive peptides: a review. Food Qual Saf 1:29–46
3. Dziuba M, Darewicz M (2007) Food proteins as precursors of bioactive peptides – classification into families. Food Sci Technol Int 13:393–404
4. Vizovišek M, Vidmar R, Drag M et al (2018) Protease specificity: towards *in vivo* imaging applications and biomarker discovery. Trends Biochem Sci 43:829–844
5. Bond JS (2019) Proteases: history, discovery, and roles in health and disease. J Biol Chem 294:1643–1651
6. Rawlings ND (2013) Proteases families, evolution and mechanism of action. In: Brix K, Stöcker W (eds) Proteases: structure and functions. Springer Vienna, Viena
7. Rawlings ND, Barrett AJ, Thomas PD et al (2018) The MEROPS database of proteolytic enzymes, their substrates and inhibitors in 2017 and a comparison with peptidases in the PANTHER database. Nucleic Acids Res 46:D624–D632
8. Schechter I, Berger A (1967) On the size of the active site in proteases. I Papain. Biochem Biophys Res Commun 425:497–502
9. Rawlings ND (2010) Peptidase inhibitors in the MEROPS database. Biochimie 92:1463–1483
10. Clemente M, Corigliano MG, Pariani SA et al (2019) Plant serine protease inhibitors: biotechnology application in agriculture and molecular farming. Int J Mol Sci 20:1345
11. Farady CJ, Craik CS (2010) Mechanisms of macromolecular protease inhibitors. ChemBioChem 11:2341–2346
12. Chang J, Chaudhuri O (2019) Beyond proteases: basement membrane mechanics and cancer invasion. J Cell Biol 218:2456–2469
13. Sapio MR, Salzmann A, Vessaz M et al (2012) Naturally occurring carboxypeptidase A6 mutations: effect on enzyme function and association with epilepsy. J Biol Chem 287:42900–42909
14. Lv Z, Chu Y, Wang Y (2015) HIV protease inhibitors: a review of molecular selectivity and toxicity. HIV/AIDS Res Palliat Care 7:95–104
15. Pejler G (2019) The emerging role of mast cell proteases in asthma. Eur Respir J 54:1–17
16. Pomun T, Wonginta P, Kubera A (2022) Malaria box compounds against *Anopheles gambiae* (*Diptera: Culicidae*) carboxypeptidase B activity to block malaria transmission. J Med Entomol 59:1355–1362
17. De Strooper B (2010) Proteases and proteolysis in Alzheimer disease: a multifactorial view on the disease process. Physiol Rev 90:465–494

[1] See Chapter "Encapsulation and Structural Modifications of Biotherapeutics".

18. Tomita T (2017) Aberrant proteolytic processing and therapeutic strategies in Alzheimer disease. Adv Biol Regul 64:33–38
19. Dai W, Zhang B, Su H et al (2020) Structure-based design of antiviral drug candidates targeting the SARS-CoV-2 main protease. Science 368:1331–1335
20. Jin Z, Du X, Xu Y et al (2020) Structure of Mpro from COVID-19 virus and discovery of its inhibitors. Nature 582:289–293
21. Cotabarren J, Lufrano D, Parisi MG et al (2020) Biotechnological, biomedical, and agronomical applications of plant protease inhibitors with high stability: a systematic review. Plant Sci 292:110398
22. Nguyen JT, Kiso Y (2015) Discovery of peptide drugs as enzyme inhibitors and activators. Pept Chem Drug Des:157–201
23. Goswami A, Van Lanen SG (2015) Enzymatic strategies and biocatalysts for amide bond formation: tricks of the trade outside of the ribosome. Mol Biosyst 11:338–353
24. Pitzer J, Steiner K (2016) Amides in nature and biocatalysis. J Biotechnol 235:32–46
25. Hamada Y, Kiso Y (2016) New directions for protease inhibitors directed drug discovery. Biopolymers 106:563–579
26. Ciulla MG, Civera M, Sattin S et al (2023) Nature-inspired and medicinally relevant short peptides. Explor Drug Sci:140–171
27. Agyei D, Tan KX, Pan S et al (2018) Peptides for biopharmaceutical applications. In: Koutsopoulos S (ed) Peptide applications in biomedicine, biotechnology and bioengineering. Woodhead Publishing, Sawston
28. Kaysser L (2019) Built to bind: biosynthetic strategies for the formation of small-molecule protease inhibitors. Nat Prod Rep 36:1654–1686
29. Wetzler M, Hamilton P (2018) Peptides as therapeutics. In: Koutsopoulos S (ed) Peptide applications in biomedicine, biotechnology and bioengineering. Woodhead Publishing, Sawston
30. Gunnoo SB, Madder A (2016) Bioconjugation-using selective chemistry to enhance the properties of proteins and peptides as therapeutics and carriers. Org Biomol Chem 14:8002–8013
31. Abbenante G, Fairlie D (2006) Protease inhibitors in the clinic. Med Chem (Los Angeles) 1:71–104
32. Lisman T, Adelmeijer J, Huskens D et al (2021) Aprotinin inhibits thrombin generation by inhibition of the intrinsic pathway but is not a direct thrombin inhibitor. TH Open 05:e363–e375
33. Boudreau RM, Johnson M, Veile R et al (2017) Impact of tranexamic acid on coagulation and inflammation in murine models of traumatic brain injury and hemorrhage. J Surg Res 215:47–54
34. Banach L, Janczewski L, Kajdanek J et al (2022) Synthesis and hemostatic activity of new amide derivatives. Molecules 27:2271
35. Junren C, Xiaofang X, Huiqiong Z et al (2021) Pharmacological activities and mechanisms of hirudin and its derivatives – a review. Front Pharmacol 12:1–23
36. Nauck M (2016) Incretin therapies: highlighting common features and differences in the modes of action of glucagon-like peptide-1 receptor agonists and dipeptidyl peptidase-4 inhibitors. Diabetes Obes Metab 18:203–216
37. Guedes EP, Hohl A, De Melo TG et al (2013) Linagliptin: pharmacology, efficacy and safety in type 2 diabetes treatment. Diabetol Metab Syndr 5:1–7
38. Mulvihill EE (2018) Dipeptidyl peptidase inhibitor therapy in type 2 diabetes: control of the incretin axis and regulation of postprandial glucose and lipid metabolism. Peptides 100:158–164
39. Sevvana M, Klose T, Rossmann MG (2021) Principles of virus structure. Encycl Virol, pp 257–277
40. Sharma A, Gupta SP (2017) Fundamentals of viruses and their proteases. In: Gupta SP (ed) Viral proteases and their inhibitors. Academic Press, London

41. Zephyr J, Kurt Yilmaz N, Schiffer CA (2021) Viral proteases: structure, mechanism and inhibition. Enzyme 50:301–333
42. Weber IT, Wang YF, Harrison RW (2021) HIV protease: historical perspective and current research. Viruses 13:839
43. Wensing AMJ, van Maarseveen NM, Nijhuis M (2010) Fifteen years of HIV protease inhibitors: raising the barrier to resistance. Antiviral Res 85:59–74
44. Lozano Jiménez YY, Sánchez Mora RM, Giraldo Quintero SE (2020) Potenciales estrategias terapéuticas basadas en péptidos para mitigar la infección por SARS-CoV-2. Ther Nova 18: 61–66
45. Pišlar A, Mitrovic A, Sabotič J et al (2020) The role of cysteine peptidases in coronavirus cell entry and replication: the therapeutic potential of cathepsin inhibitors. PLoS Pathog 16:1–23
46. Simonis A, Theobald SJ, Fätkenheuer G et al (2021) A comparative analysis of remdesivir and other repurposed antivirals against SARS-CoV-2. EMBO Mol Med 13:1–15
47. Hillary VE, Ceasar SA (2023) An update on COVID-19: SARS-CoV-2 variants, antiviral drugs, and vaccines. Heliyon 9:e13952
48. Banerjee S, Banerjee D, Singh A et al (2023) A clinical insight on new discovered molecules and repurposed drugs for the treatment of COVID-19. Vaccine 11:1–30
49. Kouroukis TC, Baldassarre FG, Haynes AE et al (2014) Bortezomib in multiple myeloma: systematic review and clinical considerations. Curr Oncol 21:573–603
50. Chen D, Frezza M, Schmitt S et al (2011) Bortezomib as the first proteasome inhibitor anticancer drug: current status and future perspectives. Curr Cancer Drug Targets 11:239–253
51. Subeha MR, Telleria CM (2020) The anti-cancer properties of the HIV protease inhibitor Nelfinavir. Cancers (Basel) 12:1–38
52. LaCasse EC, Mahoney DJ, Cheung HH et al (2008) IAP-targeted therapies for cancer. Oncogene 27:6252–6275
53. Schimmer AD, Welsh K, Pinilla C et al (2004) Small-molecule antagonists of apoptosis suppressor XIAP exhibit broad antitumor activity. Cancer Cell 5:25–35
54. Srikanth S, Chen Z (2016) Plant protease inhibitors in therapeutics-focus on cancer therapy. Front Pharmacol 7:470
55. Cid-Gallegos MS, Corzo-Ríos LJ, Jiménez-Martínez C et al (2022) Protease inhibitors from plants as therapeutic agents – a review. Plant Foods Hum Nutr 77:20–29
56. Peng J, Zhang H, Niu H et al (2020) Peptidomic analyses: the progress in enrichment and identification of endogenous peptides. TrAC Trends Anal Chem 125:115835
57. Kitts DD, Weiler K (2003) Bioactive proteins and peptides from food sources. Applications of bioprocesses used in isolation and recovery. Curr Pharm Des 9:1309–1323
58. Amigo L, Hernández-Ledesma B (2020) Current evidence on the bioavailability of food bioactive peptides. Molecules 25:4479
59. Lyapina I, Ivanov V, Fesenko I (2021) Peptidome: chaos or inevitability. Int J Mol Sci 22: 13128
60. Sandoval DA, D'Alessio DA (2015) Physiology of proglucagon peptides: role of glucagon and GLP-1 in health and disease. Physiol Rev 95:513–548
61. Hook V, Funkelstein L, Lu D et al (2008) Proteases for processing proneuropeptides into peptide neurotransmitters and hormones. Annu Rev Pharmacol Toxicol 48:393–423
62. Bosso A, Di Maro A, Cafaro V et al (2020) Enzymes as a reservoir of host defence peptides. Curr Top Med Chem 20:1310–1323
63. Roshanak S, Shahidi F, Tabatabaei Yazdi F et al (2021) Buforin I an alternative to conventional antibiotics: evaluation of the antimicrobial properties, stability, and safety. Microb Pathog 161:105301
64. Ivanov VT, Karelin AA, Philippova MM et al (1997) Hemoglobin as a source of endogenous bioactive peptides: the concept of tissue-specific peptide pool. Biopolym Pept Sci Sect 43: 171–188
65. Egger L, Ménard O (2017) Update on bioactive peptides after milk and cheese digestion. Curr Opin Food Sci 14:116–121

66. Boutrou R, Gaudichon C, Dupont D et al (2013) Sequential release of milk protein-derived bioactive peptides in the jejunum in healthy humans. Am J Clin Nutr 97:1314–1323
67. Sato K (2018) Structure, content, and bioactivity of food-derived peptides in the body. J Agric Food Chem 66:3082–3085
68. Korhonen H, Pihlanto A (2006) Bioactive peptides: production and functionality. Int Dairy J 16:945–960
69. Kleekayai T, Cermeño M, FitzGerald RJ (2021) The production of bioactive peptides from milk proteins. In: Kelly AL, Larsen LB (eds) Agents of change. Enzymes in milk and dairy producs. Springer Nature
70. Xing L, Liu R, Cao S et al (2019) Meat protein-based bioactive peptides and their potential functional activity: a review. Int J Food Sci Technol 54:1956–1966
71. Liu YF, Oey I, Bremer P et al (2018) Bioactive peptides derived from egg proteins: a review. Crit Rev Food Sci Nutr 58:2508–2530
72. Kim IS, Yang WS, Kim CH (2021) Beneficial effects of soybean-derived bioactive peptides. Int J Mol Sci 22:8570
73. Korhonen H, Pihlanto Leppala A (2003) Milk-derived bioactive peptides. In: Shortt C, O'Brien J (eds) Handbook of functional dairy products. CRC Press
74. Mora L, Toldrá F (2023) Advanced enzymatic hydrolysis of food proteins for the production of bioactive peptides. Curr Opin Food Sci 49:100973
75. Minkiewicz P, Iwaniak A, Darewicz M (2019) BIOPEP-UWM database of bioactive peptides: current opportunities. Int J Mol Sci 20:5978
76. Karami Z, Akbari-Adergani B (2019) Bioactive food derived peptides: a review on correlation between structure of bioactive peptides and their functional properties. J Food Sci Technol 56: 535
77. Meisel H (2004) Multifunctional peptides encrypted in milk proteins. Biofactors 21:55–61
78. Corrêa JAF, de Melo NT, da Rocha GF et al (2023) Bioactive antimicrobial peptides from food proteins: perspectives and challenges for controlling foodborne pathogens. Pathogens 12:477
79. Ucar Y, Ozogul Y, Ozogul F et al (2021) The impact of different levels of nisin as a biopreservative agent on the chemical, sensory and microbiological quality of vacuum-packed sea bass (Dicentrarchus labrax) fillets stored at 4 ± 2°C. Grasas Aceites 72:e401
80. Schneider JJ, Unholzer A, Schaller M et al (2005) Human defensins. J Mol Med 83:587–595
81. Dathe M, Nikolenko H, Meyer J et al (2001) Optimization of the antimicrobial activity of magainin peptides by modification of charge. FEBS Lett 501:146–150
82. Feijó Corrêa JA, Evangelista AG, Nazareth TM et al (2019) Fundamentals on the molecular mechanism of action of antimicrobial peptides. Acta Mater. https://doi.org/10.2139/ssrn.3474484
83. Li X, Zuo S, Wang B et al (2022) Antimicrobial mechanisms and clinical application prospects of antimicrobial peptides. Molecules 27:1–29
84. Häffner SM, Malmsten M (2018) Influence of self-assembly on the performance of antimicrobial peptides. Curr Opin Colloid Interface Sci 38:56–79
85. Huan Y, Kong Q, Mou H et al (2020) Antimicrobial peptides: classification, design, application and research progress in multiple fields. Front Microbiol 11:1–21
86. Sen S, Chakraborty R, Sridhar C et al (2010) Free radicals, antioxidants, diseases and phytomedicines: current status and future prospect. Int J Pharm Sci Rev Res 3:91–100
87. López-García G, Dublan-García O, Arizmendi-Cotero D et al (2022) Antioxidant and antimicrobial peptides derived from food proteins. Molecules 27:1–32
88. Wen C, Zhang J, Zhang H et al (2020) Plant protein-derived antioxidant peptides: isolation, identification, mechanism of action and application in food systems: a review. Trends Food Sci Technol 105:308–322
89. Zhu Y, Lao F, Pan X et al (2022) Food protein-derived antioxidant peptides: molecular mechanism, stability and bioavailability. Biomol Ther 12:1622
90. Xiong YL (2010) Antioxidant peptides. In: Yoshinori M, Eunice LC, Bo J (eds) Bioactive proteins and peptides as functional foods and nutraceuticals. Wiley

91. Mills KT, Stefanescu A, He J (2020) The global epidemiology of hypertension. Nat Rev Nephrol 16:223–237
92. Ahmad H, Khan H, Haque S et al (2023) Angiotensin-converting enzyme and hypertension: a systemic analysis of various ACE inhibitors, their side effects, and bioactive peptides as a putative therapy for hypertension. J Renin Angiotension Aldosterone Syst 2023:7890188
93. Cutrell S, Alhomoud IS, Mehta A et al (2023) ACE-inhibitors in hypertension: a historical perspective and current insights. Curr Hypertens Rep 25:243–250
94. Ferreira SH (1965) A bradikinin-potentiating factor (BPF) present in venom of *Bothrops jararaca*. Br J Pharmacol Chemother 24:163–169
95. Ondetti MA, Williams NJ, Sabo EF et al (1971) Angiotensin-converting enzyme inhibitors from the venom of *Bothrops jararaca*. Isolation, elucidation of structure, and synthesis. Biochemistry 10:4033–4039
96. Ondetti MA, Rubin B, Cushman DW (1977) Design of specific inhibitors of angiotensin-converting enzyme: new class of orally active antihypertensive agents. Science 196:441–444
97. Akif M, Schwager SL, Anthony CS et al (2011) Novel mechanism of inhibition of human angiotensin-I-converting enzyme (ACE) by a highly specific phosphinic tripeptide. Biochem J 436:53–59
98. Xue L, Yin R, Howell K et al (2021) Activity and bioavailability of food protein-derived angiotensin-I-converting enzyme–inhibitory peptides. Compr Rev Food Sci Food Saf 20: 1150–1187
99. He HL, Liu D, Ma CB (2013) Review on the angiotensin-I-converting enzyme (ACE) inhibitor peptides from marine proteins. Appl Biochem Biotechnol 169:738–749
100. Fitzgerald RJ, Meisel H (2000) Milk protein-derived peptide inhibitors of angiotensin-I-converting enzyme. Br J Nutr 84:33–37
101. Durak MZ, Turan NA (2020) Antihypertensive peptides in dairy products. Am J Biomed Sci Res 7. MS.ID.001139
102. L'Opez-Expósito I, Quirós A, Amigo L et al (2007) Casein hydrolysates as a source of antimicrobial, antioxidant and antihypertensive peptides. Lait 87:241–249
103. Contreras MM, Sanchez D, Sevilla MA et al (2013) Resistance of casein-derived bioactive peptides to simulated gastrointestinal digestion. Int Dairy J 32:71–78
104. Yamada A, Sakurai T, Ochi D et al (2015) Antihypertensive effect of the bovine casein-derived peptide Met-Lys-Pro. Food Chem 172:441–446
105. Otte J, Shalaby SMA, Zakora M et al (2007) Fractionation and identification of ACE-inhibitory peptides from α-lactalbumin and β-casein produced by thermolysin-catalysed hydrolysis. Int Dairy J 17:1460–1472
106. Nakamura Y, Yamamoto N, Sakai K et al (1995) Purification and characterization of angiotensin I-converting enzyme inhibitors from sour milk. J Dairy Sci 78:777–783
107. Gobbetti M, Stepaniak L, De Angelis M et al (2002) Latent bioactive peptides in milk proteins: proteolytic activation and significance in dairy processing. Crit Rev Food Sci Nutr 42:223–239
108. Foltz M, Van Buren L, Klaffke W et al (2009) Modeling of the relationship between dipeptide structure and dipeptide stability, permeability, and ACE inhibitory activity. J Food Sci 74: 243–251
109. Silva MR, Silvestre MPC, Silva VDM et al (2014) Production of ACE-inhibitory whey protein concentrate hydrolysates: use of pancreatin and papain. Int J Food Prop 17:1002–1012
110. Tavares T, Malcata FX (2013) Whey proteins as source of bioactive peptides against hypertension. In: Hernandez-Ledesma B, Chia-Chien H (eds) Bioactive food peptides in health and disease. IntechOpen, London
111. Cimino CV, Colombo ML, Liggieri C et al (2015) Partial molecular characterization of *Arctium minus* aspartylendopeptidase and preparation of bioactive Ppptides by whey protein hydrolysis. J Med Food 18:856–864
112. Bertucci JI, Liggieri CS, Colombo ML et al (2015) Application of peptidases from *Maclura pomifera* fruit for the production of active biopeptides from whey protein. LWT 64:157–163

113. Villadóniga C, Cantera AMB (2019) New ACE-inhibitory peptides derived from α-lactalbumin produced by hydrolysis with *Bromelia antiacantha* peptidases. Biocatal Agric Biotechnol 20:101258
114. Saito T, Nakamura T, Kitazawa H et al (2000) Isolation and structural analysis of antihypertensive peptides that exist naturally in *Gouda cheese*. J Dairy Sci 83:1434–1440
115. Maruyama S, Mitachi H, Awaya J et al (1987) Angiotensin I-converting enzyme inhibitory activity of the C-terminal hexapeptide of α s1-casein. Agric Biol Chem 51:2557–2561
116. Tauzin J, Miclo L, Gaillard JL (2002) Angiotensin-I-converting enzyme inhibitory peptides from tryptic hydrolysate of bovine αS2-casein. FEBS Lett 531:369–374
117. Meisel H, Walsh D, Murray B et al (2006) ACE inhibitory peptides. In: Mine Y, Shahid F (eds) Nutraceutical proteins and peptides in health and disease. CRC Press, New York
118. Maeno M, Yamamoto N, Takano T (1996) Identification of an antihypertensive peptide from casein hydrolysate produced by a proteinase from *lactobacillus helveticus* CP790. J Dairy Sci 79:1316–1321
119. Tavares T, Contreras MDM, Amorim M et al (2011) Novel whey-derived peptides with inhibitory effect against angiotensin-converting enzyme: in vitro effect and stability to gastrointestinal enzymes. Peptides 32:1013–1019
120. Pihlanto-Leppälä A, Koskinen P, Piilola K et al (2000) Angiotensin I-converting enzyme inhibitory properties of whey protein digests: concentration and characterization of active peptides. J Dairy Res 67:53–64
121. Mullally MM, Meisel H, Fitzgerald RJ (1997) Identification of a novel angiotensin-I-converting enzyme inhibitory peptides corresponding to a tryptic fragment of bovine β-lactoglobulin. FEBS Lett 402:99–101
122. Hernández-Ledesma B, Ramos M, Recio I et al (2006) Effect of β-lactoglobulin hydrolysis with thermolysin under denaturing temperatures on the release of bioactive peptides. J Chromatogr A 1116:31–37
123. Yan W, Lin G, Zhang R et al (2020) Studies on molecular mechanism between ACE and inhibitory peptides in different bioactivities by 3D-QSAR and MD simulations. J Mol Liq 304: 112702
124. Xue L, Wang X, Hu Z et al (2018) Identification and characterization of an angiotensin-converting enzyme inhibitory peptide derived from bovine casein. Peptides 99:161–168
125. Wu J, Xie D, Chen X et al (2019) Inhibitory mechanism of a substrate-type angiotensin I-converting enzyme inhibitory peptide. Process Biochem 79:97–104
126. Sangiorgio S, Vidović N, Boschin G et al (2022) Preparation, characterization and *in vitro* stability of a novel ACE-inhibitory peptide from soybean protein. Food Secur 11:1–14
127. Zhou F, Xue Z, Wang J (2010) Antihypertensive effects of silk fibroin hydrolysate by alcalase and purification of an ACE inhibitory dipeptide. J Agric Food Chem 58:6735–6740
128. Sagardia I, Roa-Ureta RH, Bald C (2013) A new QSAR model, for angiotensin I-converting enzyme inhibitory oligopeptides. Food Chem 136:1370–1376
129. Wu J, Aluko RE, Nakai S (2006) Structural requirements of angiotensin I-converting enzyme inhibitory peptides: quantitative structure-activity relationship study of di- and tripeptides. J Agric Food Chem 54:732–738
130. Geleijnse JM, Engberink MF (2010) Lactopeptides and human blood pressure. Curr Opin Lipidol 21:58–63
131. Boelsma E, Kloek J (2009) Lactotripeptides and antihypertensive effects: a critical review. Br J Nutr 101:776–786
132. Barrett JC, Acar H, Mellas MJ et al (2018) Peptides in immunoengineering. In: Koutsopoulos S (ed) Peptide applications in biomedicine, biotechnology and bioengineering. Woodhead Publishing, Sawston
133. González García M, San J, Galán J, Fidel II et al (2017) Péptidos antimicrobianos: potencialidades terapéuticas. Rev Cubana Med Trop 69:1–13
134. Nordström R, Nyström L, Andrén OCJ et al (2018) Membrane interactions of microgels as carriers of antimicrobial peptides. J Colloid Interface Sci 513:141–150

Part V
Biotherapeutic Formulation Developments

Top Med Chem
https://doi.org/10.1007/7355_2024_178

Published online: 28 December 2024

Encapsulation and Structural Modifications of Biotherapeutics

Romina Glisoni

Contents

Abstract Biotherapeutics, encompassing a broad spectrum of biological derived therapeutics such as proteins, hormones, monoclonal antibodies, peptides, and nucleic acids, have revolutionized modern medicine with their high specificity and efficacy. This chapter begins by focusing on the impact of micro and nanotechnology on human health and, in that sense, in its contribution on the encapsulation and structural modifications of biological molecules, crucial strategies that enhance their stability, bioavailability, and targeted delivery, thereby overcoming inherent challenges such as degradation, self-aggregation, and immunogenicity. The most relevant encapsulation techniques, including biodegradable-polymeric micro and nanosystems, liposomes, extracellular vesicles, and polymeric micelles are explored, evidencing their roles in protecting biomolecules from mainly the enzymatic degradation and improving their pharmacokinetic profiles. Additionally, the chapter delves into the principal structural modifications, including chemical conjugation, PEGylation, glycosylation, and fusion protein technology. These modifications are pivotal for enhancing the therapeutic index, reducing immunogenicity, and prolonging half-life of the biomolecules. The synergy between encapsulation and structural modifications is emphasized, showcasing how these approaches can be integrated to develop next-generation biotherapeutics with superior clinical

R. Glisoni (✉)
Departamento de Microbiología, Inmunología, Biotecnología y Genética, Facultad de Farmacia y Bioquímica, Cátedra de Biotecnología, Instituto de Nanobiotecnología (NANOBIOTEC) UBA-CONICET, Universidad de Buenos Aires (UBA), Buenos Aires, Argentina
e-mail: rglisoni@ffyb.uba.ar

outcomes. Recent advancements in the field highlight practical applications and the translational potential of these technologies. Finally, by providing a comprehensive overview of current methodologies and their implications, this chapter aims to offer valuable insights into the strategic enhancement of biomolecules for improved therapeutic efficacy and patient outcomes.

Keywords Delivery nanosystems, Encapsulation, Structural modifications

1 Micro and Nanotechnology: Impact on Human Health

The principles that gave rise to formal micro and nanotechnology have had a profound impact on human health since they have been used in the formulation of active ingredients, which are either insoluble, have poor permeation, or marked adverse effects at the application doses [1–4]. Accordingly, there are four main limitations in the formulation and administration of small-molecule drugs and biotherapeutics: (i) poor solubility and/or permeation, (ii) low stability in the physiological environment, (iii) severe adverse effects at the systemic level, and (iv) low acceptance by patients, which directly impacts low adherence to treatment [5–11]. Specifically detailing those main obstacles in current oncological therapy, we can highlight: a non-specific biodistribution, reduced intratumoral concentrations and resistance developed by certain tumor cells to the conventional treatment [8–10].

It is due to the abovementioned obstacles and limitations that the micro and nanotechnology-based targeting allows for planned and rational modification of the biodistribution of a small-molecule drug or a biotherapeutic after its administration and absorption [8, 12]. The selective biodistribution will be governed by the targeting ligand and not by the molecular characteristics of the substance. In this way, systemic exposure to substance would be limited [12–18]. On the other hand, the dose required to achieve the therapeutic concentration at the site of action would also be reduced, reducing the cost of treatment and guaranteeing access to said medication for a greater number of patients.

2 Molecular Targeting

Molecular targeting can be (i) passive or (ii) active and can be well exploited in various nanotechnological strategies. In the first case, the nanostructure loaded with a molecule with biological activity is not modified on its surface and accumulates preferentially in a certain site, because it is capable of crossing vascular beds with high permeability [8–10]. This effect is known as enhanced permeation and retention (EPR) effect, which is mainly capitalized in the treatment of highly vascularized malignant solid tumors [8, 9]. On the contrary, in the case of active targeting, the surface of a biomolecule or a nanocarrier must be modified through adsorption or chemical conjugation with active-ligands (e.g., aptamers, small interfering ribonucleic acid -siRNAs-, monoclonal antibodies, small peptides, sugar molecules, among others), that are selectively recognized by certain receptors overexpressed on the surface of the target cell membrane [8, 12–18]. It is important to highlight that it will be deep addressed in Section IV, titled *Structural Modifications of Biotherapeutics*, the possibility to conjugate directly a biomolecule to an active ligand and to improve the targeting of certain tissues or cells, without the need to use a nanodelivery system.

The rational and selective targeting of biomolecule and nanotechnology-based drugs could improve therapeutic success by reducing adverse effects, decreasing the frequency of administration regimens, increasing patient acceptability and adherence to treatment, and overall, substantially reducing the treatment costs [8, 12, 15–18]. Furthermore, it can increase the effectiveness of approved drugs and extend their applicability by providing means to overcome technological limitations (e.g., low bioavailability, resistance, anatomical and cellular barriers, among others) [8–11].

Among the advantages that nanostructures present over conventional treatments, it can highlight: (i) protection of the encapsulated active molecule against its eventual degradation from the moment of administration to the patient until it reaches its site of action; (ii) ability to cross biological barriers such as the skin, mucous membranes (such as the gastrointestinal or respiratory), as well as the blood-brain barrier; (iii) ability to be directed and reach the target organ, tissue, or cell group where the molecule must exert its action; (iv) ability to reach intracellular compartments, and (v) ability to perform a controlled release of the active ingredient at its site of action or absorption [12–18].

The possibility of controlling or manipulating the structure of materials at the molecular level, through micro and nanotechnology techniques, has opened up countless alternatives in the biotherapeutical field [8–18]. As mentioned above, this emerging area has been capitalized on in the resolution of various biopharmaceutical limitations of poorly soluble and/or unstable small-molecule drugs or biotherapeutic proteins and peptides [12–18]. In turn, the development of biodegradable polymeric systems with improved properties, the use of combined pharmacological therapy or the design of conjugates directed at specific molecular targets are some of the approaches that are currently being studied to obtain more efficient and specific small-molecule drugs and biotherapeutics [12–18].

Thus, the chemical versatility of the polymers allows the conjugation of several active compounds in the same skeleton or the introduction of ligands with specificity to certain cellular or tissue receptors, which markedly increases the biotherapeutic value of these macromolecules [8, 9, 12–20].

In this sense, numerous research groups have been strongly focused in recent years on the design and development of different platforms that allow the resolution of different biopharmaceutical limitations, as well as the passive and active targeting of drugs or proteins for the optimization of immunization schemes and/or disease treatments [15, 17, 19], such as chronic infectious diseases with high socioeconomic impact, including tuberculosis, different parasitic diseases (leishmaniasis, Chagas, and others), viral hepatitis, human immunodeficiency virus (HIV) [3, 5, 6, 10, 21], as well as various oncological diseases [8, 9, 12, 15, 16, 20].

Furthermore, the combination of therapeutic and diagnostic agents in a single system gives rise to an emerging concept called theragnosis [22, 23]. A steerable theragnostic system would represent a great advance in the diagnosis and treatment of chronic diseases such as cancer and would be capable of treating the disease at the same time as it could be used as a diagnostic agent for the disease [22, 23]. The main characteristic that these novel theragnostic systems must meet is their ability to preferentially direct and accumulate in the target that is expected to perform their function and thus avoid an erroneous diagnosis, such as damage to healthy tissues due to a non-specific treatment [12, 16, 22, 23].

3 Encapsulation of Biotherapeutics

Biotherapeutics, enclose a broad spectrum of molecules purified from living culture systems or from blood [24], including proteins, hormones, monoclonal antibodies, peptides and nucleic acids, among others, and have revolutionized modern medicine with their high specificity and efficacy [12, 16–18, 25–29].

Specifically, in relation to the encapsulation and structural modifications on biotherapeutics, there are crucial strategies employed in pharmaceutical research and development to enhance the stability, efficacy, and delivery of these entities, especially proteins, hormones, and small peptides.

One of the most widely used techniques to protect biotherapeutics from degradation, self-aggregation, controlled-release, and prolong its half-life is the encapsulation. This section of the chapter mainly pretends delving into aspects related to micro- and nanoencapsulation techniques [12, 16–18, 29, 30].

The encapsulation of proteins and hydrophilic macromolecules can occur especially, on the one hand, on the microscale, by the production of microparticles (MPs) [29, 30], which can be divided into (i) microspheres, which is the most used strategy on the microscale, where the biomolecule is dispersed homogeneously in multiparticulate biodegradable-polymer matrix, that modulates and controls its release according to its bioerosion rate, and (ii) microcapsules, where the active ingredient is dispersed into polymeric matrices with a protective polymeric coating on the MPs, that modulates and controls its release and, on the other hand, on the nanoscale, with a large number of nanotechnological strategies, to highlight: liposomes, extracellular vesicles (EVs), polymeric micelles (PMs), biodegradable nanoparticles (NPs), albumin-based NPs (Nab), among others [12, 16–18].

The biodegradable polymers or copolymers used in micro and nanoencapsulation techniques can entrap biotherapeutics within their structure, at the time of the controlled-drying process, protecting them from degradation and facilitating the controlled release. Likewise, these materials offer tunable properties such as degradation rate, swelling behavior, and mechanical strength, making them suitable for various applications in small-molecule drugs and, biotherapeutics delivery.

Encapsulation offers several advantages: (i) Protection: The encapsulation shields the biological agent from enzymatic degradation, pH fluctuations, self-aggregation, and other destabilizing factors, thus improving its stability and prolonging its half-life. It is important to highlight that the self-aggregation of substances, such as proteins and active compounds, can lead to the loss of their biological activity, due to a decrease in their permeation and availability, but it can also lead to toxicity. A well-known example in the biopharmaceutical field is the nephrotoxicity caused by aggregates and poly-aggregates generated by amphotericin B in biological media [1, 5–7]. The encapsulation enables targeted delivery to specific tissues or cells, minimizing off-target effects and enhancing therapeutic efficacy; (ii) Controlled release: Controlled release formulations can be achieved through encapsulation, allowing for sustained and prolonged therapeutic action, reducing the frequency of dosing; and (iii) Combination therapy: Encapsulation facilitates the co-delivery of multiple therapeutics, enabling synergistic effects and improved treatment outcomes.

In summary, the encapsulation process mostly involves enclosing the hydrophilic molecule within a protective carrier system or nanocarrier system. In the first place, one well established example in clinical on the microscale is Lupron depot® (AbbVie Inc.) [29]. Lupron depot® is a pharmaceutical product based on microspheres formed by biodegradable polyesters and a gonadotropin releasing hormone (GnRH) analog, leuprolide acetate, for subcutaneous (SC) depot-injection [29]. Leuprolide acetate is a synthetic nonapeptide that has a short plasma half-life of 2.9 h and it is a potent GnRH-receptor, used for diverse clinical applications, including the treatment of prostate cancer, endometriosis, and uterine fibroids [29].

Encapsulating hormone derivatives or hydrophilic molecules within microspheres by multiple emulsion (W/O/W) method and subsequent solvent evaporation, freeze-drying (lyophilization), or spray drying, using biodegradable copolymers such as poly(lactic-co-glycolic) acid (PLGA), is a well-used and stabilized technique and it is presented in other protein-based products for SC or intramuscular (IM) injection [29, 30]. For example, another well-known product is octreotide and PLGA microspheres, known as Sandostatin® long-acting release (LAR) (Sandoz GmbH) [30]. Sandostatin® LAR is a synthetic octapeptide, somatostatin analogue, to treat acromegaly and carcinoid tumors. To improve the hydrolysis rate of PLGA copolymer and overcome the undesired biphasic biotherapeutic release seen with earlier luteinizing hormone releasing hormone (LHRH) MPs, it was synthesized as a branched copolymer for Sandostatin® LAR, by a glucose-initiated ring-opening polymerization (ROP) of lactide and glycoside monomers [14, 30]. The multiple alcohol groups of glucose residue enabled the formation of PLGA branches, which subsequently efficiently modulated and controlled the release of the substance [30].

Another notable product of this type is triptorelin pamoate/PLGA microspheres (Trelstar® LA and Trelstar® depot, Debiopharm Group™), which uses bioabsorbable polyesters of different molecular weights (MWs) and in this sense, the release of the biotherapeutic of interest will be mediated by the perfect bioerosion rate of these materials over time, which can extend from 4 weeks (for Sandostatin® LAR) to 12 weeks (for Trelstar® depot), or in some cases up to 16–24 weeks (Lupron® depot), depending on the product, the treatment, and its doses [29–31]. Many of these pharmaceutical presentations come in the form of a lyophilized powder for subsequent reconstitution at the time of application [31].

Finally, Nutropin® and Nutropin® depot (Genentech) is a recombinant human growth hormone (rhGH, somatropin) used for various medical conditions that involve growth hormone deficiencies and it is administered via SC and IM injection [32]. Nutropin® depot was approved by U.S. Food & Drug Administration (FDA) in 1999 and is the first controlled-release system for the systemic delivery of a protein with 1 month long acting. In relation to the production method, rhGH is dissolved or dispersed in an aqueous phase and it is then emulsified into the polymer (PLGA) organic solution, forming a water-in-oil (W/O) emulsion. This step can be enhanced using high-shear mixing or ultrasonic techniques to achieve the desired droplet size. The W/O emulsion is then added to a larger aqueous phase, creating a water-in-oil-in-water (W/O/W) double emulsion [32]. This helps in forming microdroplets containing the rhGH. The organic solvent is evaporated or extracted, causing the polymer to solidify and form MPs encapsulating the rhGH. This can be achieved through techniques like solvent evaporation or solvent extraction. The formed MPs are collected by filtration or centrifugation. The collected MPs are dried using methods such as freeze-drying (lyophilization) or spray drying to remove any residual solvent and water, resulting in a stable dry powder. The size of the MPs is critical for their performance [32]. Techniques like sieving or using laser diffraction (Dynamic Light Scattering, DLS) are employed to ensure MPs fall within the desired size range for controlled release.

According to the abovementioned, depending on the MW of the polymeric material and the type of biodegradable material used for the production of MPs (e.g., polycaprolactone (PCL), poly(lactic acid) (PLA), poly(glycolic acid) (PGA) and/or its copolymers PLGA, among others), as well as its composition (e.g., final percentages of each monomer in the case of PLGA copolymers), it is possible to modulate and control the rate of release of small-molecule drugs and biotherapeutics [29–32]. Finally, slower releases can be expected from these biodegradable materials at higher MWs. Particularly, PCL has slow degradation rate compared to PLA and PGA, due to the epsilon-caprolactone monomer (ε-caprolactone), a seven-membered lactone ring, used in the production of this biodegradable and highly crystalline copolymer [10, 33]. An example of PCL-based product is Capronor® (Norplant), which is a contraceptive implant designed for long-term birth control. It consists of a biodegradable polymer matrix based on PCL (one capsule) that releases the synthetic progestin levonorgestrel over time [33]. This implant is inserted subdermally (under the skin) and can provide contraception for an extended period, usually up to 1 year [33]. This demonstrates that PCL release rates are slower and longer than those from PLA, PGA, and PLGA-based products and derivatives [10, 33].

In the second place, the most used encapsulation technique is the use of nanotechnology. The nanostructures explored to nanoencapsulate different (bio)-therapeutics are liposomes, EVs, polymeric micelles, and biodegradable nanospheres, among others [12, 17, 18]. Each type of nanostructure has its advantages and limitations and the choice for use, depending mainly on factors such as the physicochemical properties of the (bio)therapeutic entity, desired release kinetics, target site, treatment and doses, and route of administration [12, 17, 18].

It is important to highlight that the encapsulation using nanocarriers holds great promise for improving the stability, bioavailability, and therapeutic efficacy of biomolecules in future clinical applications. In this way, the use of polymeric materials for the generation of nanostructures led to the gestation of products with pharmacokinetic properties and improved efficacies, an active field of research that combines nanotechnology, pharmaceutical science, and medicine, which we can define as *nanomedicine* [34].

A survey of products approved by the FDA until 2015 showed that only 51 products could be exactly defined as *nanomedicines,* while other 77 products, were in some phase of clinical trials [34]. This category of products includes a heterogeneous set of drugs and formulation strategies intended, through particle size control, to improve the properties of the various active ingredients [34]. Within the main categories and types of nanoparticles (NPs) contained within the so-called *nanomedicines* and that are inserted as products in the bio-pharmaceutical market or advanced clinical trials, can be highlighted into four large groups [34]:

(i) ***Polymer-small-molecule drug or -biotherapeutic conjugates.*** They are mainly composed of at least one biocompatible polymer combined with chemically synthesized small-molecule drugs or biotherapeutics. Within this category are present biological medications that have been modified with the biocompatible polymer, polyethylene glycol (PEG), such as Neulasta® (Amgen), the modified version of filgrastim (PEG-filgrastim) and which constitutes a blockbuster[1] together with the immunomodulator, glatiramer acetate (Copaxone®, Teva Pharmaceuticals), for the treatment of multiple sclerosis [34, 35]. PEG-filgrastim is a long-acting artificial version of granulocyte colony stimulating factor, that stimulates the production of white blood cells and, in particular, the production of neutrophils [34, 35]. PEG, is a neutral and biocompatible polymer, which increases the hydrodynamic diameter of the protein by increasing its nominal size and protects it from the environment. Its incorporation has two main purposes: (i) on the one hand, it reduces dramatically the antigenicity of the protein since it prevents the proteolytic degradation and subsequent antigenic presentation and hides it from the immune system; and (ii) on the other hand, it increases its half-life, due to the decrease in elimination by renal filtration, a consequence of the increase in the effective hydrodynamic diameter [34, 35]. Thus, the bioconjugation using a polymer (i.e., PEG) and a biological agent, would consider a structural modification strategy on a protein *perse* (this point will be developed again later), that ultimately gives it nanometric properties and, which should follow strict quality control protocols (such as particle

[1] See Chapter "Medicinal Chemistry and Biotherapeutics: Overview and Definitions".

size, size distribution, aggregation, and Zeta-potential by DLS), as a nanomedicine itself. Although this category is formally included within *nanomedicines*, from a regulatory point of view, it is considered within the category of *biological medicines,* because the active ingredient is a biotherapeutic, and therefore they are specifically registered in this way. Of the aforementioned *nanomedicines* there are biosimilar medicines that have been registered under the specific provisions of *biologicals and biosimilars* in the European Medicines Agency and FDA. For example, there are several biosimilar versions of Neulasta® (PEG-filgrastim) [34, 35].

(ii) ***Formulations Based on Liposomes and Pegylated-Liposomes Combined with Small-Molecule Drugs or Biotherapeutics.*** Liposomes are spherical vesicles composed of lipid bilayers. They were originally described in 1960 and proposed as a drug delivery system in 1970 [12, 31, 34, 36, 37]. They can encapsulate both hydrophilic and hydrophobic biotherapeutics within their aqueous core or lipid bilayers, respectively. Liposomes offer unique advantages such as biocompatibility, versatility in size and composition, and the ability to deliver payloads to specific tissues or cells. Furthermore, it is a very solid and safe platform used in the pharmaceutical market, both for therapy, as a first line or in combination with other drugs, and in the development of nanovaccines for immunization [12, 31, 34, 36, 37]. On the other hand, liposomes can be classified in terms of composition into different types: conventional, pH-sensitive, cationic, immune-liposomes (immunological adjuvants) and those with high circulation or pegylated (PEG-coated to avoid or delay the opsonization phenomenon) [36, 37]. It can be stated, that liposomes constitute the most widely studied colloidal systems for clinical purposes, for the administration of antineoplastic, antibacterial drugs, and genetic material [36–39]. There are several pharmaceutical products based on an active ingredient formulated with liposomes. One of the exponents of this group is Doxil® (Janssen) composed of PEG-liposomal doxorubicin and used for the treatment of Kaposi's Sarcoma, metastatic ovarian cancer, and multiple myeloma. In the same direction, non-pegylated liposomal daunorubicin for intravenous (IV) administration (~100 nm in size) can be mentioned, similarly for the treatment of metastatic ovarian carcinoma and Kaposi's sarcoma associated with HIV, respectively. Likewise, it is worth mentioning intrathecally administered liposomal cytarabine in the treatment of meningeal lymphomas [36, 37]. Similarly, there are formulations based on liposomes for the treatment of fungal infections, such as liposomal amphotericin B, that aim to prevent and reduce its nephrotoxicity with the nanoencapsulation [31]. Finally, it is relevant to mention the hepatitis A virion in virosomes for SC administration and influenza antigens in virosomes (suspension of IM administration) as an influenza vaccine [38, 39]. Some of the main problems that limit industrial-scale production of liposomes are their stability, batch-to-batch reproducibility, and some difficulties in sterilization and lyophilization processes, especially in pegylated systems [31].

Similar considerations are there in the production of EVs. EVs mediate communication between cells by transferring bioactive molecules such as proteins, lipids, and RNAs [17]. They play a role in immune responses by presenting antigens and modulating immune cell function. EVs can carry disease-specific biomarkers, making them valuable for non-invasive diagnoses [17]. EVs are being explored as delivery vehicles for therapeutic agents due to their ability to target specific cells and tissues [17]. It is beyond the objectives of this chapter to delve deeper into these novel systems.

(iii) ***Formulations Based on Polymeric Micelles Combined with Small-Molecule Drugs or Biotherapeutics.*** PMs are generated by the spontaneous self-assembly in the water of amphiphilic polymeric molecules above the critical micellar concentration (CMC) [5, 10, 12–16, 19]. In general, they combine hydrophilic segments of PEG with hydrophobic segments of a different nature. Regarding the applications and advantages of the use of PMs compared to conventional micelles, the following stand out: (i) encapsulation of drugs with poor aqueous solubility in the hydrophobic core [9, 14, 15] and good adsorption of hydrophilic biomolecules with active targeting on the corona surface [12, 16]; (ii) good stability above the CMC; (iii) systems with PEG as sterically stabilized hydrophilic component that prevents or delays the opsonization phenomenon and uptake by macrophages; and (iv) lower elimination through renal filtration [5, 10, 12–16, 19]. Genexol®-PM (Samyang Biopharmaceuticals) based on methoxypoly(ethyleneglycol)(mPEG)-PLA copolymer, encapsulates paclitaxel with hydrodynamic diameters of approximately 20–50 nm [16]. Genexol®-PM is indicated for the treatment of breast cancer (phase IV) and ovarian cancer, small cell lung cancer, and pancreatic cancer [16].

In recent years, Castelli and colleagues have exhaustively explored the vehicleization of Sgc8-c-Alexa647 aptamer, with active targeting to PTK7-receptor overexpressed in a great number of tumors such as lymphoma, on the surface of PMs based on amphiphilic block copolymers of poly(ethyleneoxide)-poly(propyleneoxide) (PEO-PPO) and it was well-adsorbed on the surface of the nanomicelles, a relevant increase in the half-life of the probe, as well as notable differences in its biodistribution with respect to its free counterpart [12]. It is worth mentioning that the mean diameter of 22–27 nm PMs (SP1049C) of PEO-PPO copolymers (Pluronics® F127 and L61) that contain doxorubicin are indicated for the treatment of esophageal and stomach cancer and already found in phase III clinical trials [16, 40].

(iv) ***Biodegradable Polymer-Based Nanoparticles and Protein-Based Nanoparticles Combined with Small-Molecule Drugs or Biotherapeutics.*** NPs are spherical particles with nanoscale diameters, depending on the route of administration. Besides, solid NPs are solid colloidal particles with sizes less than 200 nm for parenteral administrations, especially IV and SC, allowing larger particle sizes in routes such as intraperitoneal (IP) and IM, and oral routes

(VO) [31, 34]. They can be produced mainly due to the same biodegradable copolymers, with which MPs are produced such as PCL, PLA, PGA and PLGA, natural polymers (chitosan), metals, lipids, or proteins such as albumin, gelatin, among others, with different compositions and biodegradation rates. These NPs or nanospheres can encapsulate biotherapeutics either homogeneously within their matrix or on their surface. They are particularly useful for sustained release formulations and targeted delivery [34].

NPs and MPs provide versatile platforms for encapsulating biotherapeutics, enabling precise control over molecule release kinetics and enhanced targeting through surface modifications. Examples include IV injectable NPs and SC or IM injectable MPs for the controlled release of hormones and/or proteins with therapeutic activity from biodegradable microspheres, generally based on slow and controlled release copolymers (PCL, PLA, PGA, PLGA of different composition of LA and GA). In the category of IV injectable NPs, there are especially albumin-based NPs (Nab) for both diagnostic and therapeutic purposes [31, 34, 41–43]. In the latter case, the product Abraxane® (Celgene) stands out, consisting of Nab loaded with paclitaxel [31, 34, 41–43]. This product deserves a special analysis due to its particularities. First, the nanoparticle is not made of an inert polymer or compound, but is obtained from human albumin. For its part, Nab provides its size (particles of approximately 130 nm) and allows paclitaxel, a highly insoluble molecule, to be formulated in high concentrations. In the absence of Nab, paclitaxel and other insoluble taxanes require excipients such as Cremophor EL or polysorbates for parenteral application [34, 41–43]. These excipients contribute to the observed toxicity for taxanes (e.g., hypersensitivity, cardiotoxicity, peripheral neuropathy, and myelosuppression), while Cremophor EL can extract plastics from polyvinyl chloride tubes sometimes resulting in fatal anaphylactic reactions. Cremophor EL micelles with paclitaxel usually show nonlinear kinetics with the absence of a dose-response curve. In this sense, Nab not only provide a physical medium to contain paclitaxel, but also provide functionality since albumin undergoes an active transport process through blood capillaries called transcytosis. Through this process Nab-Paclitaxel binds the albumin receptor gp60 and is transported to the tumor interstitium [34, 41–43]. Abraxane® has been approved for the treatment of metastatic breast cancer when the first treatment is no longer effective and the standard treatment is not appropriate; metastatic pancreatic adenocarcinoma, as a first treatment in combination with gemcitabine; and non-small cell lung cancer, as the first treatment in combination with carboplatin, when the patient does not accept surgery or radiotherapy [34, 41–43].

4 Structural Modifications of Biotherapeutics

To expand on the details discussed in the previous sections, microencapsulation and nanoencapsulation, along with potential structural modification approaches, are specifically designed to address the unique challenges of biotherapeutics, enhancing their clinical utility and therapeutic potential.

Among the most important structural modifications, we can highlight (i) *PEGylation:* PEGylation involves conjugating PEG chains to the biotherapeutic molecule. This modification increases the hydrodynamic size of the bioactive molecule, shielding it from proteolytic degradation and immune recognition, thereby extending its circulation half-life (see above, *III. Encapsulation of Biotherapeutics,* (i) *Polymer-small-molecule drug or -biotherapeutic conjugates*) [35, 44]; (ii) *Glycosylation:* Altering the glycosylation pattern of the protein of interest can modulate their stability, pharmacokinetics, and immunogenicity. Glycosylation patterns can be engineered to enhance therapeutic efficacy and reduce immunogenic responses [45]; (iii) *Site-Specific mutagenesis:* Site-specific mutagenesis allows for precise modifications of the biotherapeutic molecule, such as introducing amino acid substitutions or deletions to improve stability, binding affinity, or pharmacokinetic properties [46]; (iv) *Conjugation to active ligands:* Conjugation of the biotherapeutic to active targeting ligands, such as aptamers, antibodies, or peptides, can enhance its specificity for the intended target, improving therapeutic efficacy while reducing systemic side effects [8, 12, 14–17]; and (v) and *Fusion proteins:* Fusion proteins are engineered by combining the biotherapeutic molecule with another protein or peptide sequence. This strategy can confer novel functionalities or improve pharmacokinetic properties of the protein of origin [47].

In relation to the conjugation of biotherapeutics to active targeting ligands, such as aptamers, antibodies or peptides is a powerful strategy to enhance their specificity and efficacy. Among the most recognized methods of conjugation, it can distinguish four principal methods: (i) *Antibody conjugation:* Antibody-drug conjugates (ADCs) consist of a monoclonal antibody conjugated to a cytotoxic small-molecule drug via a linker. The antibody targets specific antigens overexpressed on the surface of cancer cells, delivering the cytotoxic payload directly to the tumor site while minimizing systemic toxicity [48]; (ii) *Bispecific antibodies:* Bispecific antibodies are engineered to simultaneously bind two different antigens. By conjugating a therapeutic payload to one arm of the bispecific antibody, it can deliver the payload specifically to cells expressing both target antigens, such as tumor cells and immune cells, enhancing therapeutic efficacy [49]; (iii) *Peptide conjugation or targeting peptides:* Short peptides with high affinity for specific receptors or proteins expressed on target cells can be conjugated to biotherapeutics to enhance their selectivity. These peptides can facilitate targeted delivery to desired tissues or cellular compartments [50]; (iv) *Cell-penetrating peptides (CPPs):* CPPs are short peptides that can traverse cellular membranes efficiently. Conjugating CPPs to biotherapeutics can enhance their intracellular delivery, overcoming biological barriers such as the cell membrane and endosomal trapping. In this sense, all the advantages are focalized in enhancing the specificity [51].

To conclude this section, it is important to remark that the conjugation of biotherapeutics to active targeting ligands improves their specificity for the intended target, reducing off-target effects[2] and enhancing therapeutic efficacy, thereby reducing systemic toxicity because targeted delivery minimizes exposure of healthy tissues to the therapeutic agent, thereby reducing systemic toxicity and improving the therapeutic index. Improved pharmacokinetics due to the conjugation can prolong the circulation half-life of biotherapeutics, enhancing their bioavailability and therapeutic duration.

Finally, focusing on the challenges and advantages of the conjugation of biotherapeutics to active ligands, it is important to take into account the following points: (i) *The linker stability:* The stability of the linker connecting the biological agent to the targeting ligand is critical to ensure efficient payload delivery and minimize premature release; (ii) *The immunogenicity*: Conjugation may introduce immune-genic epitopes, leading to immune responses against the biotherapeutic or targeting ligand; (iii) *The optimization of the design:* Designing optimal conjugates requires careful consideration of factors such as the linker chemistry, the conjugation site, and the stoichiometry to maximize the therapeutic efficacy and minimize the adverse effects [48–51].

Overall, conjugation of biomolecules to active targeting ligands offers a promising approach to enhance their specificity, efficacy, and therapeutic potential across various disease indications.

5 Final Conclusions and Perspectives

In this chapter, some of the most relevant and innovative strategies for protecting biotherapeutics were explored, through advanced encapsulation techniques, specifically microencapsulation and nanoencapsulation, and different structural modifications. These methodologies represent significant advancements in biotechnology and pharmaceuticals, offering enhanced stability, controlled release, and targeted delivery of (bio)active compounds.

Microencapsulation, with its ability to create a protective and control barrier around the biological agents, ensures prolonged stability and efficacy, which is crucial for applications in pharmaceutical industry, for extended and controlled release applications from 4 weeks to 6 months, mainly via SC and IM routes. The versatility of microencapsulation allows for the encapsulation of a wide range of agents, from peptides, hormones to vaccines and hydrophilic drugs, thereby broadening its applicability.

Nanoencapsulation takes these benefits a step further by offering improved penetration and bioavailability, especially important in medical and pharmaceutical applications. The nanoscale size of these nanoparticles allows for precise delivery to

[2] See Chapter "Medicinal Chemistry and Biotherapeutics: Overview and Definitions".

specific cells or tissues, minimizing side effects and enhancing therapeutic outcomes. Nanoencapsulation techniques have opened new avenues in targeted drug and biomolecule delivery, cancer therapy, the development of advanced diagnostic tools, and the creation of novel nanovaccines for more effective immune responses [52].

Structural modifications complement these encapsulation techniques by altering the molecular architecture of biological agents to improve their stability and functionality. By engineering these modifications, it is possible to enhance the compatibility of the agents with encapsulation materials and optimize their performance in various environments.

Collectively, these encapsulation and modification strategies provide a robust toolkit for enhancing the protection and efficacy of biotherapeutics. Future research and development in this field promise to further refine these techniques, making them more efficient and accessible for a wider range of applications. As we continue to understand and manipulate these technologies, their potential to revolutionize industries and improve human health becomes increasingly apparent. The integration of microencapsulation, nanoencapsulation, and structural modifications stands as a testament to the progress and potential of modern biotechnology.

Compliance with Ethical Standards This chapter is a review and no humans or animals were involved in the preparation of the chapter.

References

1. Glisoni RJ, Chiappetta DA, Finkielsztein LM et al (2010) Self-aggregation behaviour of novel thiosemicarbazone drug candidates with potential antiviral activity. New J Chem 34:2047–2058
2. Fuchs AG, Soraires Santacruz MC, Fusco OA et al (2021) Trypanocidal effect of Diminazene aceturate by intranasal administration. Comparison among formulations. Int J Health Sci 1:33
3. Glisoni RJ, Sosnik A (2014) Encapsulation of the antimicrobial and immunomodulator agent nitazoxanide within polymeric micelles. J Nanosci Nanotechnol 14:4670–4682
4. Dávila B, Sánchez C, Fernández M et al (2019) Selective hypoxia-cytotoxin 7-fluoro-2-aminophenazine 5,10-dioxide: preclinical studies and *in vivo* proof of concept. ChemistrySelect 4:9396–9402
5. Glisoni RJ, Chiappetta DA, Moglioni AG et al (2012) Novel 1-indanone thiosemicarbazone antiviral candidates: aqueous solubilization and physical stabilization by means of cyclodextrins. Pharm Res 29:739–755
6. Glisoni RJ, Cuestas ML, Mathet VL et al (2012) Antiviral activity against the hepatitis C virus (HCV) of 1-indanone thiosemicarbazones and their inclusion complexes with hydroxypropyl-β-cyclodextrin. Eur J Pharm Sci 47:596–603
7. Ezquerra Riega S, Cammarata A, Marino J et al (2023) Enhancement of zinc (II) phthalocyanine photophysical properties in water by sulfocalix[4] arene based nanomicelles. ChemPhotoChem 8:1–23
8. Monterrubio C, Paco S, Olaciregui NG et al (2017) Targeted drug distribution in tumor extracellular fluid of GD2-expressing neuroblastoma patient-derived xenografts using SN-38-loaded nanoparticles conjugated to the monoclonal antibody 3F8. J Control Release 255:108–119

9. Lecot N, Dávila B, Sánchez C et al (2022) Development and evaluation of 2-amino-7-fluorophenazine 5,10-dioxide polymeric micelles as antitumoral agents for 4T1 breast cancer. Polymers 14:71
10. Glisoni RJ, Quintana S, Molina Soler MA et al (2015) Chitosan-g-oligo(epsilon-caprolactone) polymeric micelles: microwave-assisted synthesis and physicochemical and cytocompatibility characterization. J Mater Chem B 3:4853–4864
11. Monterrubio C, Paco S, Vila-Ubach M et al (2015) Combined microdialysis-tumor homogenate method for the study of the steady state compartmental distribution of a hydrophobic anticancer drug in patient–derived xenografts. Pharm Res 32:2889–2900
12. Castelli R, Ibarra M, Faccio R et al (2022) T908 polymeric micelles improved the uptake of Sgc8-c aptamer-probe in bearing tumor mice: a co-association study between the probe and preformed nanostructures. Pharmaceuticals 15:15
13. Cuestas ML, Glisoni RJ, Mathet V et al (2013) Lactosylated-poly(ethylene oxide)-poly(propylene oxide) block copolymers for potential active drug targeting: synthesis and physicochemical and self-aggregation characterization. J Nanopart Res 15:1389
14. Glisoni RJ, Sosnik A (2014) Novel poly(ethylene oxide)-co-poly(propylene oxide) copolymer-glucose conjugate by microwave-assisted ring opening of a sugar lactone. Macromol Biosci 14: 1639–1651
15. Lecot N, Glisoni RJ, Oddone N et al (2020) Glucosylated polymeric micelles actively target curcumin to a breast cancer model. Adv Therap 4:2000010
16. Lecot N, Fernandez-Lomonaco M, Cerecetto H et al (2024) Indocyanine green within glycosylated polymeric micelles as potential image agents to map sentinel lymph nodes and breast cancer. RSC Pharmaceutics 1:57–67
17. Mattera V, Pereyra Gerber F, Glisoni RJ et al (2020) Extracellular vesicles containing the apotransferrin receptor as nanocarriers of transferrin. J Neurochem 155:327–338
18. Blachman A, Funez F, Saavedra S et al (2020) Targeted anti-inflammatory peptide IRW delivery in injured endothelial cells using dermatan sulfate/chitosan polyelectrolyte complex nanomaterials. Carbohydr Polym 230:115610
19. Sosnik A (2013) Temperature- and pH-sensitive polymeric micelles for drug encapsulation, release and targeting. In: Alvarez-Lorenzo C, Concheiro A (eds) Smart materials for drug delivery. Royal Society of Chemistry, London
20. Sosnik A, Chiappetta DA, Carcaboso A (2009) Drug delivery systems in HIV pharmacotherapy: what has been done and the challenges standing ahead. J Control Release 138:2–15
21. Duncan R (2006) Polymer conjugates as anticancer nanomedicines. Nat Rev Cancer 6:688–701
22. Ryu JH, Koo H, Sun IC et al (2012) Tumor-targeting multifunctional nanoparticles for theragnosis: new paradigm for cancer therapy. Adv Drug Deliv Rev 64:1447–1458
23. Fan Z, Fu PP, Yu H et al (2014) Theranostic nanomedicine for cancer detection and treatment. J Food Drug Anal 22:3–17
24. World Health Organization (2024) Biologicals. https://www.who.int/health-topics/biologicals#tab=tab_1. Accessed 18 Apr 2024
25. Simois S, Cammaratta A, Glisoni RJ et al (2024) ^{99m}Tc-AgNPs-ICG: nanoparticle for hybrid image. Salud Militar 43:e302
26. Lecot N, Rodríguez G, Stancov V et al (2022) Development of fluorescent- and radio-traceable T1307-polymeric micelles as biomedical agents for cancer diagnosis: biodistribution on 4T1 tumor-bearing mice. Braz J Pharm Sci 58:e191055
27. Hirsch DB, Martínez Álvarez L, Urtasun N et al (2020) Lactoferrin purification and whey protein isolate recovery from cheese whey using chitosan mini-spheres. Int Dairy J 109:104764
28. Hirsch DB, Baieli MF, Urtasun N et al (2018) Sulfanilic acid-modified chitosan mini-spheres and their application for lysozyme purification from egg white. Biotechnol Prog 34:387–396
29. Wilson AC, Vadakkadath Meethal S, Bowen RL et al (2007) Leuprolide acetate: a drug of diverse clinical applications. Expert Opin Investig Drugs 16:1851–1863
30. Tice T (2023) Inspirational chemistry of bioabsorbable long-acting injectables. Drug Deliv Transl Res 13:371–377

31. Dobovsek M, Glisoni RJ (2019) Formas farmacéuticas inyectables. In: Manzo RH, Nacucchio MC (eds) Tópicos de tecnología farmacéutica. Eudeba, Buenos Aires
32. Cleland JL, Duenas ET, Park A et al (2001) Development of poly-(D,L-lactide--coglycolide) microsphere formulations containing recombinant human vascular endothelial growth factor to promote local angiogenesis. J Control Release 72:13–24
33. Darney PD, Monroe SE, Klaisle CM et al (1989) Clinical evaluation of the Capronor contraceptive implant: preliminary report. Am J Obstet Gynecol 160:1292–1295
34. Glisoni RJ, Alonso L (2019) Estado actual, desarrollo y avances de la nanomedicina en el diagnóstico y tratamiento de enfermedades neoplásicas. In: Colección de propiedad industrial e intelectual: Su valor para el crecimiento y la salud. Albremática, Buenos Aires
35. Olvera AS, Albán F, Villalobos A et al (2010) Pegfilgrastim in the primary prophylaxis of chemotherapy-induced severe neutropenia. GAMO 9:3–10
36. Anselmo AC, Mitragotri S (2016) Nanoparticles in the clinic. Bioeng Transl Med 1:10–29
37. Hernández Herrero G, Moreno González A, Zaragozá García F et al (2010) MediPharm. Tratado de medicina farmacéutica. Editorial Médica Panamericana, Ciudad de México
38. Bovier PA (2008) Epaxal: a virosomal vaccine to prevent hepatitis A infection. Expert Rev Vaccines 7:1141–1150
39. Mischler R, Metcalfe IC (2002) Inflexal V a trivalent virosome subunit influenza vaccine: production. Vaccine 5:17–23
40. Sutton D, Nasongkla N, Blanco E et al (2007) Functionalized micellar systems for cancer targeted drug delivery. Pharm Res 24:1029–1046
41. Miele E, Spinelli GP, Miele E et al (2009) Albumin-bound formulation of paclitaxel (Abraxane® ABI-007) in the treatment of breast cancer. Int J Nanomedicine 4:99–105
42. Pagels RF, Prud'homme RK (2015) Polymeric nanoparticles and microparticles for the delivery of peptides, biologics, and soluble therapeutics. J Control Release 219:519–535
43. Singh R, Lillard JW (2009) Nanoparticle-based targeted drug delivery. Exp Mol Pathol 86:215–223
44. Veronese FM, Mero A (2008) The impact of PEGylation on biological therapies. BioDrugs 22: 315–329
45. Solá RJ, Griebenow K (2010) Glycosylation of therapeutic proteins: an effective strategy to optimize efficacy. BioDrugs 24:9–21
46. Carter PJ (2011) Introduction to current and future protein therapeutics: a protein engineering perspective. Exp Cell Res 317:1261–1269
47. Leader B, Baca QJ, Golan DE (2008) Protein therapeutics: a summary and pharmacological classification. Nat Rev Drug Discov 7:21–39
48. Beck A, Goetsch L, Dumontet C et al (2017) Strategies and challenges for the next generation of antibody-drug conjugates. Nat Rev Drug Discov 16:315–337
49. Labrijn AF, Janmaat ML, Reichert JM et al (2019) Bispecific antibodies: a mechanistic review of the pipeline. Nat Rev Drug Discov 18:585–608
50. Dong C, Liu Z, Wang F (2014) Peptide-based radiopharmaceuticals for targeted tumor therapy. Curr Med Chem 21:139–152
51. Guidotti G, Brambilla L, Rossi D (2017) Cell-penetrating peptides: from basic research to clinics. Trends Pharmacol Sci 38:406–424
52. Márquez PG, Wolman FJ, Glisoni RJ (2024) Nanotechnology platforms for antigen and immunostimulant delivery in vaccine formulations. Nano Trends 8:100058

Part VI
Regulatory Aspects fo Biotherapeutics

Top Med Chem
https://doi.org/10.1007/7355_2024_179

Published online: 4 January 2025

Evolution of International Standards for the Approval of Biosimilars: A Path to Flexibilities

José Ricardo Ramírez and Claudia P. Vaca G

Contents

Abstract The availability of competitors from generic versions of biotherapeutics, known internationally as biosimilars, is critical for patients who require them to have access to treatment and for the sustainability of healthcare systems. The presence of competition is conditioned by the regulatory processes and the approval of health agencies. In the last decade, technological and scientific developments for protein characterization and clinical results in the real-life usage of biosimilars have supported the adoption of flexibilities for their approval and faster entry into the market. These flexibilities refer to the extension of redundant clinical tests, especially comparative confirmatory tests, and the exemption of animal toxicological tests, as well as the acceptance of a global referenced product having wide access to full dossier information. Although the adoption of these flexibilities has had a justified rationale since 2014, they were only partially incorporated by the World Health Organization until 2022. In Latin America region, the incorporation of flexibilities is diverse and has a low impact on the configuration of competition in the market to achieve better prices. Colombia, Brazil and Argentina have adopted a pro-competitive regulation, so did England. United States of America started this discussion only in 2023..

Keywords Biosimilar, Competition, Generic, Pharmaceutical regulation

---

J. R. Ramírez and C. P. Vaca G (✉)
Think Tank "Medicines, Information and Power", Universidad Nacional de Colombia. Facultad de Ciencias. Departamento de Farmacia, Colombia, Bogotá
e-mail: joramirezco@unal.edu.co; cpvacag@unal.edu.co

1 Introduction

The increasing availability of competitors in the biological and biotechnological medicines, biotherapeutics, developed since the late 90 s and the first decade of the twenty-first century, determines the evolution of the international standards for its approval. The first standards were based on the principle of caution and the establishing of the comparability paradigm between the first entry product and the competitor who sought to be registered.

Comparability paradigm prevails and establishes that the whole development of a competitor must be compared in each of its pharmaceutical, preclinical, and clinical specifications in order to demonstrate its similarity to the reference product.

This approach gave rise to the use of the term biosimilar,[1] which became the harmonized proposal to name generic medicines of biological origin. However, this language affects the full acceptance of its use by medicinal doctors and patients. Maybe it is time to move toward the narrative of indistinguishable competing medicines.[2]

The comparability approach includes the requirement to carry out confirmatory clinical studies of efficacy and safety, especially its immunogenic expressions, to reduce uncertainty and the worries of medicinal doctors and patients about using the competing medicines.

The first health guidelines and standards were issued by the European Medicines Agency (EMA) in 2005 and, in 2009, by the World Health Organization (WHO). From 2010 to date, other health agencies, such as the U.S. Food & Drug Administration (FDA) and from Latin America, published the first regulations and technical guides for this medicine's approval.

The guidelines and standards have been updated over the years as the number of biosimilars on the market increased, and in light of scientific and technological advances in the characterization of therapeutic proteins and other biotherapeutics. The availability of a chemistry packet, manufacturing, and very robust controls (CMC) currently allow establishing comparability with greater precision [1].

In parallel, in recent decades, there has been an increase in the number of publications of studies on the effectiveness and long-term follow-up of biosimilars available in different parts of the world, as well as several systematic reviews showing that the clinical results of biosimilars are comparable to those of reference biotherapeutics, with substantially lower prices and greater access to treatments.

Below is a description of the rationale, contents, and transformations of biosimilar regulations in Europe and the United States, the WHO guidelines over time and how the exemption from confirmatory clinical trials is currently consolidated. It describes

[1] See Chapter "Development and Production of Protein-based Biotherapeutics in Mammalian Cells".

[2] Biosimilar drugs provide cost-effective yet clinically indistinguishable replications of target drugs in https://jamanetwork.com/journals/jamaoncology/article-abstract/2819238. Accessed 19 Aug 2024.

how these transformations are or are not present in the Colombian, Mexican, Brazilian, and Argentinean regulations. The speed of entry of competitors in relation to the issuance of regulations is also analyzed.

2 Evolution of the European Regulation

The EMA regulation of 2005 and the WHO guidelines of 2009 [2, 3], with nuanced differences, established that for the approval of any biosimilar, a tiered comparability exercise of the competing medicines with the reference medicine was essential, starting with a full analytical characterization, in addition to a pharmacokinetic/pharmacodynamic (PK/PD) study, plus a confirmatory comparative clinical study—with a non-inferiority or equivalence design. The requirement of these confirmatory studies delayed and increased the cost of each biosimilar that sought to enter the pharmaceutical market, with the consequent impact on the expenditure of the health system and patients' pockets.

Scientific and technologic developments for therapeutic proteins characterization, increasing experience of its usage around the world, and the challenges that healthcare systems face to guarantee access, motivated the updating of the WHO and European guidelines. Indeed, in 2014, health ministers meeting in the World Health Assembly, approved the WHA 67.21 resolution named "Access to biotherapeutic products (BTP), included biosimilars, and its quality, security and effectiveness guarantee" [4], in which the WHO was asked to update the comparability guidelines in 2014, taking into account the technological advances in biotherapeutic product characterization and considering the regulatory needs and capacities of the countries.

To fulfill the mandate, WHO initiated a process to review the scientific evidence and experience of the use of biosimilars in a way that suggested reducing clinical and non-clinical data and thus reaching consensus on the most flexible regulatory considerations to accelerate the entry of biosimilars and increase confidence in their use. The findings of the review were published in 2020 and highlighted that data on biosimilars approved since 2006 did not raise safety concerns, that effectiveness and clinical outcomes were comparable to reference, that the technological tools available to perform analytical and functional testing were multiple, and that pharmacokinetic studies were robust and could be sufficient to demonstrate comparability, while there was no need to generalize the requirement for in vivo animal studies and confirmatory clinical efficacy and safety studies [5, 6].

A new version of WHO guidelines was published in 2022 after several rounds of consultation and discussion involving experts and authorities from more than 20 countries [7]. Table 1 describes some of the main modifications to the WHO guidelines mentioned above.

The update of the WHO comparability guide implies that the international health standard suggested for the development of a biosimilar requires a greater emphasis on the exercise of physicochemical, structural, and in vitro functional testing

Table 1 Main modifications of the WHO guidelines (Adapted from [8])

WHO comparability guideline – 2009 [2]	WHO comparability guideline – 2022 [7]	Reasons for updates
Key principles for the authorization of commercialization: The development of a similar biotherapeutic product (SBP) involves an exercise or exercises of comparability *in stages*, starting with the comparison of quality characteristics of SBP and reference biotherapeutic product (RBP). The demonstration of similarity between a SBP and a RBP in quality terms is a prerequisite to reduce the required clinical and non-clinical dataset for authorization	*Characterization of reference product (RP) quality attributes must be the first step to guide the development of biosimilars*: The subsequent comparability exercise must demonstrate structural, functional, and clinical similarity Demonstration of the similarity of a biosimilar to a RP in *structural and functional aspects is a prerequisite* for establishing comparability, with a *customized clinical data* package required as needed	"Step by step" deleted to reflect the evolving approach of "step by step" to the "customized" approach based on current practices, which shows that the development of biosimilars proceeds "concurrently" rather than in a staggered manner
PK, PD, and efficiency studies: The comparability clinical exercise is a gradual procedure that must begin with PK and PD studies and continue with pivotal clinical trials. *Generally, it should be demonstrated* that the chosen SBP and RBP. in some cases, however, it could be appropriated to perform comparative PK/PD studies *Safety studies*: Pre-authorization of safety data should be obtained from a sufficient number of patients to characterize the safety profile of SBP	*PK, PD, and efficiency studies*: The clinical studies are a valuable step to confirm similarity. In general, a comparative study of bioequivalence that includes PK and/or PD comparability is required for clinical evaluation. A comparative efficacy and safety test *will not be required* if sufficient evidence of biosimilarity can be extracted from other parts of the comparability exercise *Safety studies*: Safety data should be obtained *throughout* clinical development from PK/PD studies and in clinical efficacy studies when conducted	The objective of the clinical studies was clarified, and the considerations related to the quantity and types of necessary clinical data for biosimilar evaluation were presented It was articulated that the regulatory perspective on comparative safety and efficacy studies is gradually changing from a strict, inflexible requirement to a case-by-case scenario depending on the molecule and the data presented for the demonstration of biosimilarity based on the knowledge and evidence accumulated to date
Immunogenicity: The immunogenicity of biotherapeutical products must always be researched before its authorization. In the case of chronic administration, 1 year data will generally be appropriate prior to authorization to assess the incidence of antibodies and possible clinical implications *Extrapolation of indications*: If it's been demonstrated that SBP and RBP have similar effectiveness and security for a particular clinical indication, it could be possible to extrapolate this data to other RBP indications	*Immunogenicity*: The immunogenicity studies could be not necessary for well-characterized biological substances (for example, insulin, somatropin, filgrastim, teriparatide), in which a huge bibliography and clinical experience indicate that immunogenicity does not affect the product's safety and effectiveness *Indications authorization*: The decision to authorize the requested instructions will depend on demonstration of similarity between biosimilar and RP	It is clarified that the decision to authorize the requested indications depends on an adequate demonstration of similarity between biosimilars and RP

comparability, considering that non-clinical in vivo toxicological tests are not useful in the development of biosimilars. In turn, with the scientific literature research about the effectiveness/efficiency of biosimilars, which included security and long-term immunogenicity, the robustness of the development of these molecules was confirmed, therefore it was suggested that expensive confirmatory studies of efficacy, safety, and immunogenicity would be redundant [7].

They also concluded that current data from more advanced analytical and functional studies, PK and PD comparability studies, were sufficient to demonstrate comparability – in most cases – and that in vivo animal studies would only be considered in exceptional circumstances.

In 2009, the United States, from its side, established a specific approval path for biosimilars, which is recorded in Section 351(k) of BPCIA (Biologics Price Competition and Innovation Act). This regulation is based on the comparability standard, whose final objective is to demonstrate biosimilarity in relation to a reference product through an extensive package of analytical, animal, and clinical data, such as PK/PD and comparative efficacy studies. It should be noted that, in this same regulation, the FDA is granted the discretion to establish when a requirement is unnecessary.

Since then, and as a result of the growing debate on regulatory standards, there have been certain changes in the regulation. One of these changes has been the replacement of the regulation of "animal studies" by "nonclinical testing," making it possible that in certain cases, they do not have to be carried out, as already established in the latest WHO guidelines. Another implemented change was the possibility of using PD biomarkers instead of comparative clinical studies to demonstrate the similarity in biosimilar effectiveness, as suggested in prior discussions about biosimilar registration guidelines [9]. The FDA, recognizing that few medicines have well-characterized PD biomarkers, which limits the impact of regulatory action, has encouraged the application of omics technology (metabolomics, proteomics, among others) to discover new PD biomarkers. However, as argued by Niazi [10], this change does not represent a relevant contribution to the rationalization of these standards since the evaluation of these PD biomarkers does not provide new information compared to the information already obtained by PD/PK studies [10, 11]. In the specific case of new biomarkers identified through omics technology, in addition to the significant investment of time and resources required for their discovery, the absence of differences between the two biomarkers could not be interpreted as evidence of equal efficacy. This is because, during their development, it is not necessary to establish a correlation between values and a clinical response. Furthermore, the application of these biomarkers lacks logical coherence. In this context, a discussion of FDA guidelines has been ongoing since 2023, which has resulted in further flexibility being introduced for the approval of biosimilars, particularly those previously mentioned (exemption from confirmatory clinical studies and acceptance of PK/PD studies) [11, 12].

These findings, in conjunction with the 2021 decision of the British agency aligning with this approach and the 2023 EMA statement on the acceptance of substitution without any restriction on approved products (i.e., their interchangeability without reservations), reinforce the rationale for incorporating the flexibilities that have been discussed in Latin America [13].

The adoption of these flexibilities, which are based on scientific and technological values, for the approval of biosimilars in Latin America has been gradual and largely unequal, with markedly disparate regulatory frameworks. Among the countries that have adopted some forms of flexibility are Brazil, Argentina, and Colombia [14].

Brazil was the first country in Latin America to implement a specific regulatory framework for biosimilar products, as set forth in the ANVISA (the Brazilian regulatory agency) RDC 55/2010 resolution [15]. The norm describes two distinct approval pathways: the comparability pathway and the individual development pathway. In the first instance, in addition to the provision of analytical information, nonclinical studies, animal studies, and PK/PD comparatives, it is necessary to submit comparative clinical studies (those designated as confirmatory) which demonstrate the safety and efficacy of the product in question, since no exemption has been established (Art. 46). Regarding the individual development pathway, certain flexibilities are afforded, including the possibility of reducing preclinical studies in accordance with specific criteria, such as molecule complexity (Art. 39). Additionally, phase I and phase II studies are not necessarily required to be comparative (Art. 40). While phase III studies must be comparative, there are exceptions, such as in the case of blood derivatives (Art. 41). This is contrary to the principle of comparability, which is to extend and extrapolate findings to the full range of approved indications (that is, to test the biosimilar in an indication and extend the approval of other indications of the reference product). If the first route is chosen, extrapolation of reference medicine indications is permitted; otherwise, it is not possible. In accordance with the regulatory amendment that has been under development and with the objective of harmonizing its regulation with that of the principal regulatory agencies, ANVISA initiated in 2022 a review process in accordance with RDC 55/2010. Following a series of public consultations and discussions, a new regulation for biosimilar registration by comparability was approved in May 2024 [16]. In this update, several changes were incorporated, including the introduction of an exemption for certain confirmatory comparative clinical studies if a high level of comparability, functionality, and characterization of the product has been demonstrated in previous stages. This significant step, which positions the Brazilian regulatory framework at the vanguard of the region, may serve as a model for other countries in Latin America to emulate [16].

In contrast, Argentina has also established regulatory frameworks pertaining to the registration of biosimilars, Provision 7729/2011 of the ANMAT (Argentine regulatory agency) [17]. The referenced norm stipulates that the extension of clinical and preclinical studies depends on the results of the comparability exercise conducted to date. Relevant aspects for evaluation include the nature and complexity of the active substance, information about the in vivo behavior of the pharmaceutical ingredient, the impurity profile comparative, and other factors (Art. 6). Nevertheless, it is unclear whether exemption is possible in its entirety or merely a reduction according to product characteristics.

Another regulation that stands out for its pro-competitive and innovative perspective is that of Colombia which, through Decree number 1782 of 2014 of the Ministry of Health and Social Protection, established two routes for approval of biosimilars, the comparability route and the abbreviated comparability route that anticipates what in 2024 Brazil adopted and is ahead of the British MHRA decision of 2021 and the more recent WHO decision of 2022 [18]. Colombia's regulatory innovation gave way to the possibility of exempting certain clinical and pre-clinical studies as long as the pharmaceutical ingredient was sufficiently characterized, had a defined and highly documented safety and efficacy profile, and robust pharmacovigilance (Art 9). But, despite establishing a significant flexibility based on state-of-the-art analytical tools, this abbreviated route has not had the expected impact in almost 10 years that it has been in place, since only one biosimilar (Enoxaparin sodium) has been approved under this route [19]. Despite Colombia's visionary regulatory stance, the application of this access—and competition-sensitive regulation—is not reflected in the dynamics of competitive entry into the market. A recent review of the availability of competition for three monoclonal antibody biotherapeutics considered essential by the WHO in four Latin American countries showed relevant differences in the timing of entry for each competitor (Brazil and Argentina had faster competition than either Mexico or Colombia). These delays may well be explained by the way the standard was drafted, which delayed its implementation until the technical guidelines for its application were defined (Good Manufacturing Practices for biologics Res 5,402 of 2015, immunogenicity assessment Res 4,490 of 2016, and the comparability guide Res 2,950 of 2019) and which took about 3 years to be adopted [20–22]. The delay in the entry of competition could also be explained by demands for the standard by multinational pharmaceutical companies and judicial interference with decisions of INVIMA (Colombia's regulatory agency), as occurred in other countries [23, 24].

Among the variables explaining the slow entry of competitors into the pharmaceutical market, it should not be ruled out that the national standard leaves it up to INVIMA assessors to decide whether the product can apply the abbreviated route. Most agency evaluators are risk-averse and skeptical of the scope of confirmatory testing exemptions and hold to the precautionary principle that, if applied, it would only be possible for small-molecule biologics [25, 26]. This dynamic shows the need for dialogue and discussion on the importance of benefiting health systems by incorporating the scientific developments that underpin the flexibilities.

There is another major barrier for the application of comparability regulations for the approval of biosimilars: the requirement for approval of the reference standard or innovator product in the country evaluating the biosimilar.

Access to the full dossier is a decisive factor in the comparability exercise, and also implies the purchase of the reference product at very high prices. Countries with intensively developed biotechnology capacities, such as Cuba, or African countries with limited capacities, find themselves with their hands crossed in the face of this barrier, and other countries in the global south could suffer the same limitation.

However, the WHO's 2022 guidelines provide for flexibility in the approval of biosimilars, but this has not been fully adopted. Indeed, a recent study states that the 2022 WHO's recommendation to source RPs for non-local jurisdictions has had a limited uptake after over a year of release. Most countries continue to rely on the 2009 guidelines or EMA recommendations. Given that additional flexibility is afforded by the WHO's 2022 revision compared to the 2009 guidelines, adoption rates will likely accelerate as the need for biosimilars intensifies in the coming decade [27].

On the other hand, in order to establish global regulatory comparators to support demonstrations of biosimilarity, the regulatory authorities could discuss eliminating the need for duplicative PK testing of reference products from different regions of the world to support the requirements of various health regulatory authorities when sponsors seek to use data comparing a proposed product with a non-locally approved comparator product to support a demonstration of biosimilarity. The recommended streamlining will not lower safety, efficacy, or quality standards, nor will it involve an extensive revision of the regulatory framework for biosimilars. These approaches are part of the recommendations of expert groups sensitive to medicines access.[3]

In conclusion, the availability of competition for generic versions of biotherapeutics, including biotechnology medicines is critical for patient access and for the sustainability of health systems. The presence of competition is conditioned by the regulatory and approval processes followed by health agencies and the early adoption of technological and scientific developments that support the flexibilities for approval of biosimilars (exemption from redundant clinical trials, especially confirmatory comparative studies, exemption from toxicological studies in animals, and acceptance of a global and harmonized reference product with wide access to full dossier Information). A summary of the trend in the adoption of these flexibilities at the global level is presented in Fig. 1.

---

[3]Below is the proposal of the Biosimilars Council:

Establish global regulatory comparators to support demonstration of biosimilarity. FDA should work with other regulatory authorities to establish global comparators that would eliminate the need for duplicative PK testing of reference products from different regions of the world to support the requirements of various health regulatory authorities when sponsors seek to use data comparing a proposed product with a non-U.S.-approved comparator product to support a demonstration of biosimilarity. The recommended streamlining will not lower safety, efficacy, or quality standards, nor will it involve an extensive revision of the regulatory framework for biosimilars. It does not seek to eliminate clinical efficacy studies altogether. Rather, it preserves FDA's authority to request additional evidence in exceptional circumstances when scientifically justified, unique risk-based considerations are identified [28].

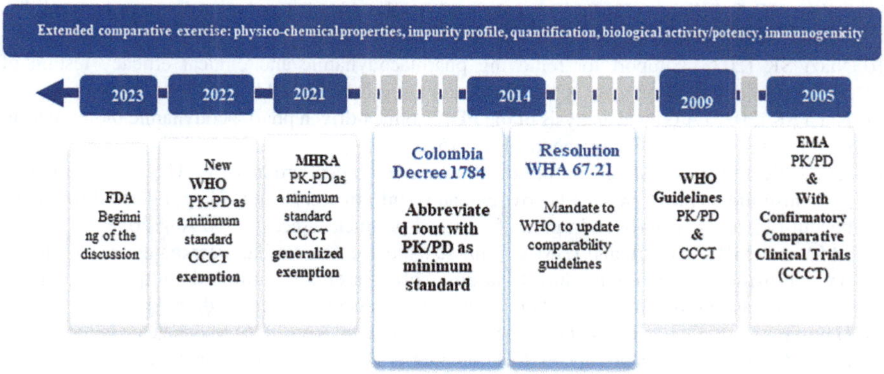

Fig. 1 Biosimilar health regulation – Colombia in an international context. Source: Self elaboration based on "Claudia Vaca. Colombia's approach to biosimilar regulation that strikes a balance between scientific rigorousness and patient access-lessons learned for vaccines" on vaccine development research forum. Faculty of Medicine. Chulalongkorn University. Bangkok. Nov 2023

Compliance with Ethical Standards This chapter is a review and no humans or animals were involved in the preparation of the chapter.

References

1. Frapaise FX (2018) The end of phase 3 clinical trials in biosimilars development? BioDrugs 32: 319–324
2. WHO Expert Committee on Biological Standardization (2009) Guidelines on evaluation of similar Biotherapeutic Products (SBPs), Annex 2, TRS No 977. https://www.who.int/publications/m/item/sbp-trs-977-Annex-2. Accessed 26 Aug 2024
3. Committee for medicinal products for human use (CHMP) (2005) Guideline on similar biological medicinal products. In: European Medicines Agency (EMA). https://www.ema.europa.eu/en/similar-biological-medicinal-products-scientific-guideline. Accessed 26 Aug 2024
4. Sixty-seventh World Health Assembly (2014) WHA67.21. Access to biotherapeutic products including similar biotherapeutic products and ensuring their quality, safety and efficacy. In: World Health Organization. http://apps.who.int/gb/ebwha/pdf_files/WHA67/A67_R21-en.pdf. Accessed 26 Aug 2024
5. WHO Expert Committee on Biological Standardization: Report of the seventy-second and seventy-third meetings (2021) Revision of the WHO Guidelines on evaluation of similar biotherapeutic products (SBPs). In: World Health Organization https://www.who.int/publications/i/item/9789240024373, Accessed 26 Aug 2024
6. Hye-Na K, Wadhwa M, Knezevic I et al (2023) WHO guidelines on biosimilars: toward improved access to safe and effective products. Ann N Y Acad Sci 1521:96–103
7. World Health Organization (2022) Guidelines on evaluation of biosimilars Replacement of Annex 2 of WHO Technical Report Series, No. 977 in https://cdn.who.int/media/docs/default-source/biologicals/annex-3%2D%2D-who-guidelines-on-evaluation-of-biosimilars%2D%2D-sj-ik-5-may-2022.pdf?sfvrsn=9b2fa6d2_1&download=true. Accessed 26 Aug 2024
8. Increasing the Efficiency of Biosimilar Development Programs-Re-evaluating the Need for Comparative Clinical Efficacy Studies in IPRP Biosimilars Working Group Workshop. https://admin.iprp.global/sites/default/files/2024-07/IPRP_BWG_Final%20IPRP%20Scientific%20Workshop%20Summary%20Report_2024_0506.pdf. Accessed 26 Aug 2024

9. Ebbers HC, Schellekens H (2019) Are we ready to close the discussion on the interchangeability of biosimilars? Drug Discov Today 24:1963–1967
10. Niazi SK (2023) Support for removing pharmacodynamic and clinical efficacy testing of biosimilars: a critical analysis. Clin Pharmacol Drug Dev 12:1134–1141
11. Niazi SK (2023) A critical analysis of the FDA's omics-driven pharmacodynamic biomarkers to establish biosimilarity. Pharmaceuticals 16:1556
12. (2023) Benefit-Risk assessment for new drug and biological products. In: U.S. Food and Drug Administration. https://www.fda.gov/regulatory-information/search-fda-guidance-documents/benefit-risk-assessment-new-drug-and-biological-products. Accessed 26 Aug 2024
13. EMA and MHRA EMA and (HMA) joint statement on biosimilar interchangeability https://www.biosim.es/posicionamiento-de-la-ema-sobre-intercambiabilidad/ https://www.ema.europa.eu/en/news/biosimilar-medicines-can-be-interchanged. Accessed 26 Aug 2024
14. Gaviria A, Vaca González CP, Gómez Muñoz C, Morales AA (2016) El debate de la regulación de medicamentos biotecnológicos: Colombia en el contexto internacional. Rev Panam Salud Pública 40:40
15. Agência Nacional de Vigilância Sanitária (2010) RDC 55/2010: dispõe sobre o registro de produtos biológicos novos e produtos biológicos e dá outras providências. Ministério da Saúde, Brasilia
16. Agência Nacional de Vigilância Sanitária (2024) Anvisa aprova novo regulamento para registro de medicamentos biossimilares. In: Agência Nacional De Vigilância Sanitária - Anvisa https://www.gov.br/anvisa/pt-br/assuntos/noticias-anvisa/2024/anvisa-aprova-novo-regulamento-para-registro-de-medicamentos-biossimilares. Accessed 26 Aug 2024
17. Administración Nacional de Medicamentos, Alimentos y Tecnología Médica ANMAT, (2011) Disposición N° 7075/11. Buenos Aires: ANMAT. https://www.argentina.gob.ar/normativa/nacional/disposici%C3%B3n-7729-2011-190137/actualizacion. Accessed 26 Aug 2024
18. Ministerio de Salud y Protección Social (2014) Decreto 1782 de 2014. Colombia. https://www.minsalud.gov.co/Normatividad_Nuevo/Decreto%201782%20de%202014.pdf. Accessed 26 Aug 2024
19. Castillo AM (2023) Análisis del conflicto en la regulación de medicamentos biológicos en Colombia. Dissertation, Universidad Nacional de Colombia
20. Ministerio de Salud y Protección social (2015) Resolución 5402 de 2015: Por la cual se expide el manual y el instrumento de verificación de las Buenas Prácticas de Manufactura de Medicamentos Biológicos. https://www.icbf.gov.co/cargues/avance/compilacion/docs/resolucion_minsaludps_5402_2015.htm. Accessed 26 Aug 2024
21. Ministerio de Salud y Protección social (2016) Resolución 4490 de 2016: Por la cual se expide la "Guía de Evaluación de la Inmunogenicidad para los Medicamentos Biológicos" y se dictan otras disposiciones. https://www.icbf.gov.co/cargues/avance/compilacion/docs/resolucion_minsaludps_4490_2016.htm. Accessed 26 Aug 2024
22. Ministerio de Salud y Protección social (2019) Resolución 2950 de 2019, Por la cual se expide la Guía para la Evaluación de la Comparabilidad de Medicamentos Biológicos y se dictan otras disposiciones. https://www.suin-juriscol.gov.co/viewDocument.asp?id=30039972. Accessed 26 Aug 2024
23. Consejo de Estado Sala de lo contencioso administrativo sección primera (2019) AUTO 11001032400020170002500 de 2019. https://www.suin-juriscol.gov.co/viewDocument.asp?id=30036504. Accessed 26 Aug 2024
24. Corte Constitucional del Ecuador (2022) Judgment No. 17460-2022-01424. https://www.corteconstitucional.gob.ec/. Accessed 26 Aug 2024
25. Vaca C, Gómez C (2020) The controversy around technical standards for similar biotherapeutics: barriers to access and competition? Pharmacoepidemiol Drug Saf 29:1518–1522
26. Vaca González CP, Rengifo JD, Gómez LM (2023) Verdades incómodas de la salud pública global. Universidad Nacional de Colombia. https://doi.org/10.5281/zenodo.10903397

27. Examining the Impact of the World Health Organization (2022) Guidelines on evaluation of biosimilars for non-local comparators in biosimilar studies on Middle East and North Africa Member States. https://www.mdpi.com/2226-4787/12/3/94. Accessed 26 Aug 2024
28. https://biosimilarscouncil.org/wp-content/uploads/2024/05/202405-BiosimilarsCouncil-Streamlining-Development-Biosimilar-Medicines.pdf. Accessed 19 Aug 2024